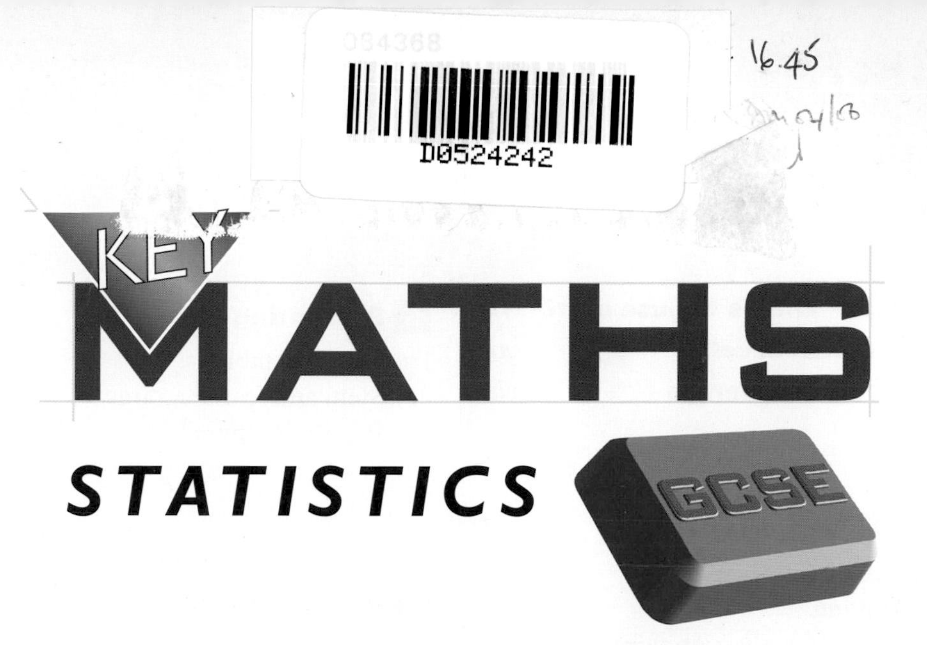

Barbara Job
Formerly of Christleton County High School, Chester

Diane Morley
Christleton County High School, Chester

AQA VERSION

Contents

First published in 2000 by:
Stanley Thornes (Publishers) Ltd

Revised AQA edition printed in 2003 by:
Nelson Thornes Ltd
Delta Place
27 Bath Road
CHELTENHAM
GL53 7TH
United Kingdom

05 06 07 / 10 9 8 7 6 5 4 3

A catalogue record for this book is available from the British Library.

ISBN 0 7487 7423 8

Illustrations by Oxford Designers & Illustrators
Page make-up by Tech Set Ltd

Printed and bound in China by Midas Printing International Ltd

Acknowledgements
The publishers thank the following for permission to reproduce copyright material:

HMSO:1, 17; John Walmsley Photolibrary: 105, 109; Martyn Chillmaid: 3, 6, 75, 167, 184, 301, 302; National Archives, Public Record Office: 1, 5; Panasonic (UK): 153; Photri-Microstock: 113 (Gary Funck); Pictor International; Spire; Stone: 95 (Martin Rogers); Topham Picturepoint: 69, 92 (Syracuse Newspapers/C W McKeen/The Image Works); TRIP: 229, 241 (C and J Walker)
All other photographs Nelson Thornes Archive.
The publishers have made every effort to contact copyright holders but apologise if any have been overlooked.

The author and publisher would like to thank the following for their additional contributions to this book:

Tim Morland for creating and generating the original material for the *Working with Excel* sections.
Kate Job for writing the Introduction for this book.
Paul Hogan and David Baker for their valuable reviews and comments on the material throughout the writing process.
Jason and Julian Morley and Brian Job for assistance throughout.

Statistics – what's the use of it?

Have you ever looked at your maths homework and asked 'What use will this ever be to me?' Statistics is one area of maths that is used all the time. You won't believe it, but you probably use it every day.

For example, take the daily press. During one year, there were over one hundred references to statistics in just one daily newspaper. These included employment, the economy, crime and medical matters. You probably think that these are boring but sometimes they may make a difference to you. For example, one report said that there will be more old age pensioners than children by the year 2008. Just how many can you help across the road?

What about the reports that GCSEs, A levels and degree courses are apparently getting easier? How do you feel about that as you work for your exams? These reports use statistics to back up their comments. Do you believe them or not? Can you afford to take facts and figures as they are reported? To understand them you need to know some statistics.

Even if you don't read newspapers, there will be situations where statistics could help more than you think. Imagine, you're about to go into your maths lesson. You haven't done your homework. The teacher is going to ask one person from the class to write the answer on the board. You're racking your brains for a way to guarantee you don't get asked. Surely statistics can't help you there ... or can it?

For a start, how great is the chance that you will be chosen? If your class has 30 students in it you have a fairly small chance of being asked. It might be worth taking the risk of not confessing and crossing your fingers. There may be a freak flu epidemic. Maybe only 10 others turn up for the lesson. Your chance of being asked is now greater.

Perhaps you should just own up and take your punishment. But maybe you can reduce the chance of being asked. The chance of being asked two questions in a row is much smaller. You could try to answer a question before someone is picked for the homework question. If you had time to do some research you could look up some of the studies in psychology. These look at how people behave in the classroom. For example, students sitting in the middle front seats tend to take part most in the lesson, i.e. they get asked and answer more questions. This might help you to work out where to sit to have the smallest chance of being asked.

This could bring up another problem however. What if you knew that research has shown that students sitting in these middle front seats are more likely to get the best grades? You may not want to move seats if you want good grades.

What if you own up? How has the teacher reacted in the past? Has she been lenient towards forgetful pupils, or is she quick to hand out detentions? What sort of punishments does she favour? What is the chance of you getting one you really don't like? You might consider the risk worth taking if the punishment is likely to be mild or non-existent. All of these possible investigations could help you make your decision. They all involve some form of statistics.

What if you had a question you couldn't find an answer to? Maybe you want to know if the time of day affects how likely you are to get a detention. How would you go about planning your own experiment to find out? What sort of data would you need to collect? Who would you ask? How would you analyse the results? How could you be sure you hadn't made a mistake? The answers to all of these questions lie with statistics.

So, what things are you interested in researching? What questions do you want to find answers to? This book takes you through the skills that you will need to get to grips with statistics. Good luck, and, by the way, maybe its better just to do your homework!

Working with *Excel*

Additional support material and activities have been provided at the end of certain chapters to show how one spreadsheet package can be used to extend statistical work, analysis and data presentation. Other packages are of course as appropriate, including *Lotus 123* or other related programs. The ones included are given as examples, which you can use or adapt accordingly.

In addition to the particular requirements of the course, the publisher has provided additional sets of data and ideas that can be obtained and downloaded by visiting our website at: www.nelsonthornes.com.

Specification coverage

Both the Foundation and Higher tiers of the GCSE statistics course from AQA are fully covered. An orange header bar indicates pages with Foundation content. A purple header bar indicates pages with Higher content. Orange and purple header bars together indicate pages with both Foundation and Higher content. In addition, orange, purple or orange and purple bars in the margin indicate the appropriate content that needs to be covered in the Summary and Questions sections.

1 Collecting data

1 Types of data

Dawn and Mary are doing a survey of the T-shirts sold in the shops in their town. Dawn is collecting data on prices and Mary is collecting data on colours.

Statistics is all about data.
There are two types of data.

Quantitative data Data that is numerical is called **quantitative data.**
The price of T-shirts is an example of quantitative data.
Age is another example.

Qualitative data The second type of data does not use numbers.
It is non-numerical data. This is called **qualitative data.**
The colour of T-shirts is an example of qualitative data.

Example Julia is collecting information about her friends' bedrooms for her Design and Technology project.
These are three of the questions she asks.

a What are the measurements of your bedroom?
b What is the colour of your carpet?
c How many electric sockets do you have?

For each question, write down if the data is qualitative or quantitative.

a The measurements could be 10 ft long and 9 ft wide.
This is numerical data so it is quantitative.

b The carpet could be green or pink.
This is non-numerical data so it is qualitative.

c The number of sockets could be 2 or 3.
This is numerical data so it is quantitative.

Exercise 1:1

1 Jennifer is collecting information for her Design and Technology project.
She is interested in the kitchens in her friends' houses.
She asks the following questions.

a What is the colour of your cooker?

b How many windows are there in your kitchen?

c What is the height of the room?

d What is the make of your fridge?

For each question, write down if the data is quantitative or qualitativc.

2 For each of these write down if the data is quantitative or qualitative.

a Don is 1.5 m tall.

b The number of my house is 7.

c Gill's favourite colour is pink.

d There are 4 bedrooms in my house.

e Jack is a British citizen.

3 You are collecting information as part of your geography coursework.

a Write four questions to collect quantitative data.

b Write four questions to collect qualitative data.

4 Dominic wants to buy a new computer. He looks on the Internet for information to answer the following questions.

a How much is a computer monitor?

b What colour are monitors available in?

c How large is the monitor?

d How much is a keyboard?

For each question, write down if the data is quantitative or qualitative.

Jennifer and Mike have done a survey. They have collected the information themselves. This is an example of **primary data**.

Primary data

Primary data is data that is collected by the person who is going to use the information.
The data can be obtained by asking questions, taking measurements, or counting.

It is possible to collect exactly what is required.
It can be very time consuming to collect primary data.

Secondary data

Secondary data is data that is not collected by the person who is going to use the information.
It is obtained from published statistics or databases.
Tables in geography books and temperature charts in newspapers are both examples of secondary data.

It can be very cheap to obtain large quantities of secondary data.
The data may not be exactly what is required.
The data may be out of date.

5 Write down whether each of these is an example of primary data or secondary data.

a Jean looked at the hospital records to find out how many babies were born each day in September.

b Joe measured the lengths of the pebbles in a sample of pebbles from the beach.

c John counted the number of yellow cars passing his school.

d Robert needs a Saturday job. He rings the local shops and supermarkets to find out how much they pay for a day's work.

e Carl is very interested in pop music. He looks at the pop charts for the last ten weeks to see which group was top of the charts.

f Anna is trying to decide if there are more hours of sunshine in Majorca or Corfu. She looks for this information in travel brochures.

g Gill draws a line, and asks her friends to find the middle without measuring. She records the results in a table.

2 Sampling

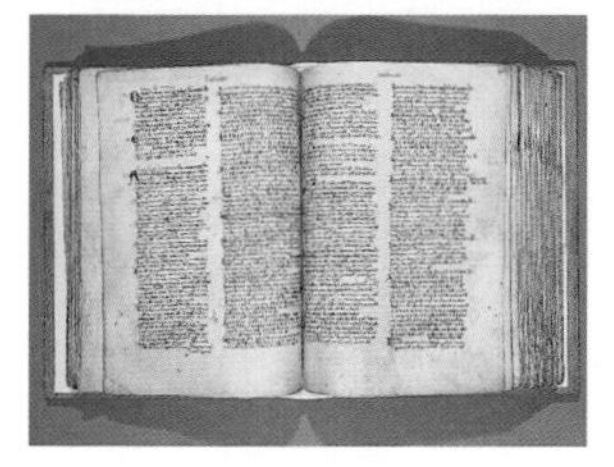

In 1085 William I wanted to know how much money he could raise in taxes. He needed to know about the people in England. William ordered a nationwide survey of his people. The results were written down in the Domesday Book.

The Domesday Book was probably the first census. These days a national census is carried out every 10 years by the government. Every household in the country completes a questionnaire. This is used as a source of secondary data. Local government, police authorities and many other organisations use the information for further planning.

Sometimes only a part of the population is questioned. This is often the case in opinion polls. These are often used before elections to see how people intend to vote.

Population The complete set of people or objects that information is collected about is called the **population**.

Sampling frame A list of all the items of the population is called the **sampling frame**.

Census When data is collected from *every* person or about *every* object, the survey is called a **census**.

Bias Anything that distorts data so that it does not fairly represent the population is called **bias**.
You need to take care to avoid bias in the way that you collect data. Otherwise your results are not reliable.

Sample Information is normally taken from a small part of a population. This is called a **sample** of the population.
It is important to choose the sample without bias so that the results will represent the whole population.

It is cheaper and quicker to take samples, than to collect information from the whole population.

The size of the sample is important. It needs to be large enough to represent the population, but small enough to be manageable.

Exercise 1:2

1 Each of the following samples does not fairly represent the population. It is a biased sample. Say what is wrong with the way that each sample is selected.

a The school librarian wants to find out how many books each pupil in the school reads. She asks the pupils in the library how many books they have read in the last week. She uses this information to work out the average number of books read each week by the pupils in her school.

b The Headteacher of a school wants to find out if the pupils enjoy school lunches. He asks ten pupils in Year 11 for their views.

c An engineer is carrying out a traffic survey. He is trying to work out how busy a particular road is. Each day, he counts the number of cars passing a particular point between 2 p.m. and 3 p.m. He uses this information to write his report.

d A local television station is keen to find out how popular their latest sports programme is. An interviewer stands inside a specialist sports shop, and asks the customers for their views. This information is used to decide if the programme is popular or not.

One of the ways to avoid bias is to use a random sample.

Random sample In a **random sample,** every member of the population has an equal chance of being selected.
Random samples need to be carefully chosen.

A tennis club with forty members has five tickets for Wimbledon.
Five people are chosen at random to receive the tickets.

Each of the forty members is given a number, 01, 02, . . . 39, 40.

The five people can be chosen at random in different ways.

Method 1 — Each number is written on a piece of paper.
The pieces of paper are put into a container and mixed up well.

To choose a random sample of five numbers, five tickets are drawn.
The people with the five chosen numbers go to Wimbledon.

Method 2 — Tables of random sampling numbers can be used to choose a sample of five numbers.

86	13	84	10
60	78	48	12
06	48	06	37
25	32	90	79
99	09	39	25

This is part of a random number table.
The smallest number is 06.
The largest number is 99.

Reading from each row of the table in order chooses the numbers.
Use the first row of the table: **86**, **13**, **84**, **10**.

The first number selected is **86** — There are only 40 members in the tennis club, so this is too large and is not used.

The second number is **13**. — This selects the member given the number 13.

The third number is **84**. — Again this is above 40, so it is too large and is not used.

The fourth number is **10**. — This selects the member given the number 10.

Continue to use the rows of the table in this way to select the rest of the sample.
The number 06 occurs twice in the table. You only select it the first time.

The chosen numbers are 13, 10, 12, 06 and 37. These people go to Wimbledon.

You can use columns instead of rows.

86	13	84	10
60	78	48	12
06	48	06	37
25	32	90	79
99	09	39	25

Using the first column gives the numbers 06, 25

Using the second column gives the numbers 13, 32, 09

The chosen numbers are 06, 25, 13, 32 and 09

This is a different random sample.

Method 3 — Scientific calculators have a random number button. This can be used, in a similar way to the random number table, to select the people going to Wimbledon.

2 The table gives the sex and heights in metres of 30 children.

Child	Height	Sex	Child	Height	Sex
01	1.43	M	16	1.25	F
02	0.98	F	17	0.89	F
03	1.24	M	18	1.62	M
04	0.87	F	19	1.20	F
05	1.10	F	20	1.53	M
06	1.15	F	21	1.60	M
07	1.29	M	22	1.23	F
08	0.94	M	23	1.44	M
09	1.00	M	24	1.30	F
10	1.21	F	25	1.00	F
11	1.53	F	26	1.54	F
12	1.43	M	27	1.12	M
13	1.27	M	28	0.98	F
14	1.24	M	29	1.06	M
15	1.42	F	30	1.25	F

a From this population of 30 children, choose a random sample of 5 children. Use this random number table to do this. Begin by using the first line. Write down the height and sex of each child in the chosen sample.

11	52	23	55
89	62	97	55
83	10	22	12
16	79	25	72
93	94	43	15

b Choose a second random sample of 5 children. Use the same random number table, but begin by using the second line. Write down the height and sex of each child in the chosen sample.

c Choose a third random sample of 5 children. Use the same random number table but begin by using the first column. Write down the height and sex of each child in the chosen sample.

d Use the random number function on your calculator to choose a third sample of 5 children.

Systematic sample

In a **systematic sample**, every member of the sample is chosen at regular intervals from a list.
A sample chosen in this way can be biased, if low or high values occur in a regular pattern.
The table lists 20 members of a bowling club.

Membership no.	Name	Membership no.	Name
01	J. Crabtree	11	J. Darnton
02	B. Stocks	12	A. Stuart
03	A. Coombes	13	D. Jones
04	J. Reeves	14	J. Wareing
05	A. Smith	15	W. Haslam
06	B. Clifford	16	A. Wardle
07	F. Hassall	17	C. Dorris
08	A. Dowle	18	B. Cook
09	C. Patel	19	F. Driver
10	D. More	20	L. Torvalds

You can use systematic sampling to choose four names from this list.
For a sample of four people, one person in five must be selected.

You can start at any point in the table between 01 and 05. This starting point can be chosen at random, or you can use a particular starting point. For example you may decide to start at 02.

Taking a starting point of 02, the sample will be the membership numbers:
02, $02 + 5 = 07$, $07 + 5 = 12$, $12 + 5 = 17$

The names chosen are: B. Stocks, F. Hassall, A. Stuart, C. Dorris.

Exercise 1:3

1 **a** Use the table given in the example above.
Start with A. Coombes and choose a systematic sample of four people.
Write down the names of the four chosen members.

b Start with J. Crabtree and choose another systematic sample of four people.
Write down the names of the four chosen members.

The two samples in the previous question were chosen systematically.
The samples produced are different however.
It is important to be aware of this variability of samples.

2 Use the table given in the example.

a Start with J. Crabtree and choose a systematic sample of five people.
Write down the names of the five chosen members.

b Start with J. Reeves and choose another systematic sample of five people.
Write down the names of the five chosen members.

Convenience sampling

A company is researching the popularity of different brands of cola.
An interviewer stands inside a supermarket and asks 100 customers for their preferences.

A sample produced in this way is called a **convenience sample**.
It can be biased because of the way the sample is chosen.

3 Write down the method of sampling used in each part below.
Choose from: random sample, systematic sample or convenience sample.

a A careers officer uses random numbers to choose a sample of twenty school leavers to interview.

b Sally is researching customer satisfaction amongst train passengers.
She asks 100 people who are waiting for trains at a station.

c Each form in year 11 has to choose one person to represent their group on the School Council. All the pupils write their name on a piece of paper. The papers are folded and placed in a box. The teacher mixes them up and then picks one out without looking.

d The librarian is interested in estimating how many books will need replacing in a central library. From an alphabetical list of all the books she chooses every 100th book to see if it needs replacing.

e Peter is researching the leisure activities of adults.
He stands in the centre of town and asks 150 adults to answer a questionnaire.

Stratified sample

A population may contain separate groups or strata. Each group needs to be fairly represented in the sample. The number from each group is proportional to the group size.
The selection is then made at random from each group.
A sample produced in this way is called a **stratified sample**.

A survey about sport is carried out among pupils in Year 11, Year 12 and Year 13.
There are 198 pupils in Year 11, 120 pupils in Year 12 and 101 pupils in Year 13.
A sample of 40 pupils is taken.

You need to decide how many pupils from each year should be included in this stratified sample of 40 pupils.

First find the total number of pupils.

Total number of pupils $= 198 + 120 + 101$
$= 419$

Now find the fraction of pupils in each year.

Year	Fraction of pupils	Number of pupils in the sample of 40
11	$\frac{198}{419}$	$\frac{198}{419} \times 40 = 19$
12	$\frac{120}{419}$	$\frac{120}{419} \times 40 = 11$
13	$\frac{101}{419}$	$\frac{101}{419} \times 40 = 10$

These numbers of pupils are rounded to the nearest whole number.

So you need: 19 pupils from Year 11.
11 pupils from Year 12.
10 pupils from Year 13.

Finally select 19 pupils in Year 11 at random.
Do the same for the pupils in Years 12 and 13.
These pupils form the stratified sample.

Exercise 1:4

1 George is collecting data for his statistics project. He decides to compare the sporting activities that Year 10 boys and girls take part in.
He begins his project by taking a stratified sample of 30 pupils. This needs to reflect the proportion of boys and girls in the year.
There are 80 boys and 102 girls in Year 10.
Calculate the number of:

a boys to be included in the sample,

b girls to be included in the sample.

Remember to give your answers to the nearest whole number.

2 Juliet lives on a housing estate. The table gives the number of people in each age group who live on her estate.

Age in years	0–19	20–39	40–59	60–79	80+
Number of people	182	88	110	72	15

For her geography project she chooses a stratified sample of 50 people which reflects these age groups.
Calculate the number of people she should include from each age group.
Give each of your answers to the nearest whole number.

3 A travel agent has 3900 customers who have children under 16 years of age and 1764 customers who do not have children under 16.
He decides to investigate the different types of holiday they take.
Find the numbers of customers with children under 16 who should be included in a stratified sample of 300 people.

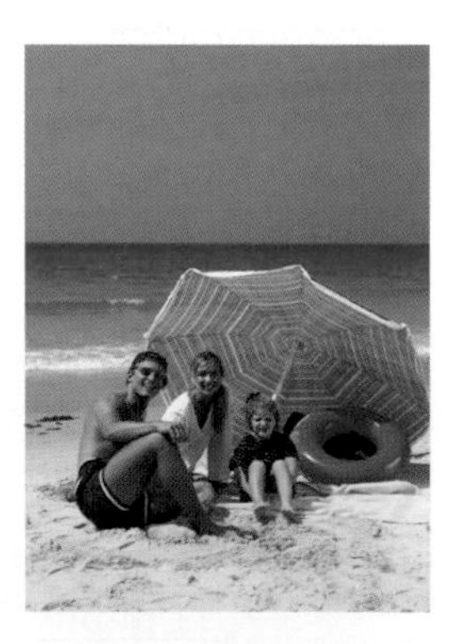

4 A town has a population of 12 567. The librarian wants the library to be used by as many people in the town as possible. She is particularly keen to help the 12% unemployed. She sends a questionnaire to 300 people to find out how the facilities can be improved. Find the number of unemployed people she should include in the stratified sample.

In this table each year has been separated into boys and girls.
You need to decide how many pupils from each year should be included in a stratified sample of 40 pupils. This sample needs to reflect both the year group and the gender.

Year	boys	girls	Number of pupils in the sample of 40 boys	girls
11	97	101	$\frac{97}{419} \times 40 = 9$	$\frac{101}{419} \times 40 = 10$
12	52	68	$\frac{52}{419} \times 40 = 5$	$\frac{68}{419} \times 40 = 6$
13	37	64	$\frac{37}{419} \times 40 = 4$	$\frac{64}{419} \times 40 = 6$

So you need: 9 boys and 10 girls from Year 11.
5 boys and 6 girls from Year 12.
4 boys and 6 girls from Year 13.

5 A headteacher has to choose a committee of 12 people to represent the staff. The table shows the staff in the school.

Choose a stratified sample of 12 to represent both the gender and type of employment.

	Men	Women
Office staff	1	6
Teaching staff	29	21
Support staff	6	6

6 A Local Authority wants to know what leisure facilities to include in a new development. As part of their survey they sample the views of young people.

Age (years)	Full-time study	Full-time employment
16–18	4311	3718
19–21	3025	5627

Choose a stratified sample of 500 to represent both the age profile and the occupation.

Quota sampling

Quota sampling is often used in market research.

The interviewer questions a certain number of people. The people have to be of a certain type. They could be of a certain age, sex or social class. The interviewer makes the choice of exactly who is asked.

Quota sampling is very cheap. It is not very reliable because it depends on the interviewer to choose the sample.

John is a part of a market research team. He uses the quota sampling method to find the views of 20 people on a new breakfast cereal. This is how he does it.

He stands in the centre of town. He asks five men aged 20–30, five men aged 30–40, five women aged 20–30 and five women aged 30–40. These twenty people form his quota sample.

Cluster sampling

The population is divided into smaller groups. These smaller groups are called clusters. One or more clusters are chosen using random sampling. This is called **cluster sampling**. The sample is then every member of the clusters chosen.

Cluster sampling is very cheap but it can be biased if the clusters are different.

Dave looks at how much time pupils in Year 9 spend on sporting activities. The six tutor groups form his clusters. He numbers the tutor groups from 1 to 6. He chooses a tutor group by throwing a dice. All of the pupils in his chosen tutor group are part of his sample.

Large-scale opinion polls often use both cluster and quota sampling.

Opinion polls

Opinion polls are another way of collecting information.
These are often used in politics to see how people perceive the performance of the government.
Examples of questions they can be asked are:

Is the Prime Minister effective?
Is the Health Service improving?
Are exams getting easier?

Large scale opinion polls often use both cluster and quota sampling.

Some methods of collecting data are
- personal interviews
- telephone surveys, including mobile phones
- postal surveys
- shoppers surveys

Exercise 1:5

Julia and her friends investigate the part-time jobs of the Sixth Form. To begin the survey, they list all the possible ways that they can choose their sample.

random sampling
systematic sampling
stratified sampling
quota sampling
cluster sampling

For questions 1–7 write down the sampling method that is used.

1 Julia has a list of all the students in the Sixth Form, arranged in alphabetical order. For her sample she chooses the first pupil and then every fifth pupil after that.

2 There are five tutor groups. They are numbered from 1 to 5. Anne rolls a dice to choose one tutor group. She questions every pupil in this tutor group.

3 Brian goes into the Sixth Form Common Room. He chooses ten girls and ten boys.

4 Derek has a list of all the students in the Sixth Form. He gives each student a number. He uses a table of random sampling numbers to choose a sample of size 20.

5 There are four school houses. Keith chooses one house at random. He asks every pupil in this house.

6 There are 55 pupils in Year 12 and 45 pupils in Year 13. Mia's sample has eleven Year 12 students and nine Year 13 students chosen at random.

7 There are 70 pupils who study Mathematics, and 30 pupils who study English. Martin's sample has 14 pupils who study Mathematics and 6 pupils who study English chosen at random.

8 Find out what 'the placebo effect' is.
It has something to do with control groups.

Example A bus company carries out a survey to find out how many times people use a bus in one week. They carry out a mobile phone survey.

a Explain why this survey is biased.

b What is the sampling frame that the bus company should be using.

c Suggest a better sampling method that the bus company could use.

a Mobile phone users tend to be younger so they do not reflect the age spread of the population.

b All the people who live in the area.

c Stratified sampling according to age.

Exercise 1:6

For each of questions 1–4.

a Write down why the survey is biased.
b What should the sampling frame be.
c Suggest a better sampling method.

1 A local travel agent carries out a survey into the type of holiday people in the area take. They go to the local airport to interview the passengers.

2 The manager of a car production line collects data on the number of faults per completed car. He looks at the last 20 cars produced on a Friday afternoon.

3 Tom is investigating how far pupils travel to school. He asks all the pupils on his bus on the way to school.

4 Polly investigates the leisure activities of people living in her village. She sends a questionnaire to all the members of a golf club who live in her village.

Control group You can use statistics to check if a new drug has any effect. A group of patients are chosen at random to form a sample. The sample is split into two groups at random. Both groups think that they are taking the new drug. Only the first group actually takes the drug. The second group just think that they are taking it. The second group is called a **control group**.
Both groups are treated identically. If more patients get better in the first group then the drug has an effect.
When only one group is compared to the control group they are called matched pairs. There may be more groups.

Control groups are used as part of an experiment.

The school nurse is part of a research team. He is asked to investigate the effect of adding vitamins to the daily diet on IQ.
He uses a **control group** as part of his research.

He uses a school with 500 pupils.
He tests and records each pupil's IQ.

Half of the pupils are chosen at random to form the control group.
The rest of the pupils form the experimental group.

The experimental group are given vitamin pills to take each day for a year.
The control group receive a placebo.
The placebo looks like the vitamin pill, but does not contain any vitamins.

The IQs of all 500 pupils are then tested and recorded at the end of the year.
The results of the two groups are then compared.

5 Julie is testing how effective two different fertilisers are. Explain how she can use 300 seedlings to test which, if any, is the more effective fertiliser.

Extraneous variables

Sometimes other things need to be considered.
Did the two groups of seedlings have:

- the same amount of sunshine,
- the same amount of liquid,
- the same room temperature . . .?

These are called **extraneous variables.**
They need to be kept the same for both the control and the experimental groups during the experiment.

6 The farmers' union is doing research into the effect of different fertilisers on wheat crops. The fertilisers are applied to whole fields of wheat.

a Explain how you would carry out the experiment.

b List the possible extraneous variables that need to be considered.

You can use sampling to estimate the size of a population.

Jack is a farmer. He has a small lake on his farm.
He wants to estimate the number of fish in the lake. He uses sampling to do this.

Jack takes a 1st sample of 40 fish. The fish are marked and returned to the lake.
Jack later takes a 2nd sample of **100** fish. In this sample, there are 25 marked fish.
It is assumed that this 2nd sample reflects the whole population of the lake.

Let the number of fish in the lake = n.
If there are n fish in the lake this means that 40 out of the n fish are marked.

The fraction of marked fish is $\frac{40}{n}$

This is assumed to be the same as the fraction of marked fish in the 2nd sample.
There are 25 marked fish out of **100** in the second sample.

The fraction of marked fish is $\frac{25}{100}$

So
$$\frac{40}{n} = \frac{25}{100}$$
$$40 \times 100 = 25 \times n$$
$$25n = 4000$$
$$n = 160$$

An estimate of the number of fish in the lake is 160.

This sampling method is called the **capture/recapture method.**
It can only be used to estimate the size of a population that is contained.

---

7 The local parish council wants to know how many fish are in the village pond. To find this out, 60 fish are caught, marked and returned. A month later, 180 fish are caught and only 15 of these are found to be marked. Estimate how many fish there are in the pond.

8 Peter has a pond in his garden. He thinks there are about 300 fish in it. He catches and marks 50 fish, which he returns to the pond. Later, he catches another 40 fish. Eight of these are marked. Estimate the number of fish in the pond.

3 Designing a questionnaire

Table 80 Working parents: hours worked (10% sample)

80. Women in couple families and lone parents in employment

	Age of youngest dependent child in family	TOTAL PERSONS	3 and under
a	b	c	d
Women in couple families in employment		**8,461**	**56**
No dependent child in family	–	4,826	14
1 or more dependent child(ren)	0–4	1,084	26
	5–10	1,247	8
	11–18	1,304	8
Male lone parents in employment		**161**	–
No dependent child in family	–	79	–
1 or more dependent child(ren)	0–4	14	–
	5–10	25	–
	11–18	43	–

This is part of the questionnaire used in the 10 Yearly National Census. It is used to obtain data about the people living in the UK.

Questionnaire

A **questionnaire** is a set of questions on a given topic. There are two ways that the questions can be used.

1 An interviewer asks questions and fills in a form.

Advantages The interviewer can make sure that the questions are correctly understood. Some errors in answers can be corrected immediately. If it is a house-to-house survey, the interviewer can return until the form is completed.

Disadvantages The interviewer must ask the question in a way which does not influence the person answering the question. This method is expensive and time-consuming.

2 People are given a form to fill in themselves.
This is called the 10 Yearly National Census.

Advantages A large group of people can be questioned in this way. Only a few people are required to distribute and collect the forms and so it is cheap. The person filling in the questionnaire has time to answer the questions.

Disadvantages If forms are posted, some of them may not be returned. This is called a non-response and should be followed up. Also, there is no one to ask for advice if a question is not clear. Errors cannot be detected immediately and some questions may be missed out. Errors cannot be detected immediately and some questions may be missed out.

When you write your own questionnaire try to obey these rules.

Rule 1

Questions must not be **biased**.
They should not make you think that a particular answer is right.
A question that does this is called a **leading question**.
Look at this question:

Normal people enjoy swimming. Do you enjoy swimming?

This question is biased. The first sentence should not be there.
It makes you think that you aren't normal if you don't enjoy swimming.

Exercise 1:7

1 Which of these questions do you think are biased? Write down their letters and explain what makes them biased.

- **a** How many hours of television did you watch last Sunday?
- **b** It is important to have a school coat with a designer label. What make is your school coat?
- **c** Most normal people enjoy pizza. Do you eat pizza?
- **d** It is very important to eat fruit. Do you eat fruit?
- **e** How long does it take you to travel to school each day?

Rule 2

Questions can give a choice of possible answers.
There are usually boxes to tick.

Example **1** Will you vote in the next General Election?

☐ Yes ☐ No ☐ Don't know

Sometimes you can use groups for the boxes.

The groups must not overlap.
There must be no gaps between the groups.
There should be at least three boxes.

Example **2** How old are you?

☐ 21–30 ☐ 31–40 ☐ 41–50 ☐ 51–60

You can also use a line that people are asked to mark.

Example **3** All children should learn to swim.

Strongly agree ——————————————— Strongly disagree

A point representing the opinion is marked somewhere on this line. The line is called an opinion scale.

Exercise 1:8

Here is a list of questions and statements. For each one, write down the letter of the style of answer you would use.

a ❑ Yes ❑ No ❑ Don't know

b ❑ Agree ❑ Disagree ❑ Don't know

c ❑ 0 ❑ 1 ❑ 2 ❑ 3 or more

d Strongly agree ——————————————— Strongly disagree

1 How many brothers and sisters do you have?

2 Mathematics is fun.

3 Is the UK a member of the European Union?

4 Everybody should be able to use a computer.

Rule 3

Questions should not upset people or embarrass them.

There are times when statisticians need to ask sensitive questions. An example of a sensitive question is "Do you pay the correct amount of tax?"

People may not give a truthful answer if they have not been paying the correct amount of tax.

To overcome this, the person answering the question uses a random method to choose one of two questions to answer. The questions could be:
Do you pay the correct amount of tax?
Do you pay too little tax?

The interviewer does not know which question the person has answered so cannot directly interpret the answer.
Probability theory is then used to estimate the proportion of people who pay the correct amount.

Exercise 1:9

Use this list for questions **1** and **2**.

a How heavy are you?

b How many children do you have?

c How old are you?

d How many books have you read this week?

e Do you enjoy sports?

f How much money do you earn?

1 Which of these questions do you think are upsetting or embarrassing? Write down their letters.

2 Write down what is wrong with each of the questions that you chose in **1**.

3 This is a part of John's questionnaire.

Toss a coin. If the coin shows a head tick the box marked Yes.
If the coin shows a tail answer this question: “Do you smoke?’

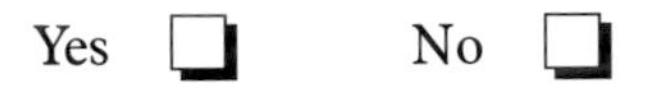

Out of 100 questionnaires, 69 have the Yes box ticked.
John expects 50 heads out of 100 tosses of the coin so $69 - 50 = 19$ people have answered Yes to the question.
John estimates that 19 out of the remaining 50 people smoke.
This is 38%.
This is another part of John's questionnaire.

Toss a coin. If the coin shows a head tick the box marked Yes.
If the coin shows a tail answer this question: “Have you ever shoplifted?’

Yes ❑ No ❑

This time 62 out of 100 tick the Yes box.
What percentage does John estmate to have shoplifted?

Rule 4 Questions should be clear, short and easily understood. If people do not know what you mean, you will not get the information that you need.

Rule 5 Questions should be relevant.
Don't ask questions that have nothing to do with your survey.

Rule 6 Don't ask questions that allow people to give many different answers. These are called open questions.
This makes it very difficult to draw diagrams to show your results.

Closed questions give a small number of possible answers.

An example of an open question is: How long did you watch TV last night?

The closed question is: How long did you watch TV last night?

❑ Less than 60 mins ❑ 60 mins to less than 120 mins ❑ 120 mins or more

Rule 7 Questions should be in a sensible order.
Don't jump from one idea to another and back again.

You now need to use your questionnaire. First you need to carry out a **pilot survey**.

You give the questionnaire to a small group.
This tests that the questions are clear and that you get the information you want.
Questions can then be changed before the questionnaire is used in the actual survey.

Always carry out a pilot survey when using your own questionnaire in a coursework task.

Not all the questionnaires may be returned. You need to follow up any **non-responses**.

Any non-returned questionnaires means the sample will be incomplete.
This could cause the sample to be biased.

It may be helpful to **check the information** obtained through a questionnaire.

This can be done in different ways. You could:

carry out a second survey using the same questionnaire and a different sample.
The results will not be identical but should be similar

carry out a post survey with the same questionnaire but a smaller sample.

1

a Write down **one** way of obtaining primary data for a statistical analysis. *(1)*

b What is the statistical term given to data obtained from published statistics or known databases? *(1)*

c Give **one** disadvantage of using published statistics. *(1)*

SEG, 1996, Paper 2

2

a What is the main difference between obtaining information by census and by sampling? *(1)*

b Why is it **necessary** to use sampling in a survey to find the lifetime of electric bulbs? *(1)*

c Why is it **usual** to use sampling for an opinion poll? *(1)*

NEAB, 1997, Paper 1

3

In a street with 30 houses, it was decided to ask a sample of five residents if they approve of the Council's proposed introduction of speed restriction humps in the road.

The map shows their location

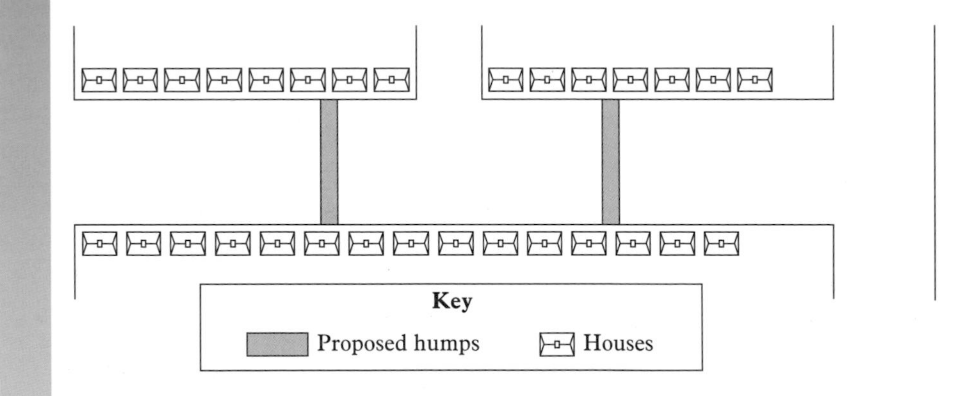

a Explain how you would choose a random sample of five for this enquiry. *(2)*

It was suggested that a systematically chosen sample could be taken.

b Explain how you would select such a sample. *(2)*

c Which of these two sampling methods would you choose?
Give a reason for your answer. *(2)*

SEG, 1996, Paper 2

4 A market research company is conducting a survey to find out whether most people had a holiday in Britain, elsewhere in Europe or in the rest of the world, last year. It also wants to know if they stayed in self-catering accommodation, hotels or went camping.

Design **two** questions that could be used in a questionnaire to find out all this information. *(4)*

NEAB, 1998, Paper 1

5 **a** Amanda wants to choose a sample of 500 adults from the town where she lives. She considers these methods of choosing her sample.

Method 1: Choose people shopping in the town centre on Saturday mornings.

Method 2: Choose names at random from the electoral register.

Method 3: Choose people living in the streets near her house.

Which method is **most** likely to produce an unbiased sample? Give a reason for your answer. *(2)*

b How would you choose a random sample of 20 children from a school with 100 pupils? *(2)*

NEAB, 1997, Paper 1

6 One of the questions from a questionnaire used in a survey is shown.

"Are you tall?"	Please tick	Yes	☐
		No	☐

Out of 100 people who took part in this survey 20 ticked 'Yes' and 30 ticked 'No'.
There was a non-response of 50.

a What proportion of those that answered this question ticked 'Yes'? *(1)*

b Give **two** reasons why the 'non-response' category cannot be ignored. *(2)*

c Rewrite the question in a more suitable form. *(2)*

SEG, 1998, Paper 4

7 Give a reason why questions A and B should be reworded before being included in a questionnaire. Rewrite each one showing exactly how you would present it in a questionnaire.

Question A: How old are you? *(3)*

Question B: The new supermarket seems to be a great success.
Do you agree? *(3)*

NEAB, 1997, Paper 2

8 The table below gives the scores, out of 100, of 20 schoolchildren in a mathematics test. The sex (M – male, F – female) and age (years) of each child are also given.

Child	Sex	Age	Score	Child	Sex	Age	Score
01	M	9	41	11	M	15	62
02	M	11	43	12	F	7	44
03	M	16	57	13	F	13	54
04	M	12	49	14	M	17	59
05	M	17	66	15	M	6	40
06	F	8	36	16	M	13	50
07	M	14	68	17	M	5	35
08	M	14	61	18	M	12	56
09	F	10	47	19	F	18	60
10	M	10	42	20	M	11	50

For this population of schoolchildren the mean score of all 20 children is 51.

Random numbers (01–20 range)				
14	07	11	17	18
13	03	14	06	18
19	08	10	07	01

a Using the random number table above, and reading from left to right, draw samples from the population of schoolchildren as follows.

i Use Line 1 of the random numbers above to select a random sample of size 4. Calculate the sample mean score. *(2)*

ii Use Line 2 of the random numbers above to select a random sample of size 4 which reflects the population sex structure. Calculate the sample mean score. *(3)*

iii Given that there are 5 children in the school population aged 5–9 years, 10 aged 10–14 years and 5 aged 15–19 years, use Line 3 of the random numbers above to select a random sample of size 4 which reflects this age structure. Calculate the sample mean score. *(3)*

b Which of the three sampling methods used in part **a** do you think is the best? Give a reason for your choice. *(2)*

NEAB, 1998, Paper 3

9 Medford is a new town, with several large housing estates occupied by young families. There is also a large amount of sheltered accommodation for the elderly.
Medford Borough Council intends to build a new leisure centre for its residents.
In order to provide the facilities that the people want, the council decides to conduct a survey by giving out a questionnaire.
The first 100 people arriving at Medford railway station are used as the sample.

a Give **one** reason why this is likely to be a biased sample. *(1)*

b How would you choose a more representative sample for this survey? *(1)*

c The following are suggested questions for the questionnaire. Comment, with a reason, on the suitability of each question.

i How old are you? *(1)*

ii Would you use a leisure centre? *(1)*

iii What is your favourite sport? *(1)*

d Write down **two** further questions as they should appear on the questionnaire. *(2)*

SEG, 1996, Paper 3

10 The number of workers on the three floors of a factory are shown.

A different type of work is done on each floor.

The owner wants to ask 50 workers what they think of their working conditions.

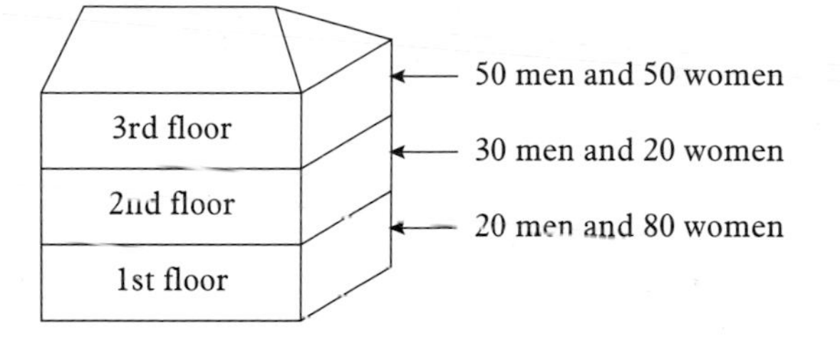

a Give **one** advantage of taking a sample of the workforce to obtain this information. *(1)*

b Explain why a **random** sample is considered to be an unsuitable way of selecting this sample. *(1)*

It was decided to obtain a **stratified** sample according to the number of workers on each floor.

c Calculate the number of workers that should be questioned from each floor. *(3)*

d How many women should be included from the 2nd floor to make the sample a fair representation of that floor? *(2)*

e Explain how you would finally make a systematic selection of the women who would represent the 2nd floor. *(2)*

SEG, 1997, Paper 1

11 In a school of 800 students it was decided to undertake a sample survey to find their television viewing patterns.

The following two methods of selecting the sample were proposed:

A: to ask for volunteers

B: to use the alphabetical list of students in the school, taking every fifth name on the list.

a Give **one** advantage and **one** disadvantage to each method of selection.

Method A

Advantage

Disadvantage

Method B

Advantage

Disadvantage *(4)*

Random Number Table		
29764	45692	65555
86193	46243	44779
58142	79482	02718
82390	87177	64325

b Using the random number table above, explain briefly

i how you would select a simple random sample of 10 students from the school. *(4)*

ii how you would select a systematic sample of size 10 from the school. *(3)*

c Give **one** reason why it would be preferable to introduce stratification into the sampling procedure. *(1)*

Part of the survey will investigate students' views on the standard of television news programmes.

d Describe **two** opinion scales that could be used. *(4)*

Data There are two types of data, **quantitative** and **qualitative**.
Quantitative data is **numerical**, e.g. weight, age, length.
Qualitative data is **non-numerical**, e.g. colour, country, sex.

Primary data is obtained by asking questions, taking measurements or counting.
Secondary data is obtained from published statistics or databases.

Population and samples The complete set of people or objects that information is collected about is called the **population**.
Information is normally taken from a small part of the population called a **sample**.
The sample must be large enough to represent the population, but small enough to be manageable.
Samples can be chosen by different methods:

1. Random sampling
2. Systematic sampling
3. Convenience sampling
4. Stratified sampling
5. Quota sampling
6. Cluster sampling

The aim is to choose the sample without **bias,** so that the sample will represent the whole population.

Control group A **control group** is used to check the reliability of the outcomes of an experiment.

The size of a contained population can be estimated using the **capture/recapture** method.

Questionnaire A questionnaire is a set of questions on a given topic. Either an interviewer asks the questions and fills in the answers or people fill in forms themselves.

Remember:

1. Questions should not be biased.
2. Questions can give a choice of possible answers.
3. Questions should not upset people or embarrass them.
4. Questions should be clear, short and easily understood.
5. Questions should be relevant.
6. Don't ask questions that allow people to give many different answers (open questions).
7. Qu estions should be in a sensible order.

Working with *Excel*

Generating random numbers for sampling using Microsoft Excel

These are the exam marks of 400 students.

77	61	66	69	60	68	80	76	63	78	74	61	72	68	59	67	55	52	72	64
75	86	62	87	67	66	71	67	75	70	65	61	66	58	71	61	74	76	65	57
80	66	78	68	72	78	70	68	81	78	74	74	67	69	74	65	70	68	73	62
76	59	67	69	69	72	76	64	65	58	61	79	68	70	60	85	62	60	52	67
64	89	83	68	85	65	70	75	72	70	72	75	69	69	66	69	70	66	63	69
58	72	69	71	58	71	84	55	66	64	76	61	63	70	61	63	64	80	66	68
74	73	72	69	80	80	69	65	69	69	78	77	76	68	67	64	55	60	78	78
77	66	78	70	71	72	74	68	66	65	83	59	70	71	71	70	79	72	73	73
57	73	70	79	54	71	71	68	82	75	67	62	72	68	66	75	84	79	84	62
61	72	71	75	83	66	62	60	58	70	63	64	77	60	69	68	71	73	70	77
67	74	69	65	68	81	67	73	78	57	75	53	70	52	80	76	63	63	73	79
70	56	63	69	64	70	63	59	64	78	69	75	58	64	70	63	72	76	59	80
67	69	82	71	88	80	65	78	71	58	80	62	76	78	81	77	60	77	68	70
67	69	71	59	66	67	75	62	75	63	74	70	53	70	76	74	75	55	67	65
76	74	62	62	78	76	71	71	73	69	67	70	65	72	83	67	69	54	70	71
70	77	86	73	64	66	69	77	72	61	74	73	65	73	61	81	63	78	77	64
71	70	67	69	60	77	82	69	75	78	73	61	55	69	63	73	72	75	71	61
79	71	76	54	80	58	64	76	67	82	55	62	63	58	70	60	74	79	75	72
66	76	68	85	61	70	64	69	69	65	87	67	71	81	75	76	69	70	64	58
76	69	70	69	64	67	69	69	76	73	46	67	77	80	73	73	58	72	72	70

Use *Excel* to select a **random sample of five marks** from the population of 400 students.

Open a new *Excel* document

Click Start. Click Programs. Click Microsoft Excel.

Click the New icon on the Toolbar.

Labelling columns

In Cell A1, type
Random Nos. between 0 and 1
Press the Enter key.
In Cell B1, type
Random Nos. between 0 and 400
Press the Enter key.
In Cell C1, type Sampling Nos
Press the Enter key.

	A	B	C	D
1	Random N	Random N	Sampling Nos.	
2				
3				
4				
5				
6				
7				
8				

Widening the columns

	A	B	C	D
1	Random Nos. between 0 and 1	Random Nos. between 0 and 400	Sampling Nos.	
2				
3				
4				
5				
6				
7				
8				

The column labels may overlap.

If so, place the cursor ✥ on the grey bar showing the cell letters, A, B, C and so on.

Move the cursor ✥ to the vertical line between A and B.

The cursor will change to ↔

Double-click. *Column A is now wide enough.*
Repeat this for Columns B and C.

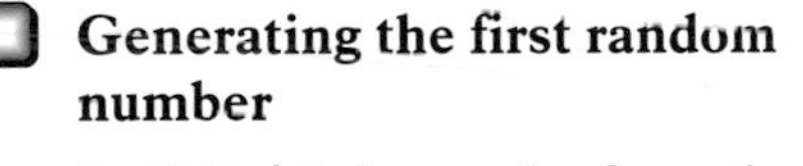

Generating the first random number

	A	B	C	D
1	Random Nos. between 0 and 1	Random Nos. between 0 and 400	Sampling Nos.	
2	0.35562092			
3				
4				
5				
6				
7				
8				

In Cell A2, insert the formula.
Type =**RAND()**

Press the **Enter** key.

The cell now displays a random number between 0 and 1.

Generating four more random numbers

	A	B	C	D
1	Random Nos. between 0 and 1	Random Nos. between 0 and 400	Sampling Nos.	
2	0.971674764			
3	0.764383063			
4	0.217391663			
5	0.214845462			
6	0.208610157			
7				
8				

Click on Cell A2 to select it.

Place the cursor ✥ over the bottom right-hand corner of cell A2.

The cursor will change to +

Hold down the left mouse button and drag over the next four cells A3, A4, A5 and A6.

When you do this, you will notice the first random number changes. This happens every time you change the spreadsheet. It is not important.

This copies the random number formula into each cell.
Each cell now displays a random number.

These random numbers are between 0 and 1.
*The sample has **400** students, so the sampling numbers must lie between 0 and **400**.*
Each random number is multiplied by **400**.

Generating the first random number between 0 and 400

In Cell B2, insert the formula:
=A2*400

Press the **Enter** key.

	A	B	C	D
1	Random Nos. between 0 and 1	Random Nos. between 0 and 400	Sampling Nos.	
2	0.684390529	273.7562114		
3	0.610707442			
4	0.969111358			
5	0.85397379			
6	0.087917335			
7				
8				

The cell now displays a random number between 0 and 400.

Generating four more random numbers between 0 and 400

Select Cell B2.

Place the cursor over the bottom right-hand corner of cell B2.

	A	B	C	D
1	Random Nos. between 0 and 1	Random Nos. between 0 and 400	Sampling Nos.	
2	0.611918718	244.7674873		
3	0.259657144	103.8628577		
4	0.304139966	121.6559864		
5	0.169932599	67.9730394		
6	0.119376906	47.75076231		
7				
8				

Hold down the left mouse button and drag over the next four cells B3, B4, B5 and B6.

This copies the multiplication formula into each cell. Each cell now contains random numbers between 0 and 400.
They do not include 0 or 400.

You need to make sure that the number 400 can also be included since there are 400 students.
You use the ROUNDUP function to do this.

Generating the sampling numbers

In Cell C2, insert the formula:
=ROUNDUP(B2,0)

	A	B	C	D
1	Random Nos. between 0 and 1	Random Nos. between 0 and 400	Sampling Nos.	
2	0.670082176	268.0328705	269	
3	0.52688287	210.7531479	211	
4	0.727989811	291.1959244	292	
5	0.799261418	319.704567	320	
6	0.757708864	303.0835457	304	
7				
8				

*The **B2** identifies the number to be rounded.*
*The **0** is the number of decimal places to round to.*

Press the **Enter** key.

Select Cell C2.

Place the cursor over the bottom right-hand corner of Cell C2.

Hold down the mouse button and drag over the next four cells C3, C4, C5 and C6.

This copies the formula into each cell.
Each cell now displays the sampling number.

Checking the sampling numbers
The third column contains random numbers from 1 to 400 inclusive.
If a number occurs more than once in your sample press the **F9** key.
This will generate a new set of random numbers.
If these are not all different, press the **F9** key again and so on.

Using the sampling numbers to give the random sample of five marks
The sampling numbers are 269, 211, 292, 320 and 304.

Use the table of exam marks to find the 269th mark, the 211th mark, the 292nd mark and so on.
For the 269th number, you need to count along the rows until you get to the 269th mark.
For the 211th number, you need to count along the rows until you get to the 211th mark.
Count along the first row, then the second row and so on.

The marks that you get are 75, 75, 70, 64 and 73.

Finding the median
Put the marks in order of size: 64, 70, 73, 75 and 75.
The median is the middle number. The median is 73.

Exercise 1:10

1 **a** Generate a new sample of size 5.
Write down the numbers and find the median.
b Generate four more samples of size 5 and write down the median of each sample.

2 **a** Generate a new sample of size 11.
Write down the numbers and find the median.
b Generate four more samples of size 11 and write down the median of each sample.

1 For each of the following statements, write down if the data given is quantitative or qualitative.

a Julia weighs 50 kg.
b Hamish is Scottish.
c The colour of my kitchen is grey.
d My Grandad is 82 years old.

2 Write down whether each of these is an example of primary or secondary data.

a Rajiv measured the lengths of the arms of the pupils in his class.
b Eileen looked in a book for the population of London.

3 Each of these samples does not fairly represent the population. It is a biased sample. Say what is wrong with the way each sample is selected.

a The PE department wants to find out if the pupils enjoy PE lessons. They ask the members of the netball and football teams.
b A car manufacturer is keen to find out how popular a new model is. Interviewers stand outside their showrooms and ask the customers for their views. This information is used to decide if the car is popular or not.

4 Mandy works at a leisure club. She is carrying out a survey into the number of hours the members spend at the club. Mandy is aware that different age groups spend different amounts of time. She chooses a stratified sample of 50 people to reflect the proportion of members in each age group. The table shows the membership of the club by age.

Age in years	18–25	26–40	41–60	61+
Number of members	220	130	50	100

Calculate the number of people she should include from each age group.

5 James wants to know how many fish there are in the village lake. He catches 80 fish, marks them and returns them to the lake. A week later, he catches 150 fish, of which 15 are found to be marked. Estimate the number of fish that are in the lake.

6 For each of these, write down which sampling method is used.

a There are 300 names on a list. For the sample, the seventh name is chosen and then every tenth name after that.
b Louise interviews 15 men aged 20–40, 15 men aged 40–60, 15 women aged 20–40 and 15 women aged 40–60. All are chosen at random.

2 Dealing with data

1 Presenting data
- Pictograms and bar-charts
- Multiple bar-charts
- Composite bar-charts
- Discrete and continuous data
- Open-ended classes
- Choropleth maps
- Level of accuracy

2 Line graphs and pie-charts
- Drawing and interpreting line graphs
- Frequency polygons
- Pie-charts
- Comparative pie-charts

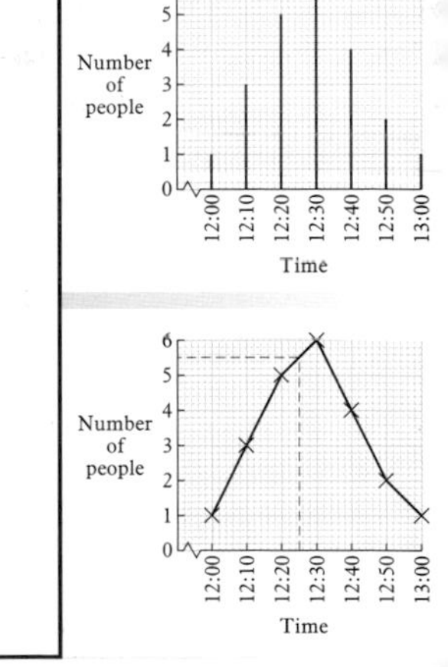

EXAM QUESTIONS

SUMMARY

ICT IN STATISTICS

TEST YOURSELF

1 Presenting data

Ian Hatem is the Headteacher of Adeney School. He is preparing a report for parents on the Key Stage 3 results in Maths, English and Science.

He has to report separately on boys' and girls' results. He has to decide which diagrams are best to make the results clear to the parents.

Here are some of the diagrams he has produced so far.

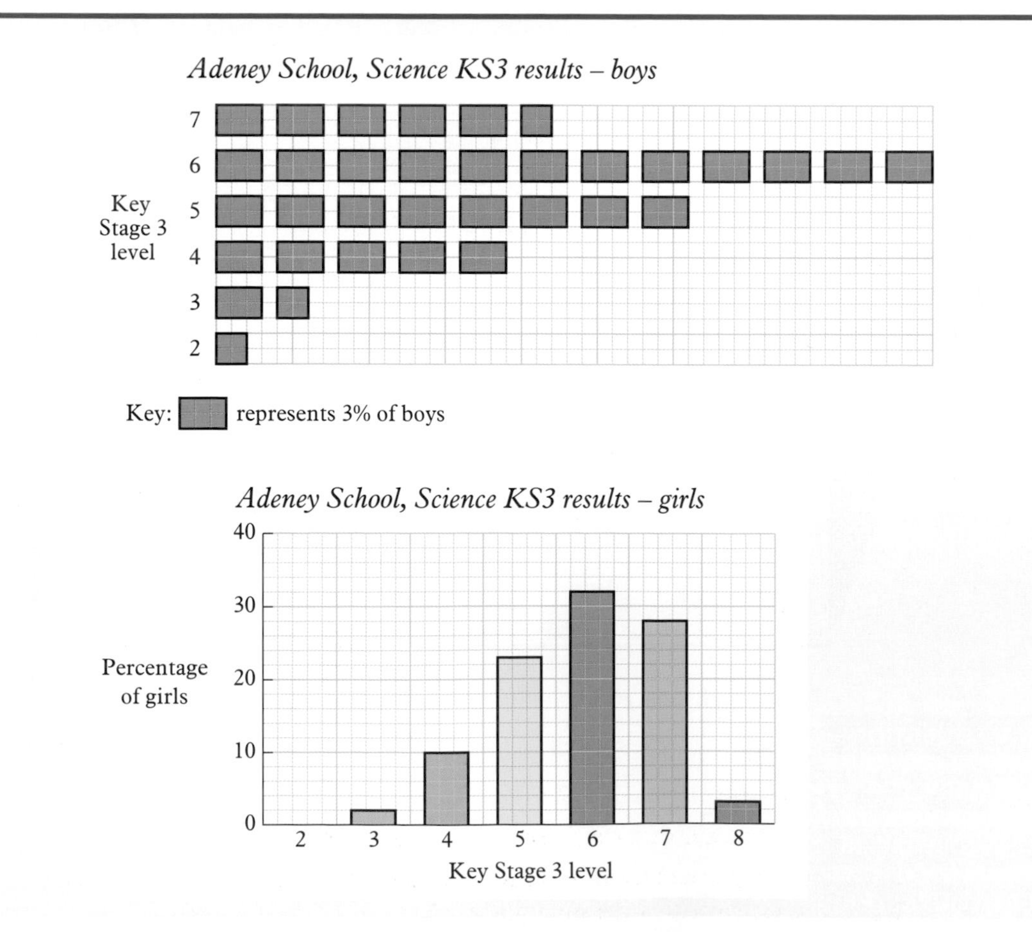

The first diagram is a pictogram.
A pictogram uses pictures to show the data. It should always have a key.
The second diagram is a bar-chart. The bars in a bar-chart can be vertical or horizontal. The bars must all be the same width. The gaps between the bars must all be the same size. The length or height of the bar represents the frequency. Both axes must be clearly labelled.

Exercise 2:1

1 Copy the table
Use the pictogram to complete the table.

Level	2	3	4	5	6	7	8
Percentage of boys							

2 Copy the table.
Use the bar-chart to complete the table.

Level	2	3	4	5	6	7	8
Percentage of girls							

3 Kate is investigating the number of cars owned by each household in her street.
The pictogram shows her results.

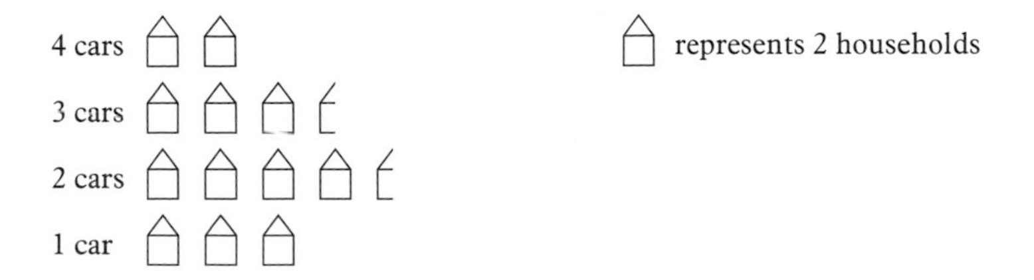

a How many households own only 1 car?

b How many households own 2 cars?

c How many households are there in Kate's street?

d Work out the percentage of households that own 3 cars.
Write your answer to 1 decimal place.

4 Sian asked pupils in Year 10 to choose their favourite sport.
These are her results.

Sport	Football	Hockey	Tennis	Rounders	Swimming	Rugby
Number of pupils	62	22	38	18	37	26

a Write down the most popular sport.
b How many pupils in Year 10 took part in the survey?
c Draw a bar-chart of the results of Sian's survey.

5 Ian asks every pupil in his tutor group their favourite car colour.
These are his results.

Colour	Red	Purple	Black	Yellow	Green	Silver	Blue
Number of pupils	3	5	2	1	3	6	4

a Write down the least popular car colour.
b Draw a bar chart of the results of his survey.
c Two pupils were absent on the day of the survey. Work out the percentage of pupils who took part in Ian's survey. Give your answer to the nearest whole number.

A multiple bar-chart is useful to compare two or more items.
This diagram shows the output of a toy factory over three years.

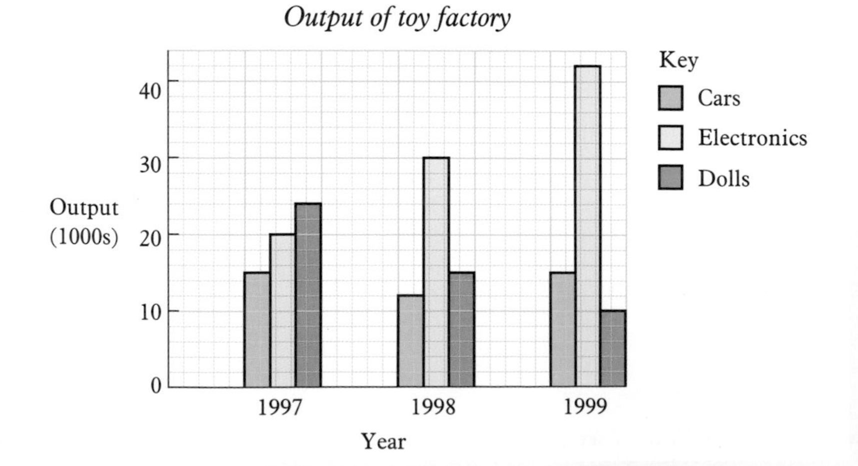

You can see the changes in production over the three years.
It is easy to see that the production of electronic toys has increased rapidly over the three years. There has been little change in the sale of cars.

6 Copy this multiple bar-chart.
Use the data you found in questions **1** and **2** to complete the chart.

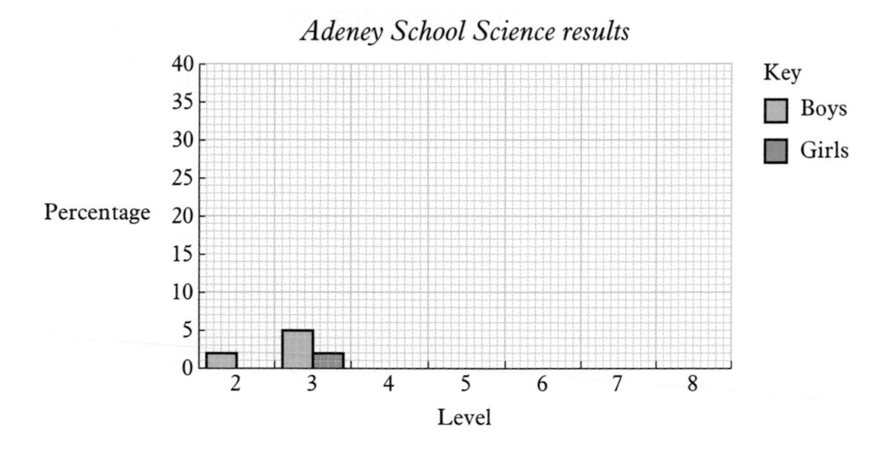

7 The table shows the percentage of their income that the Davies family spent on food, travel and entertainment over a three-year period.

Year	1997	1998	1999
Food	46%	51%	56%
Travel	31%	42%	38%
Entertainment	23%	7%	6%

Copy and complete the bar-chart to show the data.

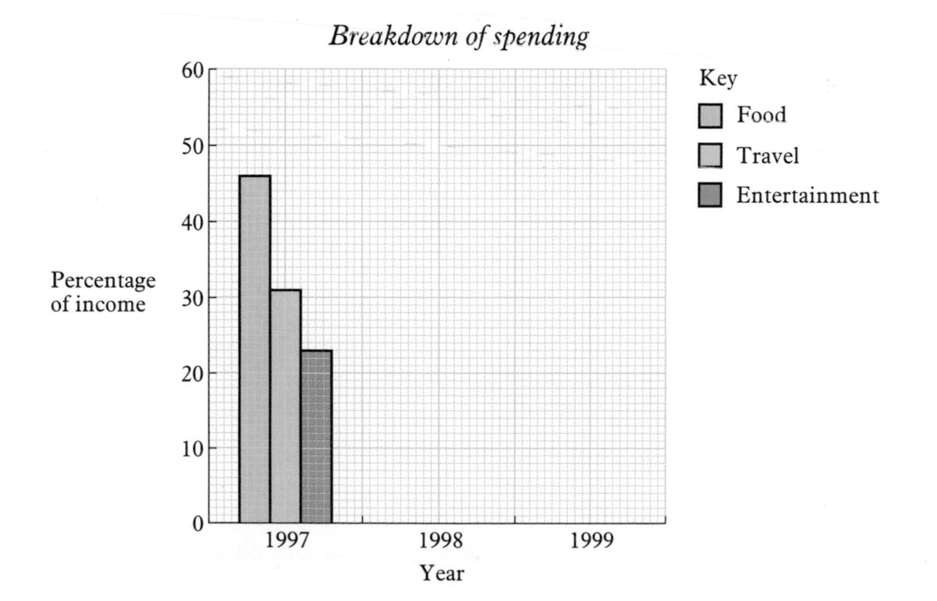

Composite bar chart

A **composite bar chart** shows how the whole is split into parts.
The length of the bar represents 100%.

Liam is drawing a chart to show the data on the expenditure of the Davies family.

He draws the bar for 1997.
The length is 5 cm.

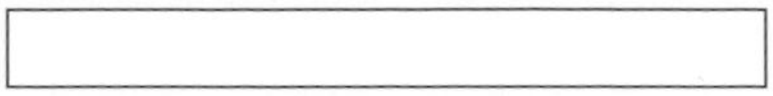

He marks off 46% for food.
46% of 5 cm = 2.3 cm.

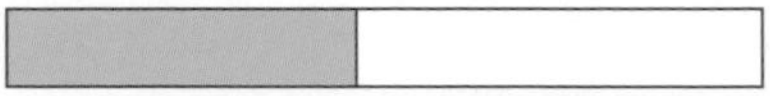

He then marks off 31% for travel.
31% of 5 cm = 1.55 cm.
This leaves 1.15 cm for entertainment.

He now draws two more bars for the other two years.

All the bars are the same length.
Each bar represents 100% of the income for that year.
Liam draws a key to show what each section represents.

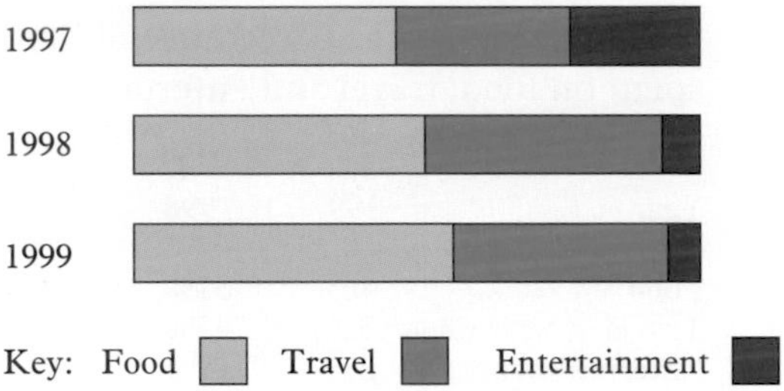

1997

1998

1999

Key: Food Travel Entertainment

8 The table gives the percentage output of a factory over three years.

Year	1997	1998	1999
Dishwashers	21%	30%	38%
Washing machines	65%	63%	55%
Tumble driers	14%	7%	7%

a The factory produced a total of 50 000 machines in 1999.
How many of these were washing machines?

b The factory produced 6132 tumble driers in 1998.
Find the total number of machines that the factory produced in 1998.

c Draw a percentage composite bar-chart to show the data.

d Does the chart show that the output of the factory changed each year? Explain your answer.

9 A survey was completed to find the ages of people using the local swimming baths. The diagram shows the results.

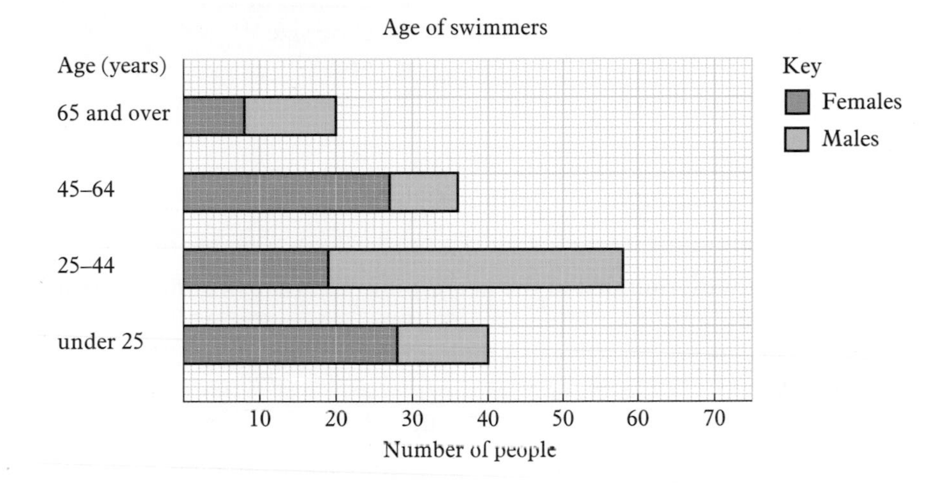

a How many people were involved in the survey?
b How many of the under 25s are female?
c What percentage of the under 25s are female?
d What is the ratio of males to females for the 45–64 age group?

10 These are the results of a survey to find the ages of people using the local library. Draw a composite bar-chart to show the data.

Age (years)	under 25	25–44	45–64	65 and over
Male	25	61	48	39
Female	30	96	71	85

There are two main types of data.

Discrete data

When data can only take certain individual values it is called **discrete data**.

The number of people is discrete data. You can't have 4.3 people!

Continuous data

When data can have any number value in a certain range it is called **continuous data**.

The length of a garden, the mass of a baby or the temperature of a swimming pool are all examples of continuous data.

Exercise 2:2

1 Write down if each of these types of data is discrete or continuous.

a temperature
b marks in an exam
c mass
d number of books
e number of cars
f shoe size
g area
h time

2 **a** Give two examples of discrete data.
b Give two examples of continuous data.

These are the times, rounded to the nearest minute, that swimmers spent in the swimming pool.

26	48	52	23	21	59	29	31	26	33
23	33	37	44	29	26	21	45	31	60
52	51	39	54	37	39	40	52	39	48

A tally-table showing every time separately would take too long.
It would also be difficult to draw a bar-chart with so many bars.
The data needs to be put into groups. The groups *must not* overlap.

Time (min)	Tally	Frequency
21–30	卌 ||||	9
31–40	卌 卌	10
41–50	||||	4
51–60	卌 ||	7

Once you group data you no longer know the size of each item of data, only the group that it is in. You have lost some of the detail of the data.
For example you know there are 10 items of data in the group 31–40. You do not know what they are.

When you use grouped data the bars of the bar-chart must touch. The width of each bar must still be the same.

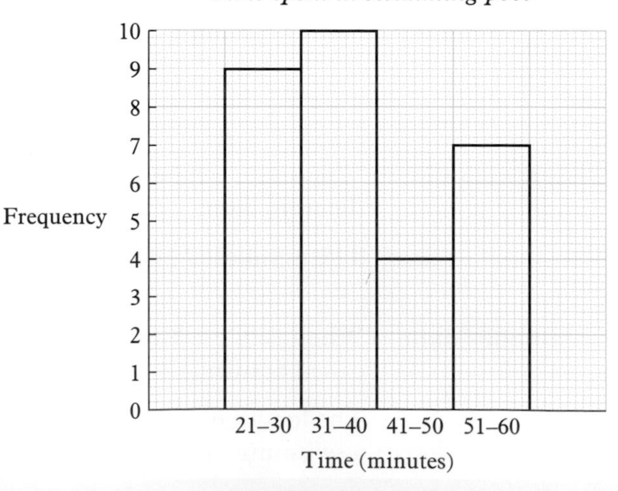

3 These are the heights, rounded to the nearest centimetre, of a test group of 50 plants.

56	82	70	69	72	37	28	96	52	88
41	42	50	40	51	56	48	79	29	30
66	90	99	49	77	66	61	64	97	84
72	43	73	76	76	22	46	49	48	53
98	45	87	88	27	48	80	73	54	79

a Copy and complete this tally-table for the data.

Height (cm)	Tally	Frequency
21–30		
31–40		
.		
.		
91–100		

b Complete a second tally-table for the data using groups 21–40, 41–60 …
c Complete a third tally-table for the data using groups 21–60, 61–100.
d Draw bar-charts for your three tables.
e Compare how well each bar-chart represents the data.

You group data so that there are not too many bars in the bar-chart.

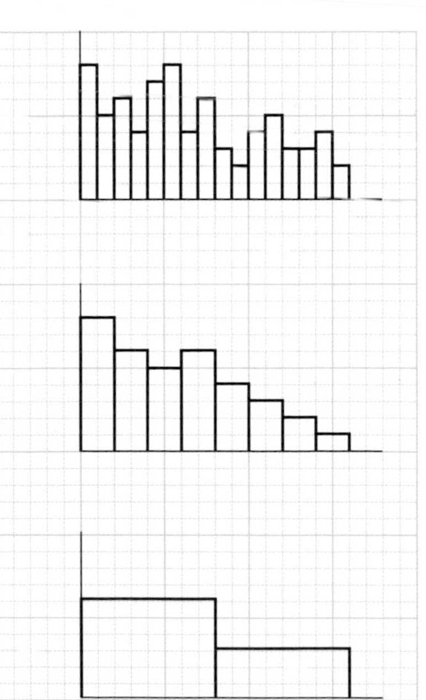

This chart has too many bars.
It is not easy to see the overall shape of the data.

This chart has about the correct number of bars.
You can clearly see the shape of the data in this chart.

There are too few bars in this chart.
The shape of the data is lost.

4 **a** Lucy asks the girls in her year group how many marks they each achieved on a Maths test.
She is going to use these groups for her data:

1–10	10–20	20–30
30–40	40–50	above 50.

Explain what is wrong with her groups.

b Jack asks the boys the same question.
These are the groups that he uses:

$1 \leqslant \text{marks} < 10$	$11 \leqslant \text{marks} \leqslant 20$	$21 \leqslant \text{marks} \leqslant 30$
$31 \leqslant \text{marks} \leqslant 40$	$41 \leqslant \text{marks} \leqslant 50$	> 51

He has made two mistakes.
Write down what they are.

Open-ended classes

Sometimes the last group does not have an upper limit.
It is left **open-ended**.
It is called an **open-ended class**.

You use open-ended classes when there are very few items at the end of the data.
These items would be spread over several classes if the same width was used.
Some of these classes could even be empty.

---

5 Lucy's class had three maths tests.
These are their total scores.

23	42	92	59	39	66	23	46	50	51
38	75	43	57	93	33	24	42	28	29
150	74	98	97	84	73	67	42	47	74
56	45	55	43	36	32	138	69	70	68
8	36	27	59	85	42	58	31	73	62

a Complete a tally-table for the data
(1) using equal groups
$1 \leqslant \text{marks} \leqslant 20$, $21 \leqslant \text{marks} \leqslant 40$, $41 \leqslant \text{marks} \leqslant 60$, …,
$141 \leqslant \text{marks} \leqslant 160$
(2) using groups

$1 \leqslant \text{marks} \leqslant 20$	$21 \leqslant \text{marks} \leqslant 40$	$41 \leqslant \text{marks} \leqslant 60$
$61 \leqslant \text{marks} \leqslant 80$	$81 \leqslant \text{marks} \leqslant 100$	$\text{marks} < 60$

b Which is the most suitable of these tally-tables to represent the data?

When you measure continuous data you have to decide on the level of accuracy.

Liam is measuring the lengths of earthworms.
He finds their lengths to the nearest centimetre.
He cannot be more accurate than this.
This means that a worm he measures as 6 cm could actually be any length in the range 5.5 cm to 6.499999… cm as these lengths would be rounded to 6 cm.

You don't use 6.499999… cm, it's too complicated. You use 6.5 instead.
You say that 6 cm can be any length in the range 5.5 cm to 6.5 cm.

Exercise 2:3

1 Tom weighs a tomato. The mass is 47 g to the nearest gram.
What range can the actual mass of the tomato lie in?

2 Pam measures a length of time as 48 minutes to the nearest minute.
What range can the actual time lie in?

3 A factory must produce packets containing 454 g to the nearest gram.
These are the masses of some packets. Which are acceptable?

a 454.3 g **c** 454.09 g **e** 454.501 g
b 453.3 g **d** 453.9 g **f** 453.999 g

4 For each of these numbers write down the range in which they can lie:

a 6 to the nearest whole number
b 8.3 to 1 decimal place
c 40 to the nearest 10
d 17 to the nearest whole number

When you collect information by measuring you need to decide on the level of accuracy.

The length of a road would be measured to the nearest metre.
The height of a plant would be measured to the nearest centimetre.
The time to solve a number puzzle would be measured in minutes.
The time to solve a mental arithmetic question would be measured in seconds.

Choropleth maps

Choropleth maps have areas of different shading. They are widely used in Geography.

A key gives the values of the variable for each shading.
The shading gets darker as the value of the variable increases.

This choropleth map shows the distribution of the elderly population in the town of Baxton.

The lightest green shading shows that the elderly are 13%–16% of the population in that area.

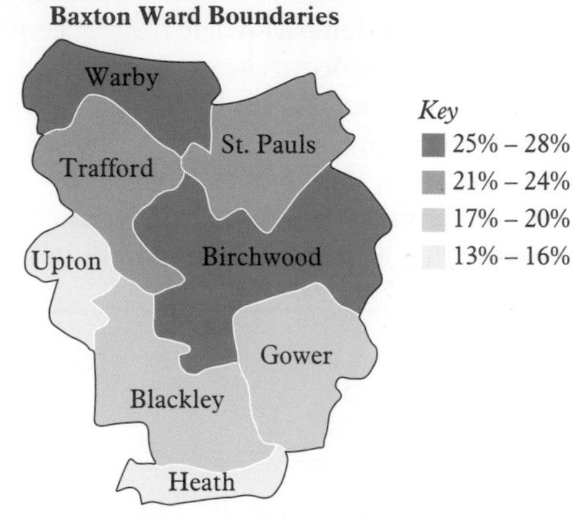

Exercise 2:4

Use this choropleth map to answer these questions.

1 In which area has the population increased?

2 Which area has the highest population change?

3 Write down the areas which have a population change of −4.3 to −5.2

4 Write down the most common population change.

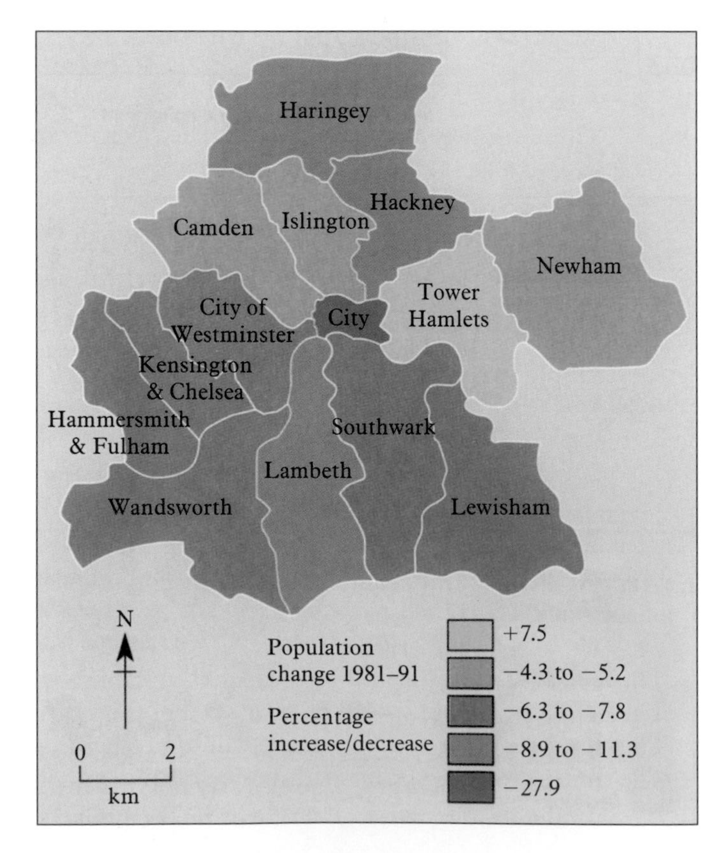

2 Line graphs and pie-charts

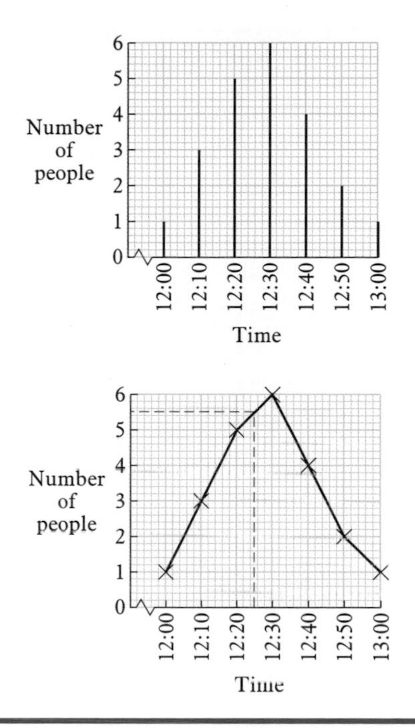

Peter is recording the number of people in a queue in a supermarket between 12:00 and 13:00.
He wants to draw a graph to show his results. He draws these two graphs.

This graph on the top is called a **vertical line graph**. This is like a bar-chart where the bars are just lines. You can draw graphs like this for discrete data.
So Peter can draw this type of graph to show the number of people in the queue.

The graph on the bottom is called a **line graph**. You join the points together with lines. You can only draw this type of graph for continuous data. The number of people is discrete data, so Peter cannot draw this type of graph.
This graph shows that there were 5.5 people in the queue at 12:25!

You can only draw a line graph for continuous data.
If you have discrete data you can draw a vertical line graph.

Exercise 2:5

1 Write down whether you would draw a line graph or a vertical line graph to show each of these:

a the number of CDs sold in a shop each day
b the temperature in a room overnight
c the profit made by a company each year for 10 years
d the number of people on a bus during a two-hour period.

2 This graph shows the results of a survey on the number of cars per household in a street.

a How many households don't have a car?

b How many households were surveyed?

c How many households had three or more cars?

d Why can't you join the points and draw a line graph for this data?

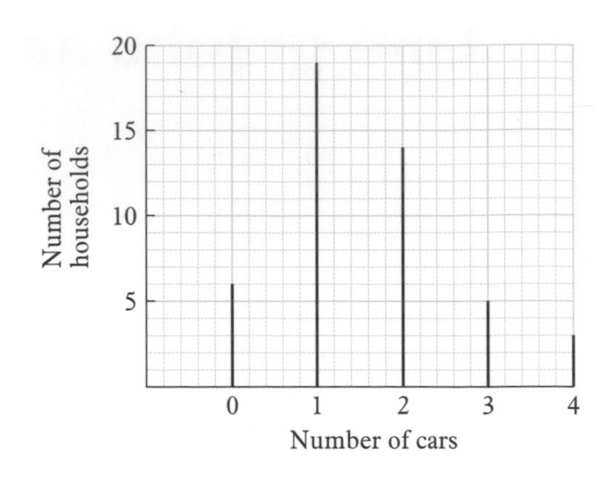

3 Here is a line graph showing the temperature in Liverpool over a 24-hour period. The temperature is recorded every hour.

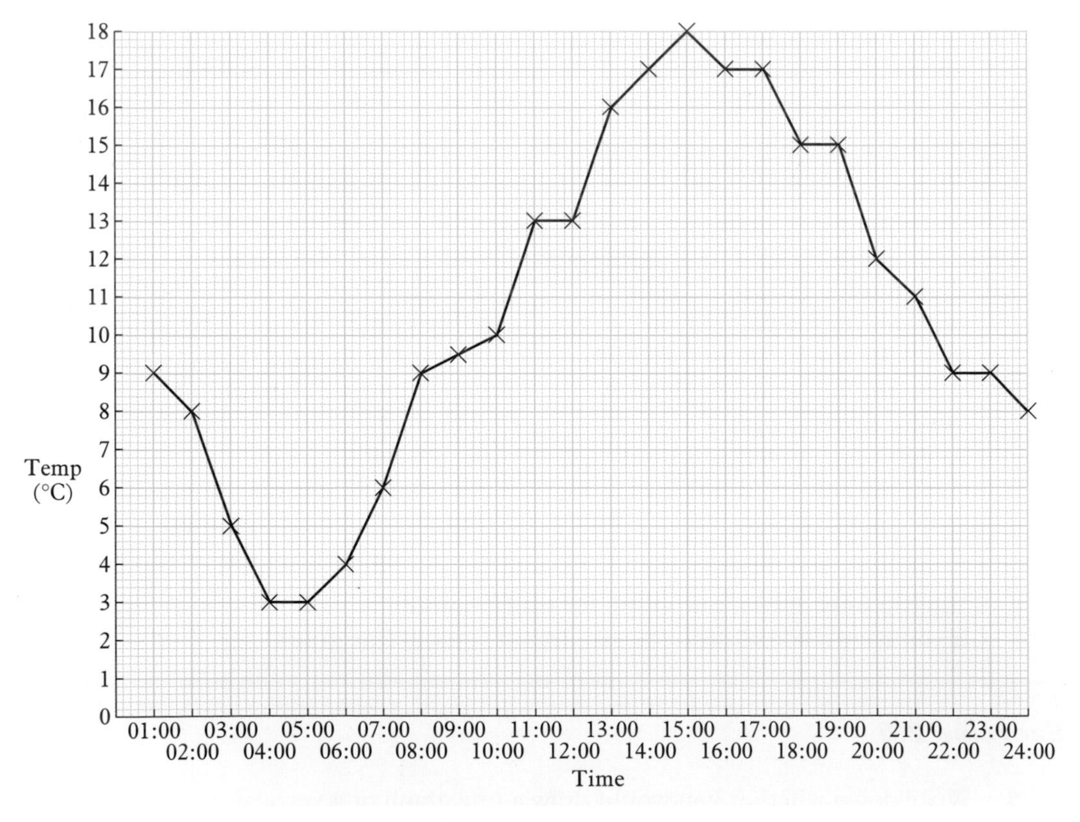

a What is the temperature at 2 p.m.?

b Between what times is the temperature falling?

c At what times is the temperature 14 °C?

d Use the graph to estimate the temperature at 12:30. Why is this answer only an estimate?

4 A garage records the number of bikes it sells over a week.
Here are the results.

Day	Number sold	Day	Number sold
Mon	6	Thurs	5
Tues	3	Fri	7
Wed	8	Sat	17

a Draw a graph to show the sales.
b What percentage of the bikes were sold on Saturday?

5 Mia uses a water-bath in the laboratory to keep a sample in a flask at a constant temperature. The heating is electric.
On Monday there was a power cut.
Mia records the temperature every hour in the flask to see what is happening to the sample.
These are her results.
12:00 34 °C means the temperature was 34 °C at time 12:00

12:00 34 °C	13:00 26 °C	14:00 20 °C	15:00 18 °C	16:00 17 °C
17:00 16 °C	18:00 14 °C	19:00 25 °C	20:00 34 °C	21:00 34 °C

a Draw a line graph to show her results.
b Estimate the time the electricity came back on.
Explain how you got your estimate.

6 The gardeners at Ness Gardens have measured the diameter of the same tree every 10 years for the last 50 years.
These are their results.

Age (years)	10	20	30	40	50
Diameter (cm)	11.5	19	28	46	68

a Draw a line graph to show this data.
b Use your graph to estimate the diameter of the tree when it was 35 years old. Why is this just an estimate?
c Estimate when the tree had a diameter of 25 cm.

Frequency polygon

Frequency polygons are often used to compare two sets of data. You join the mid-points of the groups with straight lines.

This table shows the scores of Year 11 pupils in a Geography test.

Score	1 to 10	11 to 20	21 to 30	31 to 40	41 to 50
Number of pupils	4	8	16	20	12

To draw a frequency polygon

1 First work out the mid-point of each group in the table.
The mid-point of the first group is $(1 + 10) \div 2$, the second is $(11 + 20) \div 2$ and so on.
Add a row, showing the mid-points, to the table.

Score	1 to 10	11 to 20	21 to 30	31 to 40	41 to 50
Number of pupils	4	8	16	20	12
Mid-point	5.5	15.5	25.5	35.5	45.5

2 The scale along the bottom of a frequency polygon must be like a graph scale.
It must not have labels like a bar-chart.
In this example you need a scale from 0 to 50.

3 The vertical axis shows the frequency or number of pupils.

4 Plot the mid-points and join them with straight lines.

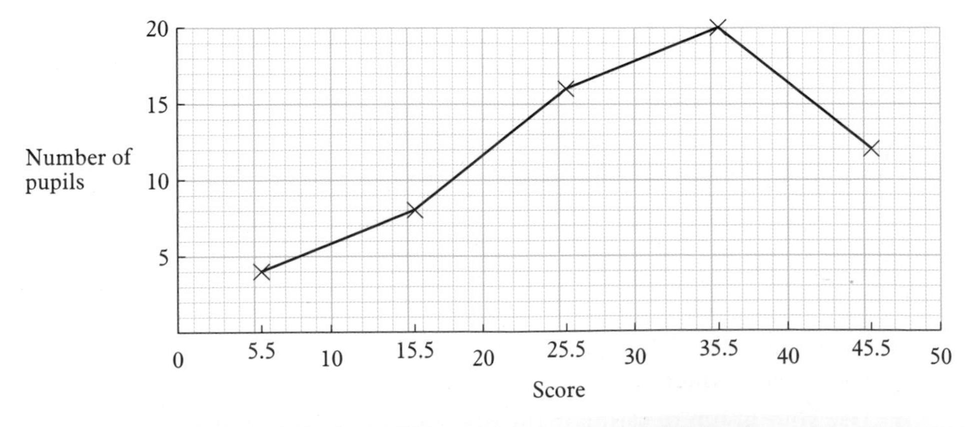

5 You can label the vertical axis 'Frequency' instead of 'Number of pupils'.

Exercise 2:6

1 Ron has asked his year group how many CDs they each have.
The table shows his results.

Number of CDs	1–20	21–40	41–60	61–80	81–100	101–120
Number of pupils	3	7	31	45	38	16

a Copy the table. Add an extra line and fill in the mid-points.
b Draw the frequency polygon of the results.

2 Jenny runs a taxi service. She asked each of her drivers how many clients they had last week. These are her results.

Number of clients	26–30	31–35	36–40	41–45	46–50	51–55	56–60
Number of drivers	7	15	22	33	18	21	11

a Copy the table. Add an extra line and fill in the mid-points.
b Draw the frequency polygon of the results.

3 The table shows the number of words per page in a sample of library books.

Number of words	1–100	101–200	201–300	301–400	401–500	501–600
Frequency	5	29	68	70	42	31

a Copy the table. Add an extra line and fill in the mid-points.
b Draw the frequency polygon for the data.

4 The table shows the number of times each year an equal number of men and women visit a local health club.

Number of visits	51–70	71–90	91–110	111–130	131–150	151–170	171–190
Men	8	13	21	38	40	29	17
Women	3	11	38	49	48	13	4

a Copy the table. Add an extra line and fill in the mid-points.
b Draw two frequency polygons to show the data.
Use the same set of axes and label each frequency polygon.
c Compare the number of visits of men and women using your frequency polygons.

Pie-chart

A **pie-chart** shows how something is divided up.
The area of the sector represents the number of items.
For single pie-charts you just use the angle of the sector.
Pie-charts are not useful for reading off accurate figures.

Example

This table shows the percentage of students gaining each level in the statistics exam.

Level	A	B	C	D	E	F	G
% of students	8	16	25	22	15	8	6

Show these results in a pie-chart.

1 Divide up the 360°.
The total is 100% so 1% = 360 ÷ 100 = 3.6°

2 Work out the angle for each level.
You will need to round the angles to the nearest degree.

Level	%	Working	Angle
A	8	$8 \times 3.6 = 28.8$	29
B	16	$16 \times 3.6 = 57.6$	58
C	25	$25 \times 3.6 = 90$	90
D	22	$22 \times 3.6 = 79.2$	79
E	15	$15 \times 3.6 = 54$	54
F	8	$8 \times 3.6 = 28.8$	29
G	6	$6 \times 3.6 = 21.6$	22

3 Check that the angles add up to 360°.
When the angles have been rounded, they may not add up to 360°.
If this is the case, add or take 1° from the biggest angle.
It will never be noticed!

4 Draw and label the pie-chart.
Always add a key.

Statistics results

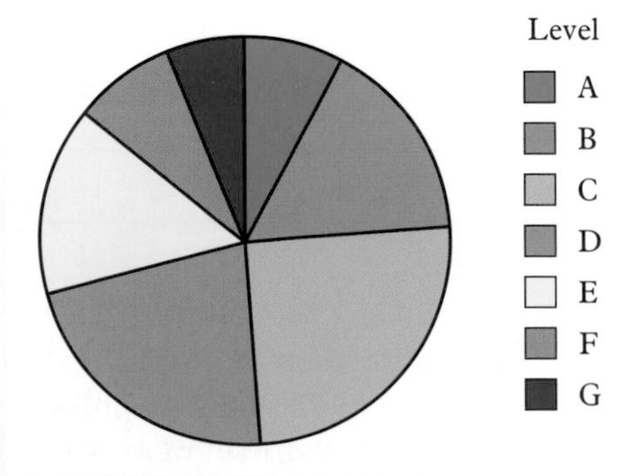

Exercise 2:7

1 The table shows the costs of a company.

a What are the total costs?
b Find the angle for £1 million.
c Draw a pie-chart to show the data.

Costs	£ million
Wages	25
Raw materials	76
Building costs	14
Energy	5

2 Here are the ingredients by weight in a pack of oatmeal biscuits:

Flour 42% Oatmeal 18% Fat 30% Sugar 10%

Draw a pie-chart to show this data.

3 The pie-chart shows the finances of a company.
a What fraction is taxes?
b What percentage is profit?
c How much was spent on costs if the total finances are £65 million?

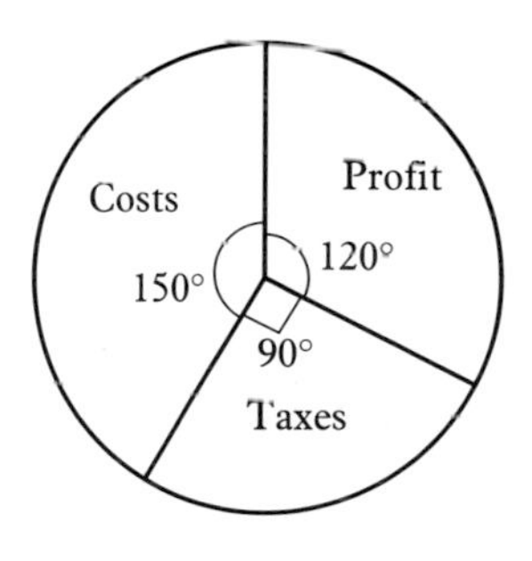

4 The volume of petrol sold by five service stations is given in the table.

Service station	Cadly	Denhall	Maxwell	Souls	Western
Sales (1000s of litres)	143	86	31	99	50

Draw a pie-chart to show this data.

5 The membership of a book club is made up of men, women and children.
The total membership is 2400.
Jacky is drawing a pie-chart to show the membership.
a She uses an angle of 150° to represent the men.
How many men are there?
b There are 800 women. What angle should Jacky use for the women?
c Draw a pie-chart to show the data.

Comparative pie-charts

You use **comparative pie-charts** to compare two sets of data. The areas of the circles must be in proportion to the two total frequencies.

The population of a village in 1920 and 1990 are given in the table.

	1920	1990
Men	257	962
Women	556	1008
Children	226	519
Total	1039	2489

Philip draws a pie-chart to show the 1920 data. He uses a radius of 2 cm.
The area of his circle is $\pi \times 2^2 = 12.57\text{ cm}^2$ to 4 sf.
This means Philip used an area of 12.57 cm^2 to represent 1039 people.
So he used $12.57 \div 1039\text{ cm}^2 = 0.012\text{ cm}^2$ to 4 sf for one person.
For 2489 people he needs to use an area of $0.012 \times 2489 = 29.87\text{ cm}^2$.
So for his pie-chart for 1990 he needs to use a radius r where $\pi r^2 = 29.87$

$$r^2 = 29.87 \div \pi$$
$$r = 3.1\text{ cm}$$

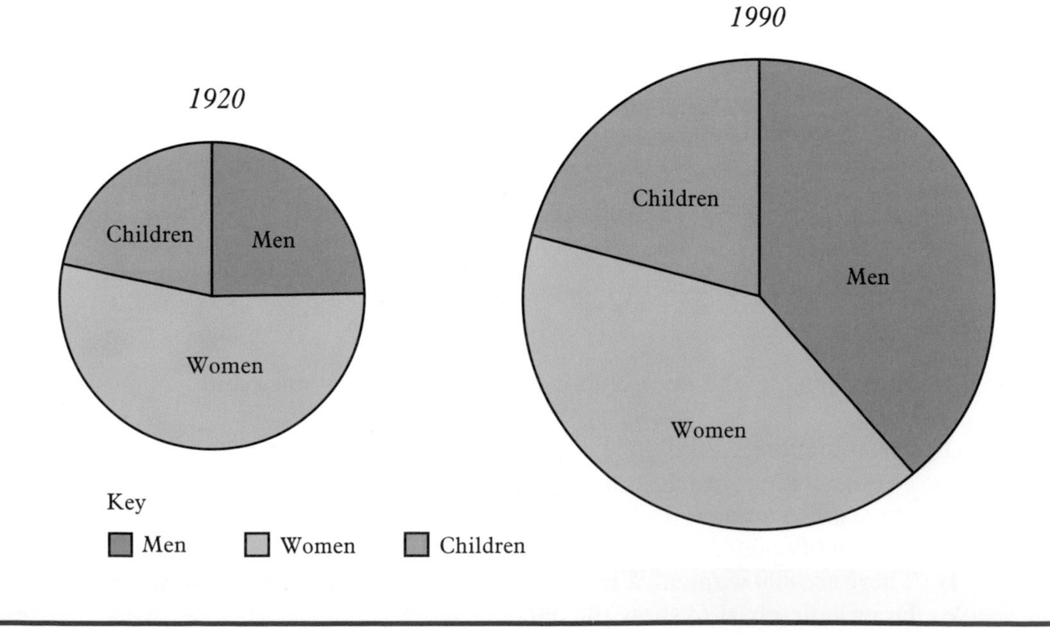

Exercise 2:8

1 Danny draws two pie-charts to compare how he spends his income.
Last year he was at college and his income was £3000.
He uses a radius of 3 cm for this pie-chart.
This year Danny earns £9800.
What radius should he use for this pie-chart?

2 The sales of a shop were £1.2 million in 1985.
In 1990 the sales were £6.1 million.
Sheila draws two pie-charts to compare the breakdown of sales.
She uses a radius of 5 cm for the 1985 sales.
What radius should she use for the 1990 sales?

3 The table shows how two families spend their income.

Jones family

Food	£45
Energy	£18
Clothing	£12
Entertainment	£9

Williams family

Food	£59
Energy	£28
Clothing	£39
Entertainment	£19

Draw comparative pie-charts, one for each family, to show this data.

4 These are the total costs of four fitted kitchens:

£3560 £6700 £9340 £12 600

Joshua is going to draw four pie-charts to compare the breakdown of the charges for each kitchen.
He uses a radius of 2 cm for the cheapest kitchen.
What radii should he use for each of the other kitchens?

5 Louise has drawn two comparative pie-charts.
One has a radius of 4 cm and represents 96 000 people.
The second has a radius of 2 cm.
How many people does the second represent?

1 **a** The pictogram shows the number of 2, 3 and 4 bedroomed houses built on a large housing estate.

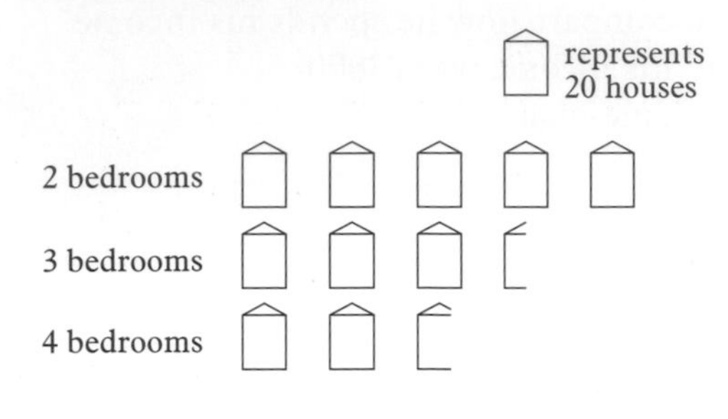

i How many 2 bedroomed houses were built? *(1)*
ii How many 3 bedroomed houses were built? *(1)*
iii There are also 90 retirement houses on the estate.
How many symbols would be needed to represent these houses on the diagram? *(2)*

b Another pictogram shows the type of aircraft used by an airline company. There are 475 aircraft represented by 19 symbols. How many aircraft does 1 symbol represent? *(2)*

NEAB, 1997, Paper 1

2 A survey was conducted to find the ages of people using the local bus service. The results are shown on the diagram.

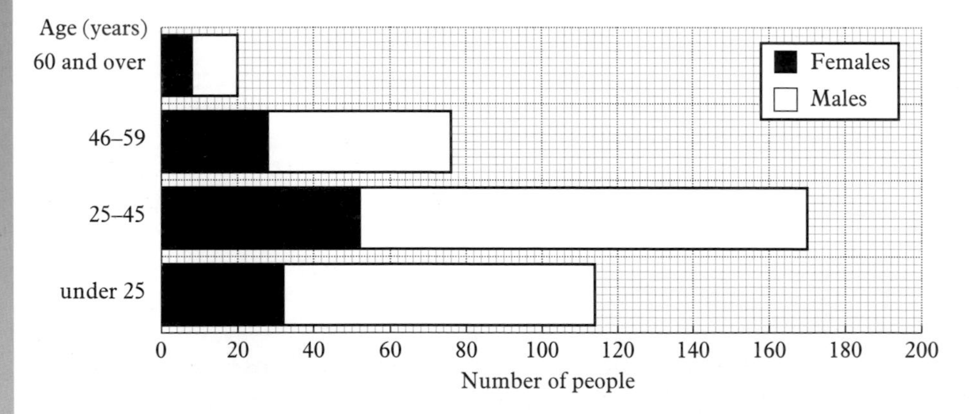

a How many of the 114 people under 25 years old were female? *(1)*
b What is the ratio of female to male for the 60 and over age group? *(1)*
c What percentage of the 46–59 age group were female? *(2)*

A marking error on the collection sheet meant that all the recorded ages should have been increased by one.

d What is the minimum number that could now be in the 46–59 age group? *(1)*
e What is the maximum number that could now be in the 25–45 age group? *(2)*

SEG, 1998, Paper 1

3 The diagram below shows how a large company spent its advertising budget in 1996.

Advertising Budget for 1996

Newspapers 55% | Television | Direct Mail 10% | Others 8%

0 20 40 60 80 100

a What percentage of the advertising budget was spent on television advertising? *(2)*

b The total advertising budget was £400 000. How much was spent on newspaper advertising? *(2)*

c The company intends to spend the same total amount on advertising in 1997.

The table shows the percentage to be spent on various forms of advertising.

Type of advertising	Percentage
Newspapers	45
Television	38
Direct mail	5
Others	12

Copy and complete the multiple bar-chart below. *(5)*

Advertising Budget for 1996 and 1997

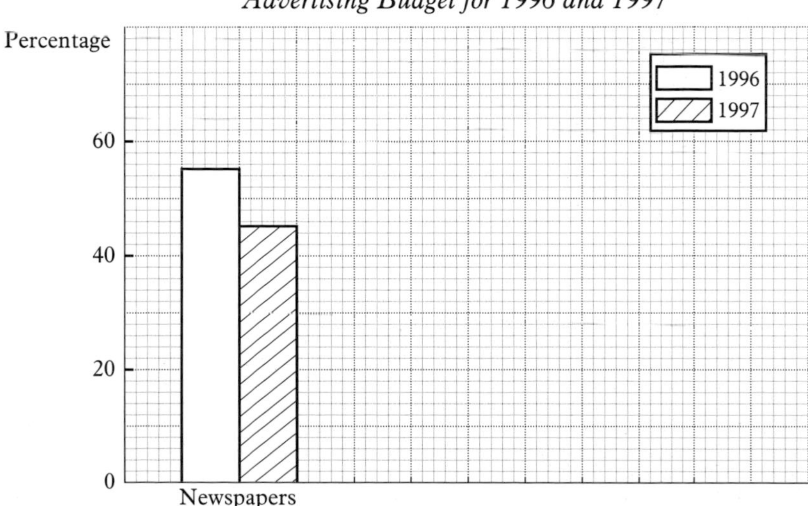

NEAB, 1997, Paper 1

4 The diagram shows the amount of revenue received by television companies, in billions of pounds.

a In which year was the revenue for the BBC the same as the total revenue for cable, satellite and digital TV? *(1)*

b Give an approximate value for the revenue of ITV/C4/C5 for 1980. *(2)*

c Describe the changes in the BBC's revenue in the period from 1980 to 1990. *(2)*

TV revenue

£ bn

0, 1, 2, 3

1980, 85, 90, 95, 2000

Total cable satellite and digital TV

ITV/C4/C5

BBC TV

Forecast

Source BBC estimates

The bar-chart shows the number of TVs per 100 people, in 1992, for ten countries.

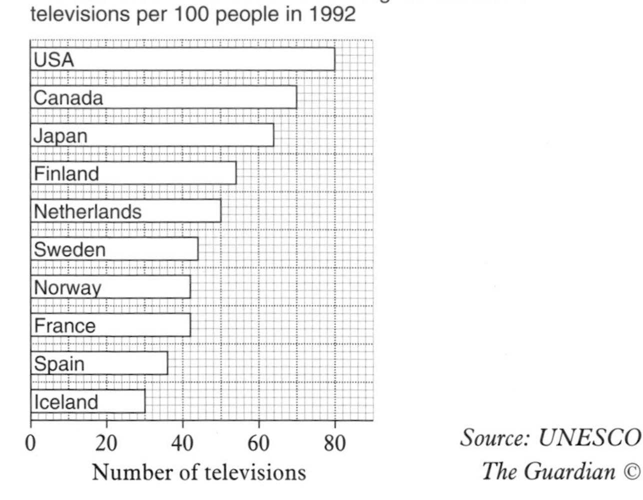

Source: UNESCO
The Guardian ©

d Copy the table below and write the names of each of the ten countries in the correct group. *(2)*

Number of televisions per 100 people (x)	Countries
$20 < x \leq 40$	
$40 < x \leq 60$	
$60 < x \leq 80$	
$80 < x \leq 100$	

e What information is lost by presenting the data in this table? *(1)*

NEAB, 1998, Paper 1

5 **a** The money spent on chilled foods in Britain, in 1995, is shown on the diagram below.

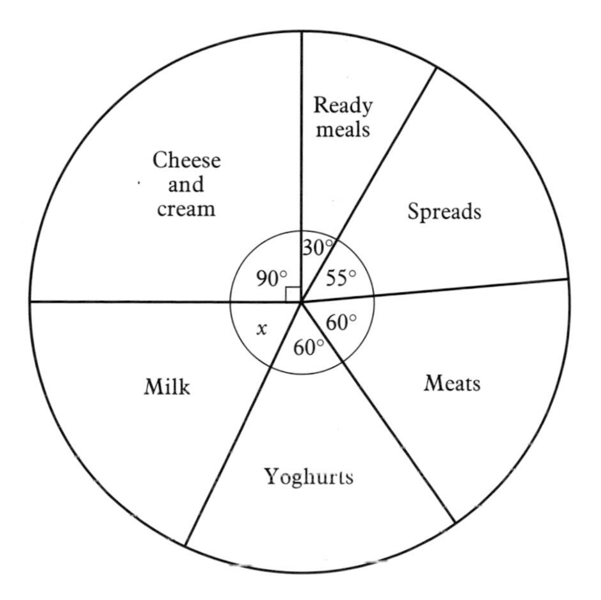

i Calculate the size of angle x.
You must show all your working. *(2)*

ii Sales of cheese and cream totalled £270 million.
How much was spent on ready meals? *(2)*

b The sales of some types of cooked meats, for the same year, are shown in the table below.

Cooked meats	Sales in millions of pounds	Angle on pie-chart
Turkey	48	
Chicken	22	
Beef	8	
Luncheon meat	12	
Total	90	

The information is to be shown on a pie-chart.
Find the missing angles and copy and complete the table.
Do NOT draw the pie-chart. *(3)*

NEAB, 1998, Paper 1

6 The number of students at a university and the numbers studying different subjects are given in the table.

Subjects	Number of students
Natural sciences	1800
Arts	2700
Social sciences	2500
Engineering	1000
Total	8000

A pie-chart of radius 5 cm is to be drawn to illustrate these data.

a Draw and label the pie-chart. *(4)*

The following comparative pie-chart represents the number of students studying the same subjects at another university.
The radius of this pie-chart is 4.6 cm.

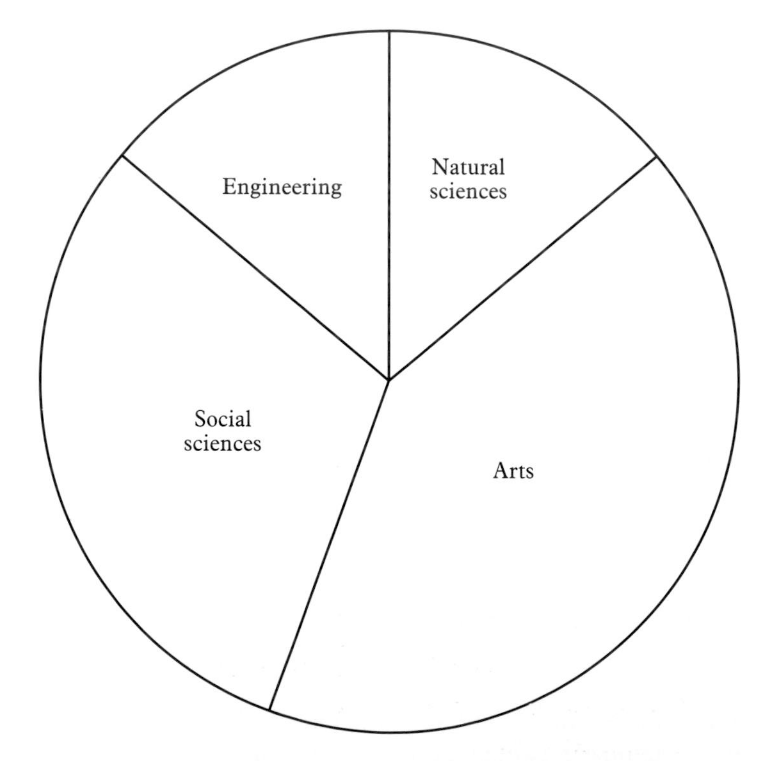

b Calculate the total number of students studying these subjects at this university. *(3)*
One difference shown by these two pie-charts is the total number of students studying these subjects.

c State another difference that is shown by these two pie-charts. *(1)*

SEG, 1998, Paper 3

7 An expert on porcelain is keen to establish values of imports and exports of porcelain to the UK in the years 1960 and 1998.

The pie-charts drawn below represent these values, in millions of pounds, for the years 1960 and 1998.

The radii of the circles are 3 cm and 4 cm, respectively.

In 1960 the total value of exports and imports was 9 million pounds.

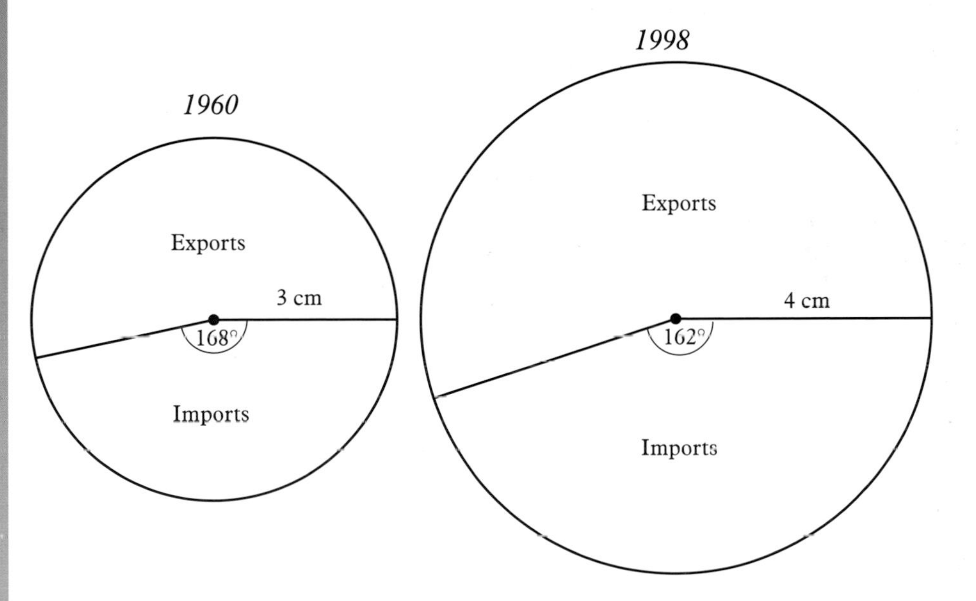

Calculate, giving your answers in millions of pounds,

i the value of exports in 1960. *(2)*

ii the total value of exports and imports in 1998. *(3)*

NEAB, 1999, Tier H (p +)

8 A doctor records information on her patients.
The variables she uses are described below.

a State whether each variable is qualitative, discrete or continuous.

i The colour of the patient's eyes *(1)*

ii the patient's weight *(1)*

iii the patient's shoe size *(1)*

iv the patient's blood group *(1)*

b Which of the above variables could be represented on a grouped frequency distribution diagram? *(1)*

NEAB, 1997, Paper 2

9 The diagram shows the percentage of teachers by age in some European countries.

a What percentage of teachers from Austria are under 30 years of age? *(1)*

b What percentage of teachers from Sweden are over 50 years of age? *(1)*

c What percentage of teachers from Germany are between 30 and 50 years of age? *(2)*

d Which of the following countries, Portugal, Austria, Ireland and UK has the greatest percengage of teachers between 30 and 50 years of age? *(2)*

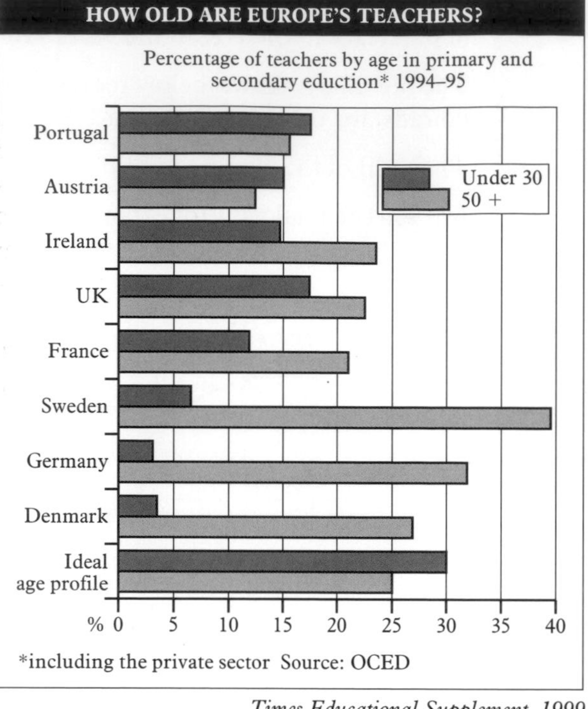

Times Educational Supplement, 1999

10 The diagram shows the income of a Sports Club in 1999.

a Use the diagram to estimate the income from

i rents,

ii subscriptions. *(3)*

b There were 92 full playing members who each paid a subscription of £57. What percentage of the subscriptions came from full playing members? Give your answer to 1 decimal place. *(3)*

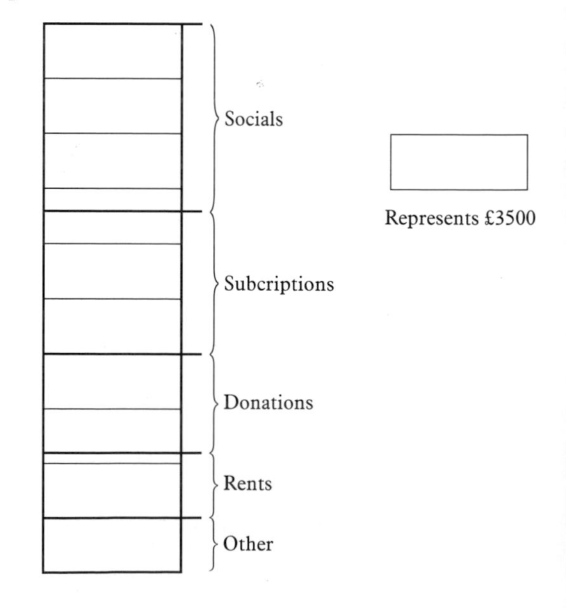

Discrete data When data can only take certain individual values it is called **discrete data**.

Continuous data When data can have any number value in a certain range it is called **continuous data**.
When you measure continuous data you have to decide on the level of accuracy.

Example Liam is measuring the length of earthworms.
He finds their lengths to the nearest centimetre.
He cannot be more accurate than this.
This means that a worm he measures as 6 cm could actually be any length in the range 5.5 cm to 6.49999… cm as these lengths would be rounded to 6 cm.
You don't use 6.4999… it's too complicated. You use 6.5 instead.
You say that 6 cm can be any length in the range 5.5 cm to 6.5 cm.

Frequency polygon **Frequency polygons** are often used to compare two sets of data. You join the mid-points of the groups with straight lines.

To draw a frequency polygon:

1. First work out the mid-point of each group in the table. Do this by adding a row to the table.
2. The scale along the bottom of a frequency polygon must be like a graph scale. It must not have labels like a bar-chart.
3. The vertical axis shows the frequency.
4. Plot the mid-points and join them with straight lines.

Pie-chart A **pie-chart** shows how something is divided up.
The area of the sector represents the number of items.
For single pie-charts you just use the angle of the sector.
Pie-charts are not useful for reading off accurate figures.

Comparative pie-charts You use **comparative pie-charts** to compare two sets of data. The areas of the circles must be in proportion to the size of the data. You need to work out the new radius.

Working with *Excel*

Drawing a bar-chart

These are the results of a survey into types of computers owned by 40 students.

Type	PC	PlayStation	Nintendo	Other
Number of students	18	5	15	3

Use *Excel* to draw a bar-chart to show this data.

Open a new *Excel* document
Click Start. Click Programs. Click Microsoft Excel.

Click the New icon on the Toolbar.

Enter the information on computer types
In Cell A1, type Type.
In Cell B1, type Number.
In Cell A2, type PC, in Cell A3, type PlayStation and so on.
In Cell B2, type 18, in Cell B3, type 5 and so on.

Select all the cells you have used
Place the cursor on Cell A1.
Hold down the left mouse button, and move the cursor to Cell B5.
Release the mouse button.
All the cells you are using are now selected.

Use the Chart Wizard button

Click the Chart Wizard icon on the Toolbar.

Drawing a bar-chart
From Chart type, choose Column.
From Chart sub-type, choose the top left diagram.
Your screen should look like the one shown.
Select the Press and hold to view sample button.
It allows you to check that your chart is turning out as intended.

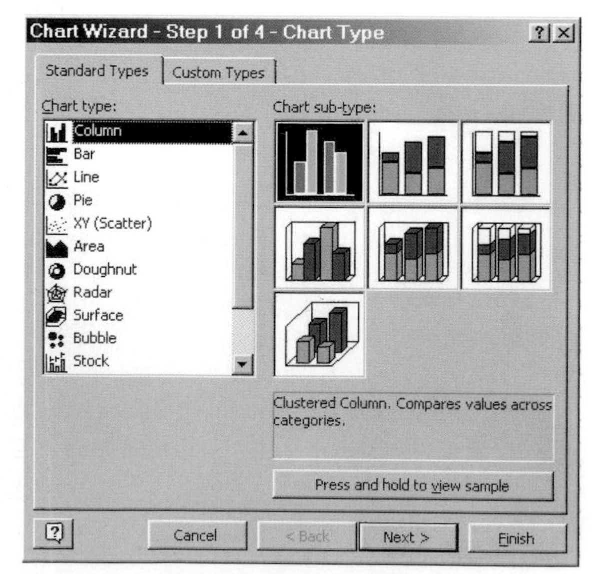

Labelling the bar-chart
Click the **Next>** button.
Repeat this until you get to the **Chart Options** window.
In the **Chart title** box, type **Survey of Computers**.
In the **Category (X) axis** box, type **Type**.
In the **Value (Y) axis** box, type **Number of Students**.
Your screen should look like the one shown.
Click **Finish**.
You can now print the bar-chart.

Drawing a pie-chart

This table shows the percentage of students gaining each level in a Statistics exam.

Level	A	B	C	D	E	F	G
% of students	8	16	25	22	15	8	6

Use *Excel* to draw a pie-chart to show this data.

Open a new *Excel* document

Enter the information on the percentage of students gaining each level in a statistics exam

Select all the cells you have used

Use the Chart Wizard button

Drawing a pie-chart
From **Chart type**, choose **Pie**.
From **Chart sub-type**, choose the top left diagram.
Your screen should look like the one shown. Click **Finish**.
You can now print the pie-chart.

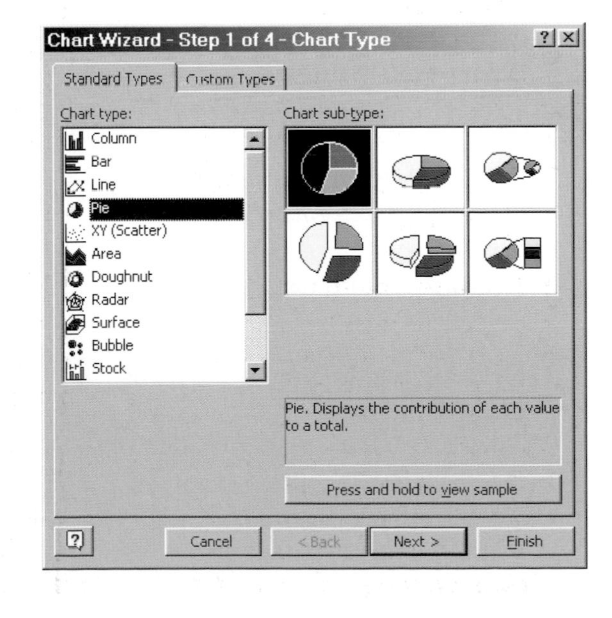

Drawing a frequency polygon

This table shows the scores of Year 11 pupils in two Geography tests.

Score	1 to 10	11 to 20	21 to 30	31 to 40	41 to 50
Number of pupils in first test	5	13	23	11	8
Number of pupils in second test	4	8	16	20	12

Use *Excel* to draw two **frequency polygons** to show this data.

 **Open a new *Excel* document**

 **Enter the information for the two Geography tests**

In Cell A1, type Score.
In Cell B1, type Mid-point.
This is needed to plot the frequency polygon.
In Cell C1, type First Test.
In Cell D1, type Second Test.
In Cell A2, type 1 to 10, in Cell A3, type 11 to 20 and so on.
In Cell B2, type 5.5, in Cell B3, type 15.5 and so on.
In Cell C2, type 5, in Cell C3, type 13 and so on.
In Cell D2, type 4, in Cell D3, type 8 and so on.

Select Cells B1 to D6

Cells A1 to A6 are not selected as this column is not used to draw the frequency polygon. The mid-points are used to draw the frequency polygon.

Use the Chart Wizard button

Drawing two frequency polygons

From Chart type, choose XY (Scatter).
From Chart sub-type, choose the bottom left diagram.
Your screen should look like the one shown.

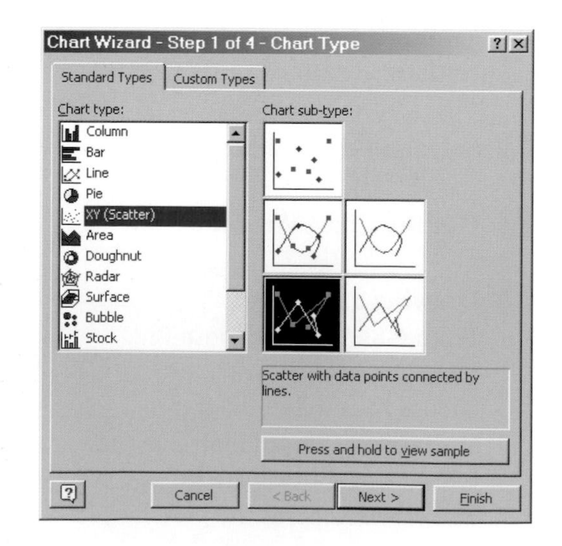

Labelling the frequency polygons

Click the **Next>** button.
Repeat this until you get to the **Chart Options** window.
In the **Chart title** box, type **Comparison of Geography Tests**.
In the **Value (X) axis** box, type **Score**.
In the **Value (Y) axis** box, type **Number of Students**.

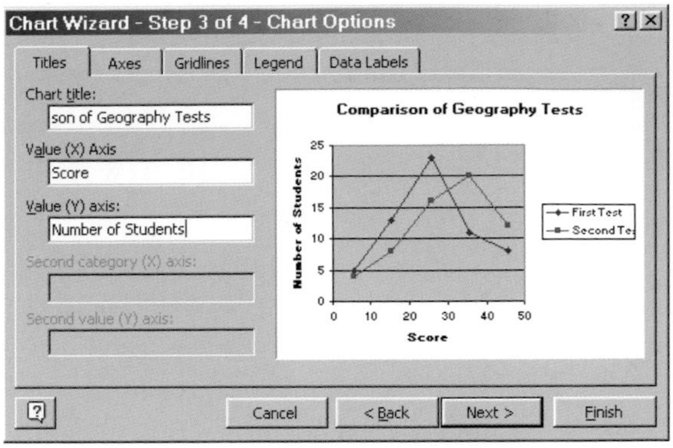

Chart Wizard - Step 3 of 4 - Chart Options
Titles | Axes | Gridlines | Legend | Data Labels
Chart title: son of Geography Tests
Value (X) Axis: Score
Value (Y) axis: Number of Students
Second category (X) axis:
Second value (Y) axis:
Comparison of Geography Tests
Number of Students
Score
First Test
Second Test
Cancel | < Back | Next > | Finish

Your screen should look like the one shown.

Click **Finish**.

You can now print the two frequency polygons.

Exercise 2:9

1 The tables show the results of a survey into the number of people living in one household. The results come from a census of all households in Great Britain.

Year 1961						
Household size (people)	1	2	3	4	5	6 or more
Percentage of households	14	29	23	18	9	7
Year 1993						
Household size (people)	1	2	3	4	5	6 or more
Percentage of households	27	35	16	15	5	2

a Use *Excel* to draw two bar-charts to show this data.
b How has the size of household changed between 1961 and 1993?

2 **a** Use *Excel* to draw two pie-charts for the data in question 1.
b Which type of chart shows the change in size of households better? Explain your answer.

3 The table gives the predicted life expectancy of males and females born in years 1901 to 2021.

Females					
Year	1901	1931	1961	1991	2021
Life expectancy (years)	49.0	61.6	73.6	78.7	82.6
Males					
Year	1901	1931	1961	1991	2021
Life expectancy (years)	45.5	57.7	67.8	73.2	77.6

a Use *Excel* to draw two frequency polygons to show this data.
b What is the difference in life expectancy for males and females over this time period?

1 A factory makes glasses by hand. The data shows the number of glasses with flaws made each day for 30 days.

4	8	2	11	18	5	9	10	4	2
0	13	5	16	10	8	1	7	5	2
14	16	9	6	4	6	9	8	0	3

a Construct a tally-chart using the groups 0–2, 3–5, 6–8, 9–11, 12–14 and so on.

b Draw a bar-chart for the data.

c These are the percentages of the total glass production for the 30 days:

wine glass 55% beer glass 28% champagne glass 17%

Draw a composite bar-chart to show this data.

2 A sample of 100 onions was weighed, to the nearest gram, and the masses tallied.
These are the results:

Mass (g)	1–50	51–100	101–150	151–200	201–250
Number	12	28	37	21	2

Draw a frequency polygon for the data.

3 Robert measured the height, to the nearest centimetre, of his sunflower plant at the end of each week. These are his results:

Week	1	2	3	4	5	6	7	8
Height (cm)	11	33	58	90	114	129	140	178

a Draw a graph to show the heights.

b Estimate the height halfway between week 4 and week 5.
Why is this just an estimate?

4 This is Cara's pie-chart showing how she spends her wages of £120 each week.

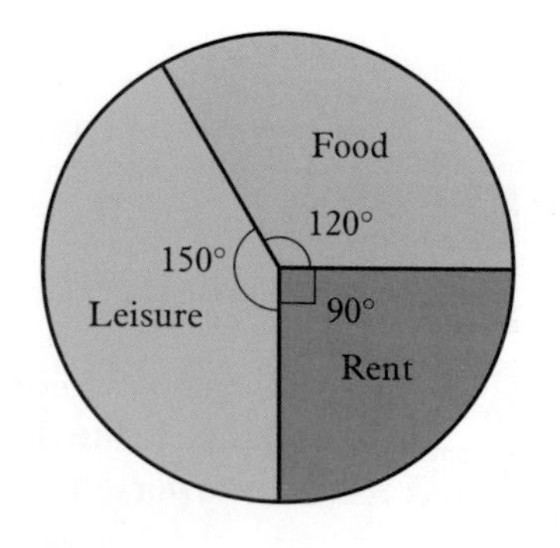

a What fraction is spent on food?

b How much does she spend on rent?

c Cara's friend Tom earns £200 each week.
Cara is going to draw another pie-chart showing how Tom spends his money.
She wants to compare the two pie-charts.
What radius should she use for Tom's pie-chart?

Example

The table shows the number of women Members of the European Parliament (MEPs) for each member state, in 1994.

State	B	DK	D	El	E	F	IRL	I
No. of women	8	7	34	4	21	27	3	10

State	L	NL	A	P	FIN	S	UK
No. of women	3	10	7	3	8	9	15

EP News, July 1999

Find:

a the mode **b** the median **c** the mean.

a The modal number of women MEPs is 3.
However, 3 does not represent the data at all well.
There are two 7s, two 8s and two 10s, but these are ignored if the mode is used to represent the data. The mode is also very small, compared with most of the data values.

b The data is placed in order of size.

3 3 3 4 7 7 8 8 9 10 10 15 21 27 34

The position of the median is $\frac{15+1}{2} = 8\text{th value}$

The eighth value is 8, so this is the median.
The value 8 is fairly representative of this data.
The median is not affected by the extreme values of 3 and 34.

c The mean, $\bar{x} = \frac{\text{total of all the data values}}{\text{total frequency}}$

$$= \frac{3+3+3+4+7+7+8+8+9+10+10+15+21+27+34}{15}$$

$= 11.3$ (1 d.p.)

This value of 11.3 does not represent the data well. This is because all values are taken into account. Thus, the mean is distorted by the large values of 27 and 34. It is about $\frac{3}{4}$ of the way through the data.

For this data, the best representative figure is the median.

Exercise 3:1

1 Rosalind is word processing her statistics coursework. She records the number of errors she makes on each page. These are the numbers of errors that she recorded.

19 6 51 14 17 20 16 13 16

a Write down the modal number of errors.
b Find the median number of errors.
c Calculate the mean number of errors to 1 d.p.
d Write down the value which best represents this data. Explain your answer.

2 Jon owns a garage. He records the number of cars that he repairs each day. These are the numbers of cars that he repaired in July.

8	2	8	9	6	11	5	8	6	9	10
6	5	10	8	5	13	9	10	5	8	15

a Write down the modal number of repairs.
b Find the median number of repairs.
c Calculate the mean number of repairs.
d Write down the value which best represents this data. Explain your answer.

Calculating the mode, median and mean for a frequency distribution

The school football team records the number of goals it scores each week. The table shows the results for the 1999 season.

Number of goals, x	Frequency, f
0	7
1	10
2	9
3	4

To find the mode you look for the value with the highest frequency.
The mode is 1.

To find the **median** you look for the middle value.
Find the total number of matches by adding together all the frequencies.
The total frequency is written as Σf

Σ means **the sum of** $\quad \Sigma f = 7 + 10 + 9 + 4 = 30$

So, the position of the median is $\frac{30 + 1}{2} = 15.5\text{th value}$

The median is the value between the 15th and 16th terms.
To find this value, use a table with running totals.

Number of goals, x	Frequency, f	Running total
0	7	7
1	10	$7 + 10 = 17$
2	9	
3	4	

By the time you have listed all the 0s and all the 1s, there are 17 terms. The 15th and 16th terms must both be 1.

So the median is $\frac{1 + 1}{2} = 1$

To find the **mean** you must first find the total of all the data values.
The total number of goals is the sum of $f \times x = \Sigma fx$
The total number of matches is the sum of all the f values $= \Sigma f$

$$\text{The mean, } \bar{x} = \frac{\text{total number of goals}}{\text{total number of matches}} = \frac{\Sigma fx}{\Sigma f}$$

Use a table to find these values. Put in an extra column to calculate $f \times x$.

Number of goals, x	Frequency, f	$f \times x = fx$
0	7	$7 \times 0 = 0$
1	10	$10 \times 1 = 10$
2	9	$9 \times 2 = 18$
3	4	$4 \times 3 = 12$
	$\Sigma f = 30$	$\Sigma fx = 40$

$$\text{Mean number of goals, } \bar{x} = \frac{\Sigma fx}{\Sigma f} = \frac{40}{30}$$
$$= 1.3 \; (1 \; d.p.)$$

It is impossible to have 1.3 goals, but the mean value gives a representative value. All the data values are considered. You must not round the mean to the nearest whole number.

Exercise 3:2

1 The table shows the number of items of mail received daily by Mrs Jones.

a Write down the modal number of items of mail.

b Copy the table. Add an extra column for the running total. Use this to find the median.

c Make a second copy of the table. Add an extra column for $f \times x$. Calculate the mean using the formula $\bar{x} = \frac{\Sigma fx}{\Sigma f}$

Give your answer to 1 decimal place.

No. of items of mail, x	Number of days, f
0	6
1	5
2	7
3	9
4	3
5	0
6	1

2 Nathan carried out a survey. He asked each member of his form how many crayons they had in their pencil case. The table shows the number of crayons.

a Write down the modal number of crayons.

b Copy the table. Add an extra column for the running total. Use this to find the median.

c Make a second copy of the table. Add an extra column for $f \times x$. Calculate the mean using the formula $\bar{x} = \frac{\Sigma fx}{\Sigma f}$

Number of crayons, x	Number of pupils, f
0	2
1	4
2	3
3	2
4	6
5	3
6	5
7	4
8	3
9	0
10	2
11	1

3 Pam has 6 people in her family. She wonders how many people are in her friends' families. She asks each of her friends and records the information in a table.

Number in family, x	2	3	4	5	6	7	8
Frequency, f	2	4	6	5	2	0	1

Calculate:

a the mode **b** the median **c** the mean.

Calculating the mode, median and mean for a grouped frequency distribution of discrete data

A local shop sells mobile phones and keeps a record of the daily sales. The table shows these sales.

Number of phones, x	Frequency, f
0–4	5
5–9	8
10–14	4
15–19	9
20–24	3

The **modal group** is 15–19 phones as it has the highest frequency.

It is impossible to find the exact value of the **median** when the data has been grouped. It can be **estimated** by using **interpolation**.

The total frequency $\Sigma f = 29$

The position of the median is $\dfrac{29 + 1}{2} = 15\text{th value}$

To find this value, use a table with running totals.

Number of phones, x	Frequency, f	Running total
0–4	5	5
5–9	8	$5 + 8 = 13$
10–14	4	$13 + 4 = 17$
15–19	9	
20–24	3	

By the time you have listed all the values up to and including 9 phones, there are 13 values. The 15th term must be in the group 10–14 since this takes you to 17.

Look at the group **10–14**.

The group is 5 numbers wide: 10, 11, 12, 13 and 14.

The frequency is 4. So the group has 4 values in it.

The 15th value is the **2**nd of these 4 values since $13 + \mathbf{2} = 15$

So the estimate would be $\dfrac{2}{4}$ of the way through this group.

$$\text{An estimate of the median} = \mathbf{10} + \frac{\mathbf{2}}{4} \times 5$$
$$= 12\tfrac{1}{2}$$

It is impossible to find the exact value of the mean when the data has been grouped. It can be estimated by using the mid-values of each group.

The mid-values are $\frac{1}{2}(0 + 4) = 2$, $\frac{1}{2}(5 + 9) = 7$, and so on.

To work out the mean, draw two extra columns.

Use the mid-values as the x values.

Number of phones	Frequency, f	Mid-values, x	$f \times x$
0–4	5	2	$5 \times 2 = 10$
5–9	8	7	$8 \times 7 = 56$
10–14	4	12	$4 \times 12 = 48$
15–19	9	17	$9 \times 17 = 153$
20–24	3	22	$3 \times 22 = 66$
	$\Sigma f = 29$		$\Sigma fx = 333$

Estimate of the mean: $\bar{x} = \frac{\Sigma fx}{\Sigma f} = \frac{333}{29}$

$= 11.5$ (1 d.p.)

Exercise 3:3

1 Julia is keen on reading. She notices that, in her recent book, the number of words in a sentence is very varied. She records the number of words in each sentence in the first chapter. The table shows her data.

Number of words, x	Frequency, f
1–5	15
6–10	26
11–15	29
16–20	12
21–25	11
26–30	5
31–35	3

a Write down the modal group.

b Copy the table. Add an extra column for the running total. Use interpolation to estimate the median to 1 decimal place.

c Make a second copy of the table. Add an extra column for the mid-values. Use this to help you estimate the mean to 1 decimal place.

2 Tina works on Saturday mornings in the local library. She has to record the number of books issued to each customer. The table shows the number of customers and how many books each of them borrowed.

Number of books, x	Number of customers, f
1–3	4
4–6	5
7–9	6
10–12	3
13–15	3

Find:

a the modal group

b an estimate of the median to 1 decimal place

c an estimate of the mean to 1 decimal place.

3 June is amazed at the number of different types of cheese. Each time she visits a shop, she records how many different cheeses they have in stock. The table shows the number of shops and how many types they stock.

Number of types, x	Number of shops, f
3–5	2
6–8	3
9–11	5
12–14	3
15–17	2

Find:

a the modal group

b an estimate of the median to 1 decimal place

c an estimate of the mean.

4 Ian has 6 different trees in his small garden. He is a keen gardener and wishes to know how many trees his neighbours have. He records the number of trees in each garden in his street.

Number of trees, x	Number of houses, f
3–7	4
8–12	9
13–17	5
18–22	3

Find:

a the modal group

b an estimate of the median to 1 decimal place

c an estimate of the mean to 1 decimal place.

Calculating the mode, median and mean for a grouped frequency distribution of continuous data

You cannot give an **exact** value to continuous data because it is impossible to measure it exactly. Continuous data always has to be given to a chosen accuracy. The groups for continuous data need some adjusting when working out means and medians.

Measurement of time, weight, height and speed are often given to the **nearest unit**. The groups for these are usually written like this:

21–30	
31–40 ⟵	This means 30.5 up to, but not including 40.5.
41–50	You use 30.5 and 40.5 to work out the mid-value.
51–60	You use 30.5 as the beginning of the median calculation.

Another way of writing groups is:

100–	
200– ⟵	This means 200 up to, but not including, 300.
300–	You use 200 and 300 to work out the mid-value.
400–	You use 200 as the beginning of the median calculation.

Special care is needed with age, if it is counted in **completed** years.

21–30	
31–40 ⟵	This means 31 up to, but not including, 41.
41–50	You use 31 and 41 to work out the mid-value.
51–60	You use 31 as the beginning of the median calculation.

Sometimes algebra is used in a table to show the size of the groups.

$10 < x \leqslant 20$	10 is not included in the group but 20 is included.
	You use 10 and 20 to work out the mid-value.
	You use 10 as the beginning of the median calculation.
$10 \leqslant x < 20$	10 is included in the group but 20 is not included.
	You still use 10 and 20 to work out the mid-value.
	You still use 10 as the beginning of the median calculation.

Example The table shows the times in minutes, to the nearest minute, spent by people travelling to work.

Time in minutes, x	0–14	15–29	30–44	45–59
Number of people, f	10	14	12	7

Estimate the median time spent travelling to work.

The median is the $\frac{44}{2} = 22$nd value

An estimate for the median $= 14.5 + \frac{12}{14} \times (29\frac{1}{2} - 14\frac{1}{2})$

$= 27.4$ minutes to 1 d.p.

Exercise 3:4

1 Dillon collected apples from the tree in his garden. He weighed each one and recorded the weight, to the nearest g, in a table.

Weight in grams, x	Number of apples, f
121–140	8
141–160	6
161–180	9
181–200	5
201–220	4

a Write down the modal group.

b Estimate the mean weight, to the nearest gram, of Dillon's apples.

c Estimate the median weight, to the nearest gram, of an apple in Dillon's garden.

2 Carla measures the heights of pupils in her class in cm. The table gives her measurements.

Height in cm, x	$110 \leqslant x < 120$	$120 \leqslant x < 130$	$130 \leqslant x < 140$	$140 \leqslant x < 150$	$150 \leqslant x < 160$
Number of students, f	10	6	5	4	3

For the pupils in Carla's class:

a write down the modal group

b estimate the mean height to the nearest centimetre

c estimate the median height to the nearest centimetre.

2 Calculations involving the mean

Carly's Mum goes to a night club with Carly and her friends. She says that she has increased the average age there!

Adding one more item of data can change the mean value.

Carly goes to a night club with three of her friends. Their average age is 17 years and 3 months. Carly's Mum is 43 years and 6 months old. To find the average age of the group when Carly's Mum is included you need to find the total of all the ages.

For Carly and her three friends:

The mean $= \dfrac{\text{total age}}{4}$

So total age = 4 × 17 years 3 months = 4 × 17.25 = 69 years

When Carly's Mum joins in:

Total age = 69 + 43.5 = 112.5

This total age includes Carly and her three friends and her Mum.

This is 5 people.

The new mean, $\bar{x} = \dfrac{112.5}{5}$

= 22.5 years

= 22 years 6 months

Carly's Mum has increased the mean age from 17 years 3 months to 22 years 6 months.

Exercise 3:5

1 Pete has seven Shetland ponies. They have a mean height of 116 cm. Pete buys an eighth pony. The height of this pony is 128 cm. Find the mean height of all eight ponies.

2 The mean height of seven pupils is 123 cm. One pupil of height 147 cm leaves the group. Find the mean height of the remaining six pupils.

Example There are **12** children in Phil's group. Their mean mark in a Maths test is **76%**. In Paul's group there are only 8 children. Their mean mark is 84%. Find the overall mean mark for the 20 children.

The mean $= \dfrac{\text{total of all data values}}{\text{total frequency}}$

Total of marks for Phil's group $= \mathbf{12 \times 76} = \mathbf{912}$

Total of marks for Paul's group $= 8 \times 84 = 672$

Total of all data values $= \mathbf{912} + 672 = 1584$

Number of children in Phil's group $= \mathbf{12}$

Number of children in Paul's group $= 8$

Total frequency $= \mathbf{12} + 8 = 20$

New mean $= \dfrac{1584}{20} = 79.2\%$

---

3 Don delivers pint bottles of milk to two streets. For the first street of 10 houses, the mean number of bottles of milk that he delivers is 3.1
For the second street of 6 houses, the mean number of bottles of milk that he delivers is 2.5
Find the mean number of bottles of milk he delivers per household for the two streets together.

4 Nigel has scored a mean of 18 runs in the last 5 cricket matches. His mean score must be 20 or more for him to be chosen for the school team. Find the number of runs that he must make in the next match if he is to be chosen for the school team.

5 Sally manages a factory making glass animals.
There are four levels of wages for people working in the factory.
These wages are given in the table.

Type of work	Annual wage (£)
Cleaner	8 000
Office worker	11 000
Glassblower	17 000
Manager	32 000

a Find the mean of these four wages.

b Does the mean that you found in **a** represent the wages paid to the staff fairly? Explain your answer.

It is sometimes important to calculate a mean as a **weighted mean**. The table gives the wages, and the number of people who earn each wage, in Sally's factory.

Type of work	Annual wage (£)	Number of people
Cleaner	8 000	5
Office worker	11 000	4
Glassblower	17 000	15
Manager	32 000	1

To find the mean of all the wages you first multiply each annual wage by the number of people being paid that wage. Then divide by the total number of people.

$$\text{Mean} = \frac{8000 \times 5 + 11\,000 \times 4 + 17\,000 \times 15 + 32\,000 \times 1}{5 + 4 + 15 + 1} = £14\,840$$

This mean takes into account the number of people getting each wage. It is called a weighted mean. The numbers of people for each salary are called the weightings.

---

6 These are the wages paid to the people working in a shop.

Type of work	Annual wage (£)	Weighting
Cleaner	7 500	3
Salesperson	13 000	8
Office staff	18 000	3
Managers	26 000	2

Find the weighted mean for the wages paid to the staff of the shop.

Weightings can be expressed as **percentages.**

Example In a Physics examination, there are two papers. Paper 1 counts for 40% of the final mark. Paper 2 counts for 60% of the final mark. Glenda scores 82 marks out of 100 for Paper 1 and 78 marks out of 100 for Paper 2. Find her overall percentage mark.

The weighted mean mark $= \frac{\Sigma wx}{\Sigma w}$ where w is the weighting given to each value of x.

Use a table with an extra column to calculate $w \times x$

Mark, x	Weighting, w	$w \times x$
82	40	$40 \times 82 = 3280$
78	60	$60 \times 78 = 4680$
	$\Sigma w = 100$	$\Sigma wx = 7960$

$$\text{The weighted mean mark} = \frac{\Sigma wx}{\Sigma w} = \frac{7960}{100} = 79.6\%$$

Exercise 3:6

1 Peter sat the same exam in Physics as in the example above. He scored 34 marks out of 100 for Paper 1 and 42 marks out of 100 for Paper 2. Use the weightings from the example to find his overall percentage mark.

2 In a French examination, there are three papers. Paper 1 counts for 30% of the final mark, Paper 2 counts for 35% and Paper 3 also counts for 35%. On Paper 1, Ceri scores 80 marks out of 100. On Paper 2, she scores 68 marks out of 100. On Paper 3, she scores 74 marks out of 100. Find her overall percentage mark.

3 A farmer sells bags of potatoes in 5 kg, 10 kg and 15 kg bags. He sells different quantities of each size. 15% of his total sales are 5 kg bags. 55% of his total sales are 10 kg bags. 30% of his total sales are 15 kg bags. Find the average weight of a bag of potatoes that the farmer sells. Give your answer to the nearest kilogram.

Weightings can also be given as **ratios**.
The same formula can still be used.

Example The prices of theatre tickets are £15, £20 and £30. The tickets are sold in the ratio of 2:3:1 respectively. Find the average price of a ticket.

The mean price of a ticket $= \dfrac{\Sigma wx}{\Sigma w}$ where w is the weighting given to each value of x.

Use a table with an extra column to calculate $w \times x$.

Price, x	Weighting, w	$w \times x$
15	2	30
20	3	60
30	1	30
	$\Sigma w = 6$	$\Sigma wx = 120$

$$\text{The mean price of a ticket} = \frac{\Sigma wx}{\Sigma w} = \frac{120}{6}$$
$$= £20$$

4 The prices of theatre tickets are £15, £20 and £30. On a certain evening the tickets were sold in the ratio of 3:4:2 respectively. Find the average price of a ticket. Give your answer to the nearest penny.

5 Judy works in a local café. She serves coffee, tea and hot chocolate in the ratio of 4:3:2 respectively. A cup of coffee is £1.10, a cup of tea costs £0.95 and a mug of hot chocolate costs £1.50. Find the average price of a drink that she serves. Give your answer to the nearest penny.

6 Jules is decorating his house. He buys tins of paint at different prices. Some tins cost £12, others cost £9 and £5. He buys these tins in the ratio of 1:1:2 respectively. Find the average cost of a tin of paint he buys, to the nearest penny.

7 A German examination has three papers. The final mark is a weighted average reflecting the time spent on each paper. Paper 1 lasts 1 hour, Paper 2 is $1\frac{1}{2}$ hours and Paper 3 is 2 hours.

a Joanna scores 40 on Paper 1, 60 on Paper 2 and 70 on Paper 3. Calculate her weighted mean mark.

b Jerry scores 32 on Paper 1, 40 on Paper 2 and 61 on Paper 3. Calculate his weighted mean mark. Give your answer to one decimal place.

Sometimes there is a quick way to find the mean.

Assumed mean Kate needs to find the mean of 307, 325, 315, 309, 322, 318
She guesses what the mean will be.

Kate thinks the mean will be 318.
This guess is called the **assumed mean**.

Kate finds the difference between the assumed mean and each data value.
These are $-11, 7, -3, -9, 4, 0$ because $307 - 318 = -11$, $325 - 318 = 7$ etc.
She now needs to find the **mean** of these differences.

Add the differences $(-11) + 7 + (-3) + (-9) + 4 + 0 = -12$
Divide by the number of differences $-12 \div 6 = -2$

Now Kate adds the answer to the assumed mean to find the actual mean.

$$318 + (-2) = 316$$

The mean of the data values is 316.

You can check this by doing the full calculation

$$\text{Mean} = \frac{307 + 325 + 315 + 309 + 322 + 318}{6} = \frac{1896}{6} = 316$$

Exercise 3:7

1 Saleem is finding the mean of 48, 41, 56, 59, 58, 45, 50
He chooses 50 as the assumed mean.

a Write down the difference between the assumed mean and each item of data.

b Add all the differences and divide by 7.

c Find the actual mean.

d Check your answer by doing the full calculation.

2 Find the mean of 127, 114, 125, 110, 121, 132, 119, 128.
Use 120 as the assumed mean.

3 Find the mean of 464, 511, 500, 489, 519, 497, 517, 495.
Use 500 as the assumed mean.

4 Find the mean of 714, 719, 720, 721, 714, 717, 716, 712, 722, 715.
Use 720 as the assumed mean.

For each set of data in questions **5** to **10**

a choose an assumed mean

b find the mean using your assumed mean.

5 229, 240, 273, 255, 236, 262, 280, 225

6 400, 404, 398, 403, 405, 396

7 818, 829, 828, 824, 835, 812, 822

8 2003, 2009, 2004, 2002, 2005, 2007

9 2410, 2380, 2420, 2390, 2440, 2360, 2420, 2350, 2430

Geometric mean

The **geometric mean** of two numbers is the square root of their product.

The geometric mean of 2 and 32 is $\sqrt{2 \times 32} = \sqrt{64} = 8$

The gemoetric mean of three numbers is the cube root of their product.

The geometric mean of 5, 9 and 12 is
$\sqrt[3]{5 \times 9 \times 12} = \sqrt[3]{540} = 8.1$ to 2sf.

---

10 Find the geometric mean of

a 2 and 8

b 3 and 14

c 4, 10 and 25

d 3, 7 and 11

e 12, 15 and 19

f 23, 28 and 34

The interest rates for bank accounts may change from year to year.
The geometric mean can be used to calculate an equivalent single rate over two or more years.

Example A bank pays interest on new accounts at a rate of 10% for the first year. The rate for the second year is 4%.
Calculate the equivalent single rate for these two years.

To find the value of the account at the end of the two years

- Multiply the balance by 1.10
  This gives the balance at the end of the first year
- Multiply the balance at the end of the first year by 1.04
  This gives the balance at the end of the second year.

The geometric mean of 1.10 and 1.04 $= \sqrt{(1.10 \times 1.04)}$
$= 1.069\,579\ldots$

The equivalent single rate is 6.96% to 3 sf.

11 A bank paid interest at rates of 5%, 7% and 3% for three consecutive years. Calculate the single rate that would pay the same over these three years.

12 The value of a car decreases by 38% in its first year and by 29% in its second year. Calculate the equivalent single rate for these two years.

13 **a** The value of a Share Group's portfolio increased at rates of 9% and 7% for two consecutive years. Calculate the equivalent rate for the two years.

b The value of a second Share Group's portfolio increased at rates of 11% and 6% for the same two years. Calculate the equivalent rate for the two years.

c Which of the two groups performed better over the two years?

14 The table shows the percentage increase in the value of two identical houses in different parts of a city over three years.

	1998	1999	2000
Brompton	8%	10%	5%
Cornwell	6%	10%	13%

a Find the equivalent single percentage increase for each house over the three years.

b Write down the area where the price increased by the highest percentage overall.

3 Estimation

Danial is collecting leaves. He needs to find out the average length of a leaf. Does he have to measure all the leaves?

Population

A **population** is a collection of individuals or items. For a large population it is not practical to consider every item to work out the mean.

Sample

Information is normally taken from a small part of a population. This is called a **sample** of the population.
If the population is very large, and a sample is chosen at random, the sample will be representative of the whole population.

The **population mean** is called μ. This is the Greek letter m and it is read as 'mu'.
An **estimate** of the population mean is the mean of the sample.
The **sample mean** is called $\bar{x}$.
$\bar{x}$ gives a better estimate of the population mean as the sample size gets bigger.
Usually you should use a sample of 50 to 100 items.

Danial collects 100 leaves from a tree in his garden. He records the length of each leaf to the nearest millimetre.

5.3	5.1	4.8	4.6	4.7	4.9	4.3	4.8	5.0	4.4	4.8	5.0	5.5	4.9	4.6
4.1	4.8	4.5	4.4	4.3	4.7	4.3	4.7	4.7	4.8	4.7	4.7	5.0	4.4	4.7
5.1	4.5	4.8	4.5	4.9	4.2	4.6	4.8	5.1	4.8	4.7	4.8	4.1	4.6	4.5
5.0	4.4	4.6	4.5	5.0	4.8	5.1	4.5	5.1	5.1	4.1	5.3	5.0	5.1	4.2
4.5	4.7	4.7	4.9	4.2	4.4	4.6	5.3	4.6	4.9	5.1	4.3	4.9	5.3	4.3
5.0	5.4	4.8	5.5	5.0	4.9	4.3	5.2	4.4	5.4	5.2	4.6	4.2	5.2	5.4
4.7	5.1	4.4	5.2	5.0	4.6	5.2	4.8	4.9	5.2					

Danial calculates the mean of his sample $\overline{x}$.

He uses a frequency table and the formula $\overline{x} = \frac{\Sigma fx}{\Sigma f}$

Length of leaf, x cm	Number of leaves, f	$f \times x$
4.1	3	$3 \times 4.1 = 12.3$
4.2	4	$4 \times 4.2 = 16.8$
4.3	6	$6 \times 4.3 = 25.8$
4.4	7	$7 \times 4.4 = 30.8$
4.5	7	$7 \times 4.5 = 31.5$
4.6	9	$9 \times 4.6 = 41.4$
4.7	11	$11 \times 4.7 = 51.7$
4.8	12	$12 \times 4.8 = 57.6$
4.9	8	$8 \times 4.9 = 39.2$
5.0	9	$9 \times 5.0 = 45.0$
5.1	9	$9 \times 5.1 = 45.9$
5.2	6	$6 \times 5.2 = 31.2$
5.3	4	$4 \times 5.3 = 21.2$
5.4	3	$3 \times 5.4 = 16.2$
5.5	2	$2 \times 5.5 = 11.0$
	$\Sigma f = 100$	$\Sigma fx = 477.6$

So the sample mean, $\overline{x} = \frac{\Sigma fx}{\Sigma f} = \frac{477.6}{100} = 4.8$ cm (to 1 d.p.)

The sample mean, $\overline{x} = 4.8$ cm

The sample mean, $\overline{x}$, is used as an estimate of the population mean, μ.

So an estimate of the population mean, μ, is 4.8 cm.

Danial's estimate of the mean length of the leaves on the tree is 4.8 cm.

If several samples of size 100 are taken then the means calculated will be spread over a range of values.
If several samples of size 400 are taken then the means will be spread over half this range.

In general, the range of the means is proportional to $\frac{1}{\sqrt{\text{sample size}}}$.

Exercise 3:8

1 Tricia is investigating earthworms. She has collected 50 worms and measured the length of each one. These are her results in cm:

5	11	16	10	14	7	9	14	6	11
10	16	15	9	8	13	15	14	7	12
9	14	17	13	18	15	15	10	7	10
16	11	9	7	12	7	15	9	12	11
15	16	13	9	12	7	14	10	14	8

a Draw a frequency table for Tricia's data.
b Calculate the sample mean, $\bar{x}$.
c Write down an estimate of the population mean, μ.
d How can Tricia improve her estimate?

2 Tom chooses a random sample of 100 sentences from a newspaper. He counts the number of words in each sentence. The frequency table shows his data.

Number of words	Frequency	Number of words	Frequency
3	1	9	23
4	0	10	11
5	7	11	12
6	9	12	4
7	12	13	2
8	18	14	1

a Calculate the mean number of words per sentence, $\bar{x}$.
b Write down an estimate of the mean number of words per sentence for the whole newspaper, μ.

Julia wears a size 6 shoe. She wants to find the proportion of girls in her school who wear the same size. There are 600 girls altogether. She cannot ask all 600 girls.

Julia chooses a sample of 60 girls at random.
An estimate of the **population proportion** is the **sample proportion**.

Julia records the shoe size for a random sample of 60 students. These are her results:

4	6	6	6	7	4	6	7	6	5	4	6	6	5	5
7	6	6	8	7	5	5	6	5	6	4	7	7	7	5
6	6	5	6	7	6	5	6	8	7	3	7	6	7	6
7	3	5	6	6	7	6	4	4	8	6	7	3	5	7

Julia now finds the proportion of the girls in the sample who wear size 6.

Number of girls in the sample $= 60$
Number of girls who wear size 6 shoes $= 22$

$$\text{Sample proportion} = \frac{\text{number of girls who wear size 6}}{\text{number of girls in the sample}} = \frac{22}{60}$$

The sample proportion is used as an estimate of the population proportion.

An estimate of the proportion of girls in the school who wear her size $= \dfrac{22}{60}$

The bigger the sample, the better the estimate of the population proportion.

Like the means, the range of the sample proportions changes as the size of the sample changes.

Again it is proportional to $\dfrac{1}{\sqrt{\text{sample size}}}$.

---

3 Use Julia's sample of 60 students from the example above.

a For the sample, calculate the proportion of girls who wear size 8 shoes.

b Estimate the proportion of girls in Julia's school who wear size 8 shoes.

4 Alan investigates the proportion of boys who keep pets. There are 200 boys altogether. He asks a random sample of 40 boys if they have any pets. He finds that 25 boys have at least one pet.

a Calculate the proportion of the sample who have at least one pet.

b Write down an estimate of the proportion of boys in the school who have at least one pet.

5 Jonty investigates the punctuality of trains arriving at his local station. In a sample of 100 trains, he finds that 20 are late, 10 are early and 70 are on time.

a Use Jonty's sample to calculate the proportion of trains that are late.

b Use the sample proportion to estimate the proportion of all the trains that are late.

4 Control charts

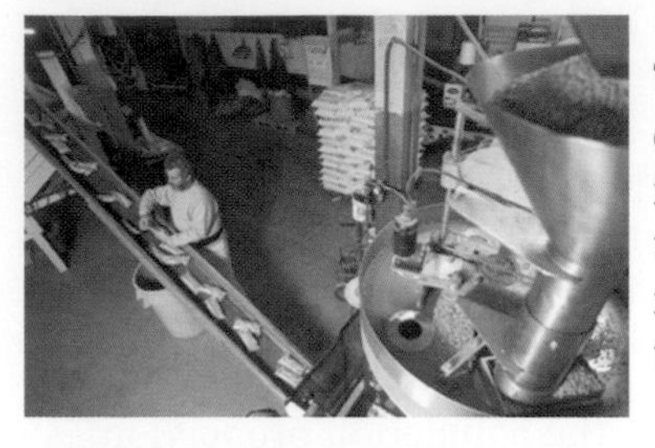

This machine is filling packets of chickpeas. The weight of the contents must be strictly controlled. It is illegal to sell an underweight packet and too many overweight packets will cut down the company's profits.

You cannot expect the weight of the contents of each packet to be exactly the same. The weight will vary. The weight is called a **variable**.

Control chart

A **control chart** is a statistical tool that monitors a process and warns us when something is going wrong.
You can use the mean, median or the range to draw a control chart.

This is a chart for the mean.

It has a horizontal line drawn on the axes. This solid line is drawn at the target value of the mean.

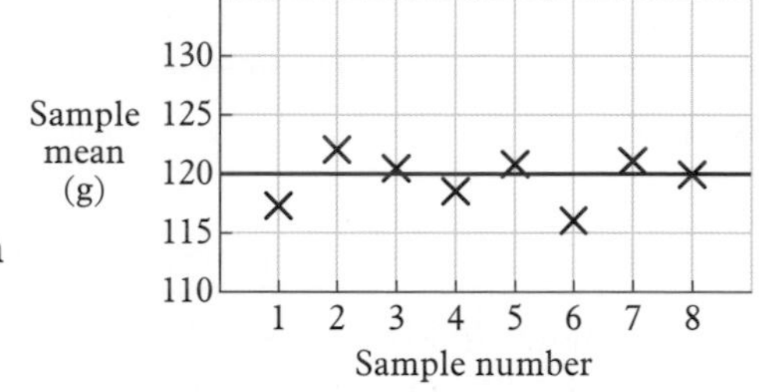

Small samples are taken from the population at regular time intervals.
The mean of each sample is worked out and these means are plotted in time order.

Sample means do vary but it is expected that all sample means will lie in a band.

If the band is narrow then the process is consistent.

A wider band shows more variability in the process.

If one or more of the sample means fall outside this band it means that a fault may have developed in the process.

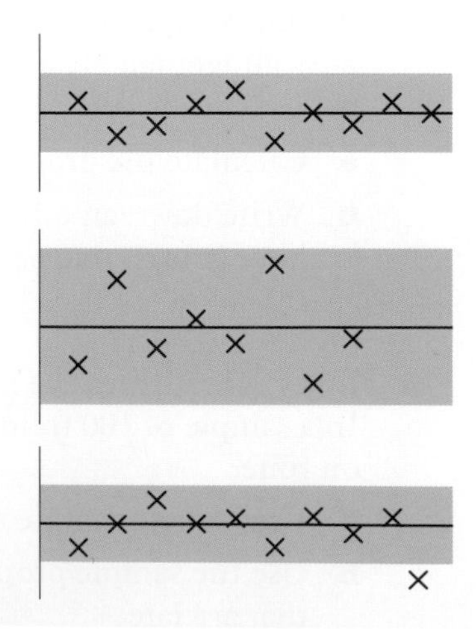

Look at these charts:

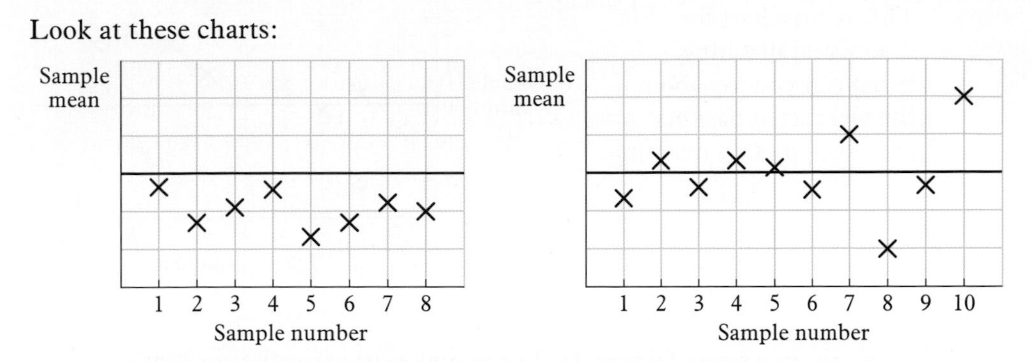

Both these charts show that it is probable that something is wrong with the process. This is a warning to have the process checked.

The sample median and the sample range could also be plotted in this way. Any unexpected trend or variability of the points are warnings of possible problems in the process.

Exercise 3:9

1 Karen is testing a machine that fills packets of crisps.
When the machine is working properly it produces packets with a mean weight of 30 g.
She takes a random sample of 4 packets every hour from the machine and works out the mean weight. This is her data.

sample number	mean weight (g)	sample number	mean weight (g)
1	31.4	6	31.2
2	29.0	7	30.6
3	29.8	8	28.9
4	30.5	9	30.9
5	28.7	10	30.0

a Copy this chart.
b Plot the sample means.
c Does the machine seem to be working properly? Explain your answer.

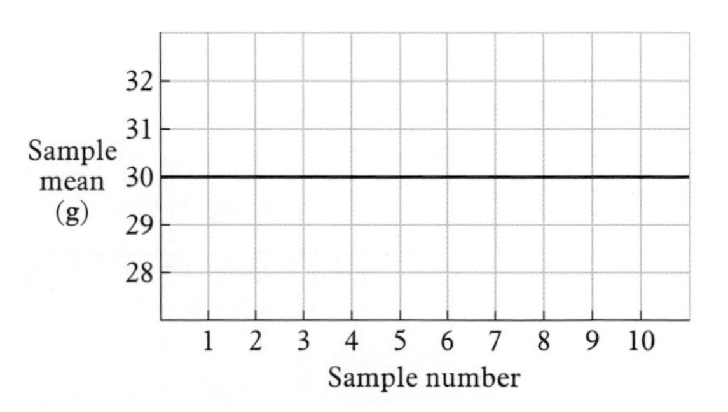

d Karen tests a second machine. This is her chart for the second machine. What can you say about the weights of packets produced by this machine?

Sample mean (g)
32
31
30
29
28
1 2 3 4 5 6 7 8 9 10
Sample number

2 Ethan works in a sweet factory. He has to make sure that the correct number of sweets go into each bag of assorted sweets.
There are four machines which fill the bags with assorted sweets.
Ethan randomly samples the contents of the bags for each machine at regular intervals.
These are his charts for the four machines. He has plotted the mean number of sweets for each sample.
What can you say about each machine?

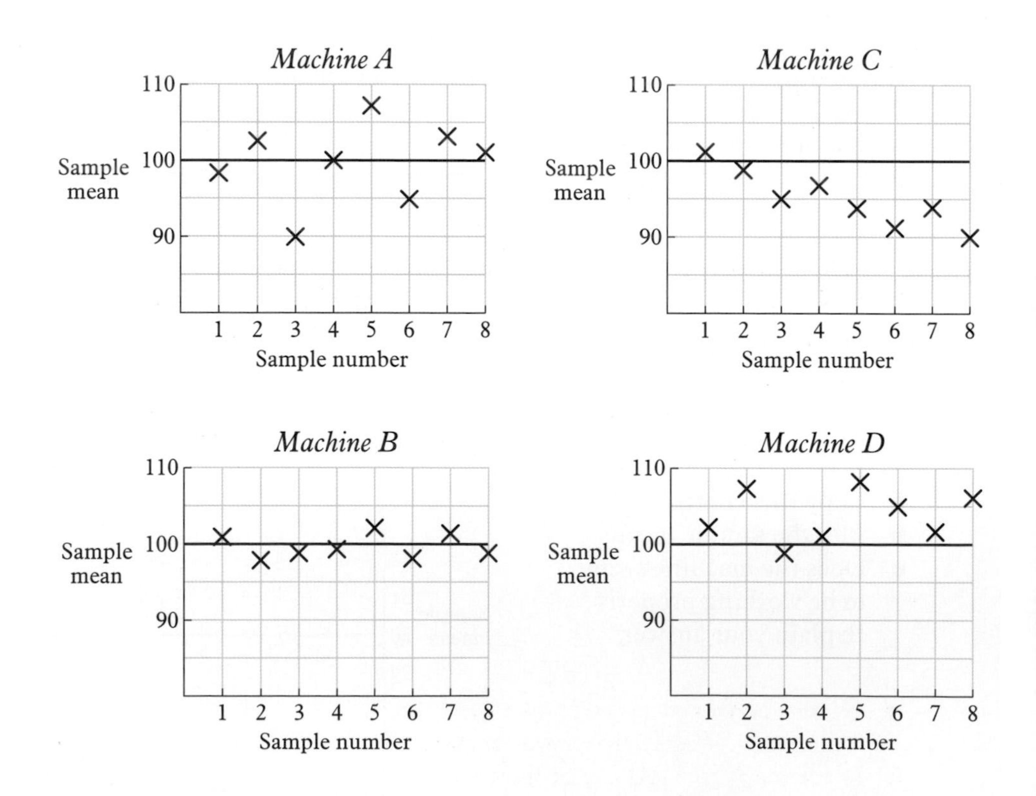

3 A factory makes pistons.
The diameter of the piston is very important.

Niamh tests the machine by taking a random sample of six pistons every hour and working out the mean diameter. The table shows her results.

sample number	sample mean (mm)	sample number	sample mean (mm)
1	27.30	6	27.29
2	27.35	7	27.33
3	27.24	8	27.62
4	27.35	9	27.34
5	27.26	10	27.36

a Draw a control chart to show this data.
The target value of the mean is 27.32 mm.

b At which point do you think the process was halted and the machine reset?
Explain your answer.

4 A factory produces resistors for electrical circuits.
Ben is testing the machine that makes the 100-ohm resistors.
Ben chooses 5 resistors at random every half-hour and measures the resistance. This is his data.

sample number	sample mean (ohm)	sample number	sample mean (ohm)
1	99.4	7	99.0
2	100.1	8	100.1
3	100.2	9	99.0
4	101.0	10	97.6
5	99.9	11	98.4
6	100.5	12	97.2

a Draw a control chart to show this data.
The target value for the mean is 100 ohm.

b Is the machine working properly?
Explain your answer.

1 For ten days Ian recorded the weight of peanuts, in grams, that his parrots ate. The mean weight was 46.4 grams and the readings were:

45, 45, 47, 47, 46, 48, 46, 46, 46, 48.

Ian then recorded the next twenty days' consumption. His results are shown in the frequency table below.

Weight (g)	45	46	47	48	49	50
Frequency	0	1	1	3	8	7

a Copy and complete the table below to show the frequency distribution of the combined sample of 30 readings. *(2)*

Weight (g)	Combined frequency	
45		
46		
47		
48		
49		
50		
TOTAL	30	

b Calculate the mean weight of the 30 observations. *(3)*

c Give **one** reason why your calculated value is likely to be a better estimate of the mean weight per day than the 46.4 grams previously calculated. *(1)*

SEG, 1997, Paper 2

2 Fred went fishing each week.
Each week he recorded the number of fish caught.
After several weeks he calculated the following averages.

The **mean** number of fish caught per week was 9.3
The **modal** number of fish caught per week was 12.
The **median** number of fish caught per week was 10.

The next week he did not catch any fish.
This had never happened before.
Fred recalculated his averages.
Which of these averages could not have been affected? *(1)*
Give a reason for your answer. *(1)*

SEG, 1997, Paper 2

3 The price, in pounds, of eleven cars sold by a salesman in the last month are shown.

4500	4300	7500	3900	4000	3900
4050	3800	4175	4200	4275	

a Which of the averages, the mean, the median or the mode, best represents these data? Give a reason for your answer. *(2)*

b What type of data describes the following variables:

i the price of a car; *(1)*

ii the time taken to complete the sale of a car? *(1)*

SEG, 1998, Paper 2

4 The number of goals scored by Burntwood Town Football Club in their first 20 games are shown.

4	1	1	2	4	5	0	0	1	1
2	2	4	1	2	1	5	3	3	1

a Copy and complete the frequency table below. *(4)*

Score	Tally	Frequency

b What is the modal score? *(1)*

c What is the median score? *(1)*

SEG, 1998, Paper 2

5 The sign on a lift says

> **Maximum total weight 600 kg**
>
> **Maximum number of people 8**
>
> **Average weight of occupants 75 kg**

a To which average does the sign refer, the mean, the median or the mode? *(1)*

b Explain how you made your choice. *(1)*

NEAB, 1998, Paper 2

6 The figures below show the number of days each pupil, in a class of 30 children, was absent during 4 school weeks.

4	0	2	5	2	1	6	0	3	1
1	0	0	0	5	7	5	5	0	2
3	0	0	2	1	1	1	0	0	1

a Copy and complete the following table, using tallies to obtain the frequencies. *(4)*

Number of days absent	Tally	Frequency
0		
1		
2		
3		
4		
5		
6		
7		

b Write down the modal number of days absent. *(1)*

c The total of the number of days absent is 58. Use this total to calculate the mean number of days absent per pupil. *(2)*

d The school has 10 classes. Gemma estimates that if there are 58 absences in 1 class there are about 580 absences in 10 classes. Give **two** reasons why her estimate may not be very reliable. *(2)*

NEAB, 1997, Paper 1

7 The table shows the number of greenfly counted on each of a sample of 30 plants.

Number of greenfly	Number of plants
25	2
26	5
27	10
28	10
29	1
30	2

a Calculate the mean number of greenfly per plant for this sample. *(3)*

A second sample of 10 plants was selected.
The mean number of greenfly per plant for this second sample was 8.5
The two samples were then combined to form one sample of 40 plants.

b Calculate the mean number of greenfly per plant for the combined sample of 40 plants, giving your answer correct to one significant figure. *(4)*

A similar survey was conducted the following year.
For this survey the mean number of greenfly per plant was 25, correct to two significant figures.

c What is the smallest possible calculated value for this mean? *(1)*

SEG, June 1998, Paper 1

8 The weekly wages (in pounds) of the twenty workers in a factory are shown below.

85	90	90	90	85	85	120	85	85	85
90	120	160	220	85	90	120	335	160	120

a Copy and complete the frequency distribution below. *(2)*

Wage (in £)	Tally	Frequency
85		
90		
120		
160		
220		
335		

b The shop steward says that the average wage of the workers is £85. Which average is he referring to, the mean, the median or the mode? *(1)*

c The manager says that the average wage of the workers is £90. Which average is he referring to, the mean, the median or the mode? *(1)*

d The twenty workers were asked to fill in this questionnaire.

> Throw a coin
>
> If it shows a HEAD, tick the YES box below.
>
> If it shows a TAIL, answer the question "DO YOU EVER SMOKE AT WORK?"
>
> YES ☐ NO ☐

How many HEADS would you expect? *(1)*

e If all the workers smoke at work how many forms will have the YES box ticked? *(2)*

f If no workers smoke at work how many forms would you expect to have the YES box ticked? *(2)*

g When the forms are returned, 16 have the YES box ticked. Estimate the number of workers who smoke at work. *(2)*

h What is the advantage of using a questionnaire like this? *(1)*

NEAB, 1996, Paper 2

9 The frequency table below has two values missing.

x	1	3	4	6	
f	2	4	6		3

a The range of x is 7. $\Sigma f = 20$.
Use this information to complete the table. *(3)*

b What is the modal value of x? *(1)*

c Find the mean value of x. *(4)*

20 June 2000

Choosing a representative figure

There are three different types of average. These can be calculated for both discrete and continuous data.
The **mode** is the most common or the most popular data value.
The **median** is the middle value when the data is written in order of size.

The **mean** can be calculated using the formula $\bar{x} = \frac{\Sigma fx}{\Sigma f}$

For grouped data, the median and the mean can only be estimated.

Example

The table shows the lifetime, in hours, of a sample of light bulbs.

Lifetime (hours)	Frequency, f	Mid-point, x	$f \times x$	Running total
201–400	50	300.5	15 025	50
401–600	125	500.5	62 562.5	175
601–800	80	700.5	56 040	
801–1000	44	900.5	39 622	
	$\Sigma f = 299$		$\Sigma fx = 173\,249.5$	

Modal group = 401–600 hours

Position of the median $= \frac{299 + 1}{2} = 150$th term

Estimate of the median $= 400.5 + \frac{100}{125} \times 200 = 560.5$ hours

Estimate of the mean, $\bar{x} = \frac{\Sigma fx}{\Sigma f} = \frac{173\,249.5}{299} = 579$ hours (nearest hour)

It is sometimes important to calculate the mean as a **weighted mean.**

The weighted mean $= \frac{\Sigma wx}{\Sigma w}$ where w is the weighting given to each value of x.

Weightings can be given as percentages or ratios.

Estimation

A population is a collection of individuals or items.
A random sample can be taken from the population.

(1) The sample mean is the best estimate of the population mean.
(2) The sample proportion is the best estimate of the population proportion.

The accuracy of these estimates improves as the sample size is increased.

Control chart

A control chart is a statistical tool that monitors a process and warns us when something is going wrong.

Working with *Excel*

Calculating a mean

Jon owns a garage. He records the number of cars that he repairs each day.
These are the numbers of cars that he repaired in July.

8	2	8	9	6	11	5	8	6	9	10
6	5	10	8	5	13	9	10	5	8	15

Use *Excel* to calculate the **mean** of this data.

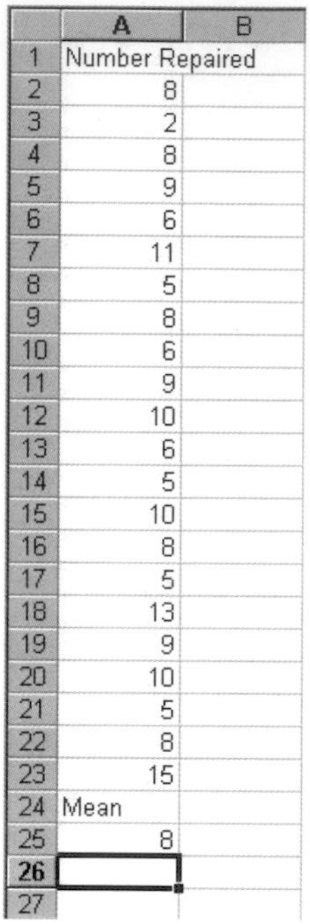

	A	B
1	Number Repaired	
2	8	
3	2	
4	8	
5	9	
6	6	
7	11	
8	5	
9	8	
10	6	
11	9	
12	10	
13	6	
14	5	
15	10	
16	8	
17	5	
18	13	
19	9	
20	10	
21	5	
22	8	
23	15	
24	Mean	
25	8	
26		
27		

- **Open a new *Excel* document**

- **Enter the information on car repairs**
  In Cell A1, type **Number Repaired**.
  In Cell A2, type 8, in Cell A3, type 2 and so on.

- **Calculate the mean**
  In Cell A24, type **Mean**.
  In Cell A25, insert the formula: **=AVERAGE(A2:A23)**
  Remember to press the **Enter** key.
  (A2:A23) *identifies the cells A2 to A23.*
  Cell A25 now displays the mean of the data.

Calculating a mean from a frequency distribution

Pam has six people in her family. She wonders how many people are in her friends' families. She asks each of her friends and records the information in a table.

Number in family, x	2	3	4	5	6	7	8
Frequency, f	2	4	6	5	2	0	1

Use *Excel* to calculate the mean of this data.

- **Open a new *Excel* document**

- **Enter the information on family size**
  In Cell A1, type **Number in Family, x.** In Cell B1, type **Frequency, f.**
  In Cell C1, type **Number in Family × Frequency**.
  In Cell A2, type 2, in Cell A3, type 3 and so on.
  In Cell B2, type 2, in Cell B3, type 4 and so on.

Calculate the mean

In Cell C2, insert the formula: **=A2*B2**
Select Cell C2.
Place the cursor over the bottom right-hand corner of Cell C2 and drag down to Cell C8.
This copies the formula into each cell.

In Cell B9, enter **Sum**.
In Cell B10, insert the formula:
=SUM(B2:B8)
(B2:B8) *identifies the cells B2 to B8.*
Cell B10 displays Σf.

	A	B	C	D
1	Number in Family, x	Frequency, f	Number in Family x Frequency	
2	2	2	4	
3	3	4	12	
4	4	6	24	
5	5	5	25	
6	6	2	12	
7	7	0	0	
8	8	1	8	
9		Sum	Sum	
10		20	85	
11				
12			Mean	
13			4.25	
14				
15				

In Cell C9, enter **Sum**.
In Cell C10, insert the formula:
=SUM(C2:C8)
(C2:C8) *identifies the cells C2 to C8.*
Cell C10 displays Σfx.

In Cell C12, enter **Mean**.
In Cell C13, insert the formula: **=C10/B10**
Cell C13 displays the mean, $\bar{x}$.

Exercise 3:10

1 The table shows the attendance for six days at an art exhibition and a pottery exhibition.

	Mon	Tues	Wed	Thurs	Fri	Sat
Art	48	61	69	53	39	78
Pottery	52	45	68	59	37	81

a Use *Excel* to find the mean attendance for each exhibition.
b Which exhibition was the most popular?
Give a reason for your answer.

2 Pasreen measured the mid-day temperature for the month of June.
This is her data.

Temperature (°C)	16	17	18	19	20	21	22	23
Number of days	3	6	2	5	1	6	3	4

Use *Excel* to find the mean daily temperature.

1 In Rob's class of 20 pupils the mean height is 125 cm. A new pupil joins the class. His height is 144 cm. Find the new mean height of the class of 21 pupils. Give your answer to the nearest centimetre.

2 The prices of tickets for a pop concert are £8, £15 and £25. The tickets are sold in the ratio of 3:5:7. Find the average price of a ticket. Give your answer to the nearest penny.

3 Kirsty enjoys listening to pop music. She is interested in how many musicians there are in each pop-group.
She records the information in a table.

Number of musicians, x	2	3	4	5	6
Number of pop-groups, f	4	5	7	2	3

a Write down the modal pop-group size.
b Calculate the median pop-group size.
c Calculate the mean pop-group size. Give your answer to 1 d.p.

4 Chris lives on a busy road. One Sunday afternoon, he records the number of cars that pass his house in each 10-minute time interval.

Number of cars, x	0–4	5–9	10–14	15–19
Number of 10-minute time intervals, f	1	5	4	3

a Write down the modal group.
b Calculate an estimate of the median number of cars that pass in 10 minutes.
c Calculate an estimate of the mean, giving your answer to 1 d.p.

5 The table shows the weekly number of hours that sixth form pupils spend at their part-time jobs.

Time (hours), x	$4 \leqslant x < 6$	$6 \leqslant x < 8$	$8 \leqslant x < 10$	$10 \leqslant x < 12$	$12 \leqslant x < 14$
Number of sixth formers, f	3	15	18	7	5

a Write down the modal group.
b Estimate the mean number of hours worked each week. Give your answer to 2 d.p.
c Estimate the median number of hours worked each week. Give your answer to the nearest tenth of an hour.

4 Measures of spread

1 Range
Finding the range of sets of numbers
Upper and lower quartiles
Interquartile range
Semi-interquartile range

2 Cumulative frequency
Cumulative frequency
Using cumulative frequency to find the median and the interquartile range
Cumulative frequency diagrams
Step polygons
Using cumulative frequency diagrams to find the median and interquartile range
Percentiles and deciles

3 Standard deviation
Deviation from the mean
Variance and standard deviation
Alternative formula for the standard deviation
Changing values of a set of data; the effect on the mean and the standard deviation
Finding the standard deviation from a frequency table
Estimating the mean and standard deviation of grouped data

EXAM QUESTIONS
SUMMARY
ICT IN STATISTICS
TEST YOURSELF

1 Range

John and Peter each enter a team for the school high jump competition. The heights jumped by John's team varied between 64 cm and 105 cm. The heights jumped by Peter's team varied between 63 cm and 151 cm. Does this mean that all of Peter's team are better at the high jump than all of John's team?

Range

The **range** of a set of data is the biggest value take away the smallest value.

Example

These are the temperatures for the first five days of July in London.

29 °C   19 °C   25 °C   24 °C   20 °C

What is the range of temperatures?

Range = 29 °C − 19 °C = 10 °C

The range can help you to compare two sets of data.

These are the temperatures for the first five days of July in Chester.

23 °C   19 °C   22 °C   20 °C   21 °C

Range = 23 °C − 19 °C = 4 °C

The smaller value of the range shows that temperatures in Chester are less spread out than for London.
The smaller range means the values are more *consistent*.

Exercise 4:1

1 Write down the range for each of these sets of data.

a	14	18	11	27	21	19	30	25
b	2.6	3.7	3.0	5.1	4.6	1.9	2.2	4.6
c	324	563	724	378	178	452	934	277

2 Geoff has six people working in his shop.
Their weekly wages are £156, £182, £193, £189, £178 and £172.
  a Find the range of the wages.
  b Geoff leaves to open a new shop so he employs a seventh person as the manager. Her wage is £310. Find the new range of the wages.

The range can be greatly affected by one very high or one very low value.
The interquartile range measures the spread of the middle half of the data. It is not affected by extreme values.

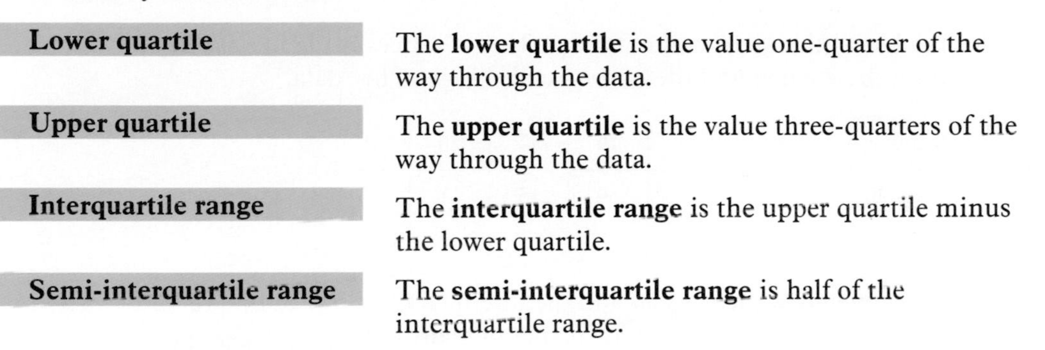

Lower quartile	The **lower quartile** is the value one-quarter of the way through the data.
Upper quartile	The **upper quartile** is the value three-quarters of the way through the data.
Interquartile range	The **interquartile range** is the upper quartile minus the lower quartile.
Semi-interquartile range	The **semi-interquartile range** is half of the interquartile range.

To find the quartiles
1. List the data in order of size, smallest first.
2. Find the position of the median.
3. Look at the data to the left of this position.
   The lower quartile is the median of this data.
4. Look at the data to the right of this position.
   The upper quartile is the median of this data.

Look at these two sets of data. The second has a 2 as an extra value.

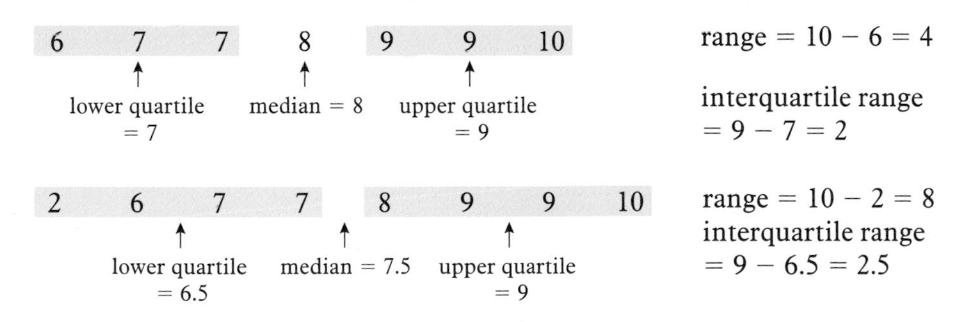

The extra number 2 has a big effect on the range.
It has influenced the interquartile range but not distorted it too much.
The interquartile range is a more reliable measure of spread than the range.
The semi-interquartile range is half of 2.5 It is 1.25
It is used in the same way as the interquartile range to look at the spread of data.

For questions **3–6** find:

a the median **b** the interquartile range **c** the semi-interquartile range.

3 5 6 7 7 9 10 12 13 15

4 13 15 17 17 18 24 27 29

5 3.2 1.8 4.6 2.9 3.1 4.0 5.7 7.2 5.0

6 64 72 49 56 61 40 84 69

7 Jan has to check the contents of packets of toffees made in a factory.
Find the median and the interquartile range of her data.

Number of toffees	15	16	17	18	19
Frequency	7	9	13	11	5

8 The table gives the median and the interquartile range of the weights in grams of strawberries grown from two types of plant.

	Median	Interquartile range
Red Glory	905	129
Sweet Giant	897	351

a Use the medians and the interquartile ranges to compare the two types of strawberry plant.

b Which type of plant would you use if you wanted to grow a plant for a competition for producing the most weight?
Explain your answer.

9 Kate has two machines in her factory that cut lengths of ribbon from a large roll.
Each length of ribbon should be at least 80 cm long.
Kate has tested the two machines by taking a sample of 100 lengths from each machine.
These are her results:

	Median	Semi-interquartile range
Machine A	81	3.2
Machine B	83	1.7

Kate has to get rid of one machine. Which machine should she keep?
Explain your answer.

2 Cumulative frequency

Have you ever wondered how exam grades are worked out? How many marks does it take to pass or to get a grade A? Examiners use cumulative frequency to answer questions like:
'How many pupils scored more than 73%?'

Cumulative frequency

Cumulative frequency is a running total of the frequencies.

Example A group of people were asked how many times they had used a bus that week. The table shows the results.
How many people used the bus five times or less?

Number of times	Frequency	Number of times	Cumulative frequency
3	4	≤3	4
4	15	≤4	19 (4 + 15)
5	21	≤5	40 (19 + 21)
6	16	≤6	56 (40 + 16)
7	4	≤7	60 (56 + 4)

The number of people that used the bus five times or less is 40.

Exercise 4:2

1 The table shows the number of aces served in a set of tennis by 100 players.

Number of aces	Frequency	Number of aces	Cumulative frequency
0	14	0	
1	26	up to 1	
2	37	up to 2	
3	11		
4	9		
5	3		

a Copy the table. Fill it in.
b Write down how many players served:
(1) two aces or less (2) less than four aces (3) more than three aces.

2 The table shows the number of visits people had made to the local library in the past three months.

Number of visits	Frequency	Number of visits	Cumulative frequency
Less than 3	21	Less than 3	
3 to less than 6	19	Less than 6	
6 to less than 9	14	Less than 9	
9 to less than 12	7		
12 to less than 15	4		

a Copy the table. Fill it in.
b How many people were involved in the survey?
c How many people visited the library less than nine times?
d How many people visited the library six times or more?

3 Mark is a window cleaner. He recorded how much money he made each week for 60 weeks. These are his results:

Money in £, w	Frequency	Money in £	Cumulative frequency
$81 \leqslant w < 100$	7	<100	
$100 \leqslant w < 120$	10	<120	
$120 \leqslant w < 140$	5	<140	
$140 \leqslant w < 160$	14		
$160 \leqslant w < 180$	19		
$180 \leqslant w < 200$	5		

a Copy the table. Fill it in.
b How many weeks did he take home less than £120?
c How many weeks did he take home £160 or more?

You can use cumulative frequency to find the median and interquartile range.

Number of visits	Frequency	Number of visits	Cumulative frequency
3	4	$\leqslant 3$	4
4	15	$\leqslant 4$	19
5	21	$\leqslant 5$	40
6	16	$\leqslant 6$	56
7	4	$\leqslant 7$	60

The median is the value half-way through the data.
The total cumulative frequency is 60 so the median is in position $60 \div 2 = 30$.
The first two rows of the table only go as far as the 19th value.

The third row of the table goes as far as the 40th value.
The 30th value must be in the third row of the table.
So the median number of visits is 5.

The lower quartile is one quarter of the way through the data.
The lower quartile is in position $60 \div 4 = 15$.
The 15th value is in the second row.
So the lower quartile of the number of visits is 4.

The upper quartile is three quarters of the way through the data.
The upper quartile is in position $15 \times 3 = 45$.
The 45th value is in the fourth row.
So the upper quartile of the number of visits is 6.

The interquartile range is $6 - 4 = 2$.

4 Find the median and the interquartile range for the data in question 1.

5 The table shows the number of GCSEs passed by Year 11 boys at Stanthorne High last year.

Number of GCSEs	Frequency	Number of GCSEs	Cumulative frequency
1	5		
2	7		
3	11		
4	16		
5	21		
6	28		
7	19		
8	8		
9	5		

a Copy the table. Fill it in.
b Write down how many pupils passed:
(1) less than 5 (2) 5 or more (3) 8 or less GCSEs.
c Find the median number of GCSEs passed.
d Find the interquartile range.
e These are the corresponding values for Year 11 girls.
Median = 7 Interquartile range = 4
Compare the results of the boys and the girls.

Cumulative frequency diagram

A **cumulative frequency diagram** shows how the cumulative frequency changes as the data values increase. The data is shown on a continuous scale on the horizontal axis. The cumulative frequency is shown on the vertical axis.

You plot the upper ends of each group against the cumulative frequency. You then join the points with straight lines or a curve. If you use straight lines the diagram is called a cumulative frequency polygon. If you use a curve then the diagram is called a cumulative frequency curve. Some questions allow you to choose which to use.

Example The table shows the ages of people in a survey.

Age (years)	Cumulative frequency
<20	3
<25	11
<30	28
<35	40
<40	46
<45	50

a Draw a cumulative frequency curve for the data.
b Estimate the numbers of people in the survey who are:
(1) less than 28 years old (2) at least 36 years old.

a The values given in the age column are all at the upper ends of their groups. You plot these against the values in the cumulative frequency column. The points to plot are (20, 3), (25, 11) and so on.

b (1) Find 28 on the age axis, move up to the graph and across. Read the value off the cumulative frequency scale.
This gives approximately 20 people.

(2) Repeat the process for 36 but subtract the answer from 50.
This gives approximately 8 people.

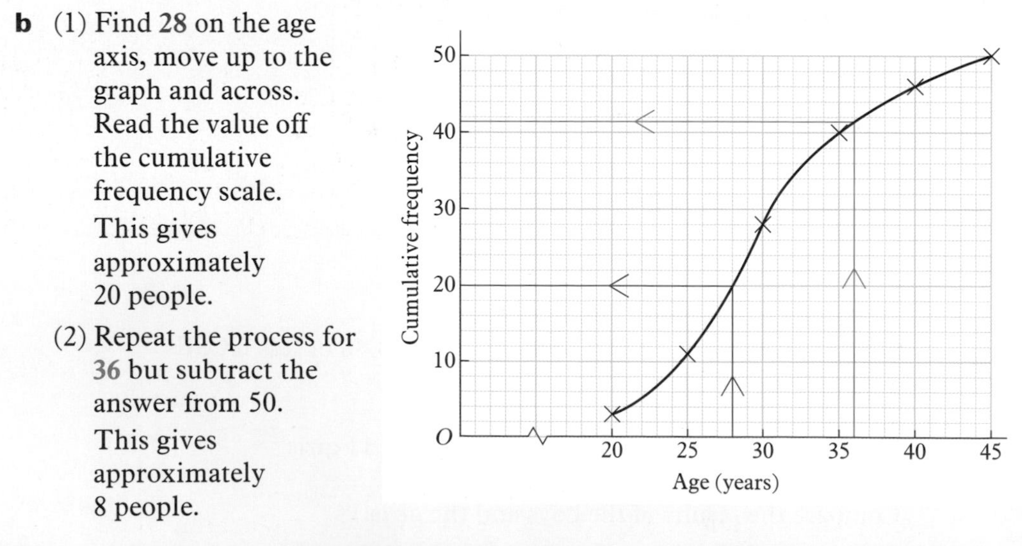

Exercise 4:3

1 **a** Draw a cumulative frequency polygon for the data in question **3** of Exercise 4:2.

b Use your polygon to estimate how many weeks Mark took home:
(1) less than £150 (2) more than £130.

2 Eighty car owners were selected at random and asked the age and value of their car. These are the results:

Age (years)	Frequency
$0 < \text{age} \leqslant 3$	15
$3 < \text{age} \leqslant 6$	36
$6 < \text{age} \leqslant 9$	14
$9 < \text{age} \leqslant 12$	8
$12 < \text{age} \leqslant 15$	5
$15 < \text{age} \leqslant 18$	2

Value (£1000s)	Frequency
$0 < \text{value} \leqslant 4$	8
$4 < \text{value} \leqslant 8$	14
$8 < \text{value} \leqslant 12$	35
$12 < \text{value} \leqslant 16$	18
$16 < \text{value} \leqslant 20$	4
$20 < \text{value} \leqslant 24$	1

a Draw a cumulative frequency polygon for the data on the age of the cars.

b Use your polygon to estimate how many cars were:
(1) less than 10 years old (2) more than 4 years old.

c Draw a cumulative frequency curve for the data on the value of the cars.

d Use your curve to estimate how many cars were valued at:
(1) more than £9000 (2) less than £5000.

3 Sarah organised a balloon competition for charity. The balloons were all released from the same place. A card was attached to each balloon and the finder of the balloon wrote on the card where it had been found. Sarah made this table to show the distances, in miles, travelled by the balloons.

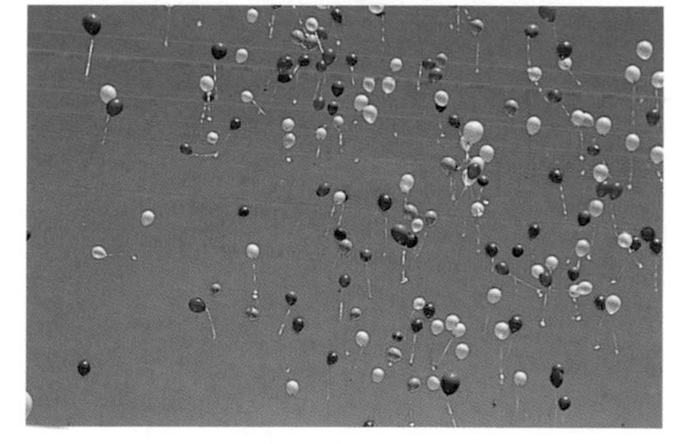

Up to 100	3
100–199	9
200–299	18
300–399	11
400–499	5
500–599	4

a Draw a cumulative frequency diagram to show this data.

b Use your diagram to estimate the number of balloons that travelled:
(1) less than 250 miles (2) more than 350 miles.

4 Michael asks 150 students in his school how long they spent watching TV over the weekend. The frequency polygon shows his data.

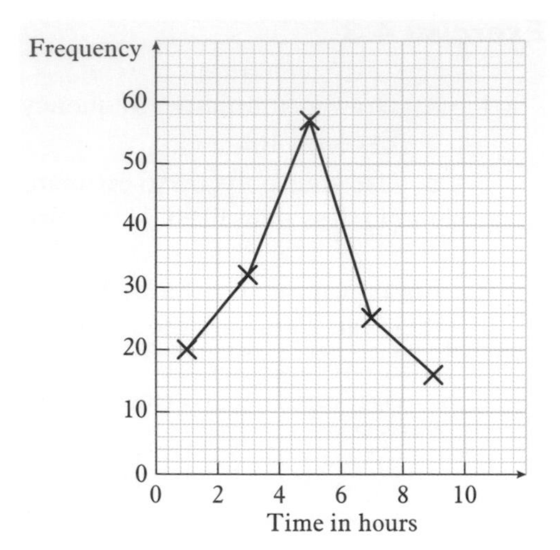

a Copy this table

Time, t, hours	Frequency
$0 \leq t < 2$	20
$2 \leq t < 4$	
$4 \leq t < 6$	
$6 \leq t < 8$	
$8 \leq t < 10$	

b Use the frequency polygon to complete the table.

c Draw a cumulative frequency diagram to show this data.

5 Sara collects data on the distances, to the nearest mile, that people travel to work. The cumulative frequency polygon shows her data.

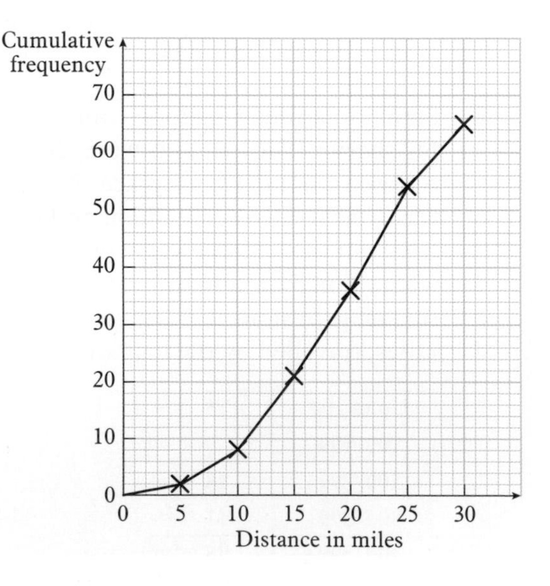

a Copy this table

Distance, m, miles	Frequency
$0 \leq m < 5$	
$5 \leq m < 10$	
$10 \leq m < 15$	
$15 \leq m < 20$	
$20 \leq m < 25$	
$25 \leq m < 30$	

b Use the cumulative frequency polygon to complete the table.

c Draw a frequency polygon to show Sara's data.

If you are dealing with discrete data you draw a step polygon to show the cumulative frequency.

Penny has recorded the number of sweets inside 58 bags of assorted sweets. The table shows her data.

Number of sweets	Cumulative frequency
10	5
11	17
12	32
13	51
14	51
15	58

To draw a step polygon for the data:
You plot the points (10, 5), (11, 17) ... as before.
You use vertical lines to show the increase in the cumulative frequency at each of the values. The diagram looks like a series of steps.

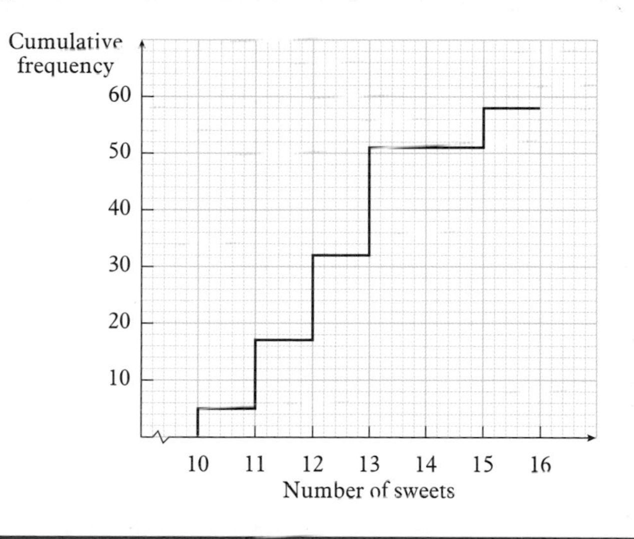

6 Draw a step polygon to show this data.

Number of TV sets per household	1	2	3	4	5
Number of households (frequency)	3	14	21	5	1

7 The table shows the number of eggs laid by Rita's chickens.

3	8	6	8	3	9	6	4	5	9
7	9	7	5	6	6	4	1	8	5
7	3	5	9	6	4	8	6	7	7
2	6	4	8	2	5	4	6	2	7

a Draw a cumulative frequency table for the data.
b Draw a step polygon to show the data.

You can also estimate the median and quartiles using a cumulative frequency diagram.

To get an estimate of the **median**:
(1) divide the total cumulative frequency by 2
(2) find this point on the cumulative frequency axis
(3) draw a line across to the curve and down to the horizontal axis
(4) read off the estimate of the median.

To get an estimate of the **lower quartile**:
(1) divide the total cumulative frequency by 4
(2) find this point on the cumulative frequency axis
(3) draw lines as you did for the median
(4) read off the lower quartile.

To get an estimate of the **upper quartile**:
(1) divide the total cumulative frequency by 4 and multiply by 3
(2) find this point on the cumulative frequency axis
(3) draw lines as you did for the median
(4) read off the upper quartile.

This is the cumulative frequency curve from page 112.

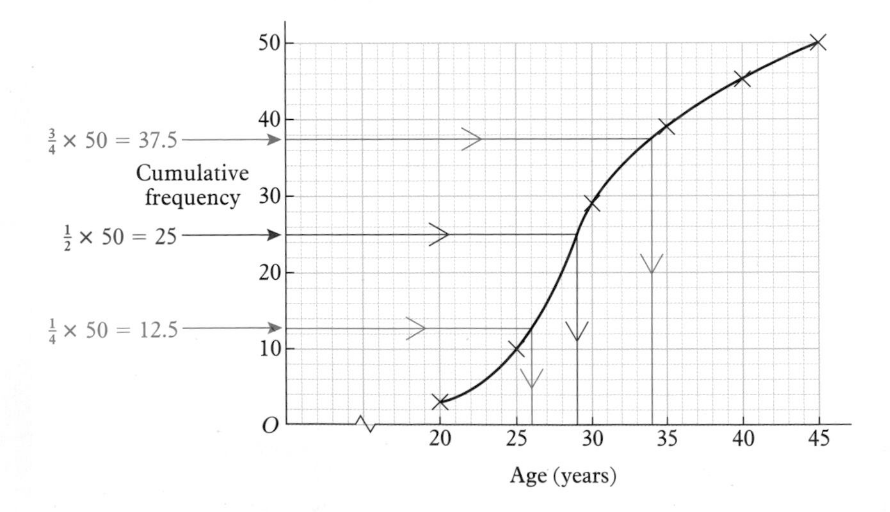

Median = 29 years **Lower quartile** = 26 years **Upper quartile** = 34 years
Interquartile range = 34 − 26 = 8 years

Exercise 4:4

1 The table shows the length of time, in minutes, that customers queued at the express check-out in a supermarket.

Time, t (mins)	Number of customers
$0 < t \leqslant 1$	5
$1 < t \leqslant 2$	13
$2 < t \leqslant 3$	23
$3 < t \leqslant 4$	36
$4 < t \leqslant 5$	46
$5 < t \leqslant 6$	36
$6 < t \leqslant 7$	30
$7 < t \leqslant 8$	11

a Draw a cumulative frequency polygon for this data.
b Estimate the median time.
c Estimate the lower and upper quartiles.
d Find the interquartile range.

2 The table shows the amount of money, in pounds (£), that families pay for mortgages each month.

Payment, p (£)	Number of families	Payment, p (£)	Number of families
$0 < p \leqslant 50$	7	$200 < p \leqslant 250$	32
$50 < p \leqslant 100$	15	$250 < p \leqslant 300$	29
$100 < p \leqslant 150$	21	$300 < p \leqslant 350$	20
$150 < p \leqslant 200$	29	$350 < p \leqslant 400$	7

a Draw a cumulative frequency curve for this data.
b Estimate the median payment.
c Estimate the lower and upper quartiles.
d Find the interquartile range.

The **percentiles** can also give information on the spread of data.

To find the percentiles you take the total cumulative frequency as 100%.
To find the 10th percentile you find 10% of the total cumulative frequency.
This cumulative frequency curve shows the times that customers spent in a shop.
The total cumulative frequency = 60
10% of 60 = 6
Find 6 on the cumulative frequency axis.

Draw lines as you did for the median.
Read off the 10th percentile.
Similarly, for the 45th percentile you find 45% of the total cumulative frequency.
45% of 60 = 27
Find 27 on the vertical axis and move to the horizontal axis as before.

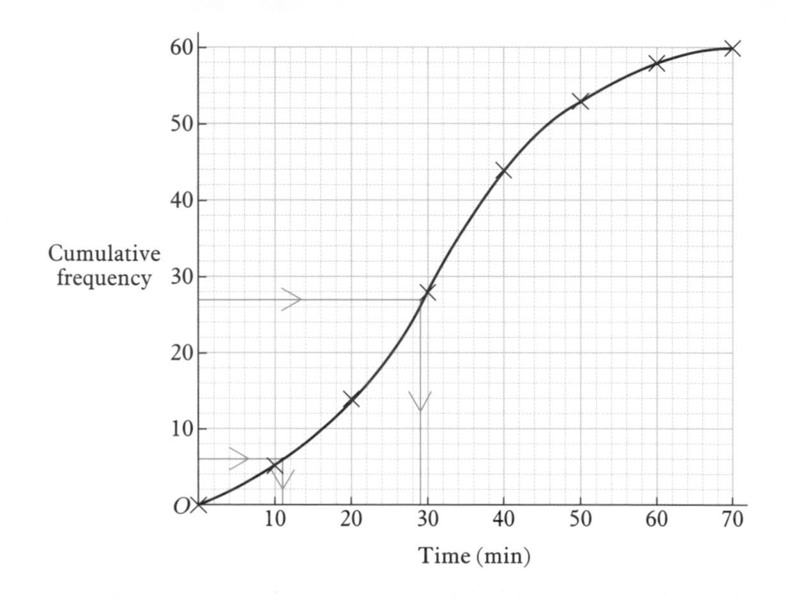

The 10th percentile is 11.   The 45th percentile is 29.

There are special names for the percentiles that are multiples of ten.
The 10th percentile is also called the 1st decile.
The 20th percentile is also called the 2nd decile.
The 30th percentile is also called the 3rd decile … and so on.

---

3 Use the graph that you have drawn for question **1**.
  a Find the 40th percentile.
  b Find the 9th decile.

4 Use the graph that you have drawn for question **2**.
  a Find the second decile.
  b Find the eighth decile.
  c How many families pay an amount of money that lies within the middle 60% of the mortgage payments?

You can also find percentiles and deciles using a frequency table.

Trish lives near the local bus stop. One day she records the number of people waiting in the queue for each bus. The table shows her data.

Number of people	3	4	5	6	7	8	9	10	11
Frequency	4	6	3	8	0	7	5	9	8
Cumulative frequency	4	10	13	21	21	28	33	42	50

To find the 4th **decile** you find $\frac{4}{10}$ of the total frequency.

The total frequency = 50 $\quad \frac{4}{10} \times 50 = 20$

The 4th decile is the 20th value.
Look at the cumulative frequency. The 20th value = 6 people.
The 4th decile is 6 people.

To find the 82nd **percentile** you find $\frac{82}{100}$ of the total frequency.

The total frequency = 50 $\quad \frac{82}{100} \times 50 = 41$

The 82nd percentile is the 41st value.
Look at the cumulative frequency. The 41st value = 10 people.
The 82nd percentile is 10 people.

5 Jamie helps at the school shop. He has recorded the number of items bought by each pupil during one lunch break.

Number of items	1	2	3	4	5	6	7	8
Number of pupils	2	5	25	12	2	1	2	1

a Copy the table. Add a row showing the cumulative frequencies.
b Find: (1) the 54th percentile (2) the 78th percentile.
c Find: (1) the 1st decile (2) the 9th decile.
d Find the range between the 1st and the 9th decile.
e Write down one advantage of using this interdecile range as a measure of spread.

6 Mary has asked each pupil in Year 11 how many people live in their house. This is her data:

Number of people	2	3	4	5	6	7	8	9
Frequency	11	35	58	60	21	10	3	2

a Copy the table. Add a row showing the cumulative frequencies.
b Find: (1) the 1st decile (2) the 8th decile.
c Find: (1) the 15th percentile (2) the 64th percentile.

You can use an interpercentile range as a measure of spread.

10th–90th interpercentile range	This is the 90th percentile minus the 10th percentile. It uses the middle 80% of the data. This gets rid of any extreme values.
20th–80th interpercentile range	This is the 80th percentile minus the 20th percentile. It uses the middle 60% of the data. This gets rid of values in the top 20% and the bottom 20% of the data.
25th–75th interpercentile range	This is the 75th percentile minus the 25th percentile. It uses the middle 50% of the data. This is the interquartile range.

Example Sam has investigated the times taken by two groups to solve a puzzle. These are his results.

	10th percentile	90th percentile	mean
Group A	5 mins	46 mins	27.0 mins
Group B	11 mins	37 mins	26.9 mins

a Work out the 10th–90th interpercentile range for each group.
b Use the means and interpercentile ranges to compare the two groups.

a Group A 10th–90th interpercentile range = 46 − 5 = 41 mins
Group B 10th–90th interpercentile range = 37 − 11 = 26 mins

b Both groups take the same time on average but Group B has a smaller interpercentile range. This means that the times for Group B are more consistent.

7 Jake investigates the life span of cats.
He collects information from two areas – one in the country and one in the town.
This is his data:

	20th percentile	80th percentile	median
Country	7 years	16 years	12 years
City	4 years	16 years	7 years

a Compare the lifespans of the two groups of cats.
b Would it have made any difference to your answer if Jake had used the range instead of the interpercentile range. Give reasons for your answer.

3 Standard deviation

Roger is considering a change of career. He wants to earn more money. He is looking at the salaries for different types of job.
He has the mean wage for each type of job.
He is interested in how close the rest of the wages are to the mean.

You need a way of finding out how far each value in the data is from the mean.

Deviation from the mean

The difference between a particular value and the mean is the **deviation from the mean** for that value.
If you use x for the value and $\bar{x}$ for the mean, then the deviation from the mean is $x - \bar{x}$.

Exercise 4:5

1 **a** Find the mean of: 3, 10, 13, 5, 14.
b Copy the table.
Fill in the missing values.

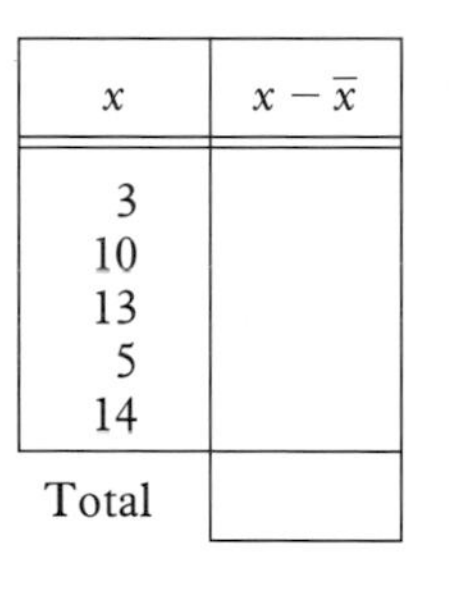

x	$x - \bar{x}$
3	
10	
13	
5	
14	
Total	

2 Copy these tables. For each one:
(1) find $\bar{x}$
(2) fill in the missing values.

a

x	$x - \bar{x}$
18	
12	
15	
11	
19	
Total	

b

x	$x - \bar{x}$
25	
48	
33	
29	
20	
Total	

c

x	$x - \bar{x}$
131	
214	
177	
160	
158	
Total	

3 Look at the totals of $(x - \bar{x})$ that you worked out in question 2.
Explain why $\Sigma(x - \bar{x})$ can't be used to measure the spread of the data.

Modulus of a number

$|x|$ means the **modulus** or size of the number x
$|-3|$ and $|3|$ have the same modulus or size.
$|-3| = 3$ and $|3| = 3$

4 Write down the sizes of each of these.

a $|4|$ **c** $|\frac{3}{8}|$ **e** $|-1.756|$

b $|-3|$ **d** $|-2.5|$ **f** $|-1\frac{3}{4}|$

Mean deviation

There is a measure of spread that uses the deviations from the mean. It is called the **mean deviation**.

Find the size of the deviation from the mean of each data value. $|x - \bar{x}|$

Find the sum of these. $\Sigma|x - \bar{x}|$

Divide by the number of data values to get an average. $\dfrac{\Sigma|x - \bar{x}|}{n}$

This is called the **mean deviation**.

Example Find the mean deviation of: 13, 20, 22, 11, 23 and 19.
Use a table to show your working.

$\bar{x} = 108 \div 6 = 18$

x	$x - \bar{x}$	$\lvert x - \bar{x}\rvert$
13	−5	5
20	2	2
22	4	4
11	−7	7
23	5	5
19	1	1
$\Sigma x = 108$		$\Sigma\lvert x - \bar{x}\rvert = 24$

The mean deviation $= 24 \div 6 = 4$.

5 Find the mean deviation from the mean of each of these sets of data.

a 19, 26, 18, 23, 24

b 10, 7, 8, 4, 5, 11, 9, 10

c 15, 18, 14, 11, 8, 7, 9, 8, 17, 18

d 51, 52, 41, 48

e 13, 17, 12, 11, 19, 16, 9, 7, 11, 5

f 12, 15, 10, 14, 15, 12

This is another measure of spread that uses the deviations from the mean.

Square the deviations from the mean to get rid of negative values. $(x - \bar{x})^2$

Find the sum of these values. $\Sigma(x - \bar{x})^2$

Divide by the number of values to get an average. This is called the variance. $\frac{\Sigma(x - \bar{x})^2}{n}$

Find the square root of the variance so that the units of spread are the same as the original units. $\sqrt{\frac{\Sigma(x - \bar{x})^2}{n}}$

Standard deviation

The standard deviation, s, of a set of data is given by the formula:

$$s = \sqrt{\frac{\Sigma(x - \bar{x})^2}{n}}$$

The higher the standard deviation, the more spread out the data is.

Example

Find the mean and standard deviation of: 12, 9, 11, 5, 7 and 10.

You will need a table to show your working.

	x	$x - \bar{x}$	$(x - \bar{x})^2$
	12	3	9
	9	0	0
	11	2	4
	5	−4	16
	7	−2	4
	10	1	1
Totals	54		34

$\bar{x} = 54 \div 6 = 9$

$$s = \sqrt{\frac{34}{6}} = 2.38047\ldots$$

The mean is 9 and the standard deviation is 2.38 to 3 s.f.
An advantage of the standard deviation over the interquartile range is that it uses all the data.

Exercise 4:6

1 Find the mean and standard deviation of each of these sets of data.

a 34, 39, 30, 25
b 15, 19, 9, 21, 26
c 54, 72, 33, 49, 83, 45
d 42, 41, 47, 46, 44
e 98, 87, 93, 86, 88, 90, 95
f 5, 1, 9, 20, 6, 11, 4, 5, 6, 3

2 These are the amounts of money that Laura's school raised for charity per month over a six-month period:

£124 £178 £67 £89 £52 £156

a Find the mean amount raised per month.
b Find the standard deviation of the amount raised. Give your answer to the nearest penny.
c James' school also raised money for charity over the same six-month period. The mean of the money raised each month was £105 and the standard deviation was £6.42
Compare the money raised by the two schools.

An alternative formula for the standard deviation is $s = \sqrt{\frac{\Sigma x^2}{n} - \left(\frac{\Sigma x}{n}\right)^2}$

This formula always gives the same results as the first formula but is much easier to work with, especially when the mean is not a whole number.

Example Find the mean and standard deviation of: 27, 34, 21, 37, 35.

Only two columns are needed to show your working.

	x	x^2
	27	729
	34	1156
	21	441
	37	1369
	35	1225
Totals	$\Sigma x = 154$	$\Sigma x^2 = 4920$

$\bar{x} = \frac{\Sigma x}{n} = \frac{154}{5} = 30.8$ $\quad s = \sqrt{\frac{4920}{5} - \left(\frac{154}{5}\right)^2} = 5.946427\ldots$

The mean is 30.8 and the standard deviation is 5.95 to 3 s.f.

---

3 Use the alternative formula to find the standard deviations of the data given in question **1**.

4 These are the scores of two players over six rounds of golf:

Martin 87, 71, 90, 82, 84, 82
Jane 77, 91, 85, 90, 70, 67

a Find the mean and standard deviation of the scores of each player using the new formula.
b Use your answers to part **a** to compare the two players.

5 **a** Find the mean, median and standard deviation of this data: 35, 41, 32, 33, 32, 37, 36.
b Sam adds 4 to each item of data. The numbers are: 39, 45, 36, 37, 36, 41, 40. Find the mean, median and standard deviation of this new data.
c Use your answers to parts **a** and **b** to explain the effect on the mean, median and standard deviation of adding 4 to each item of data.

If you change a set of data by adding the same number A to each value you can work out what happens to the measures of average and spread.

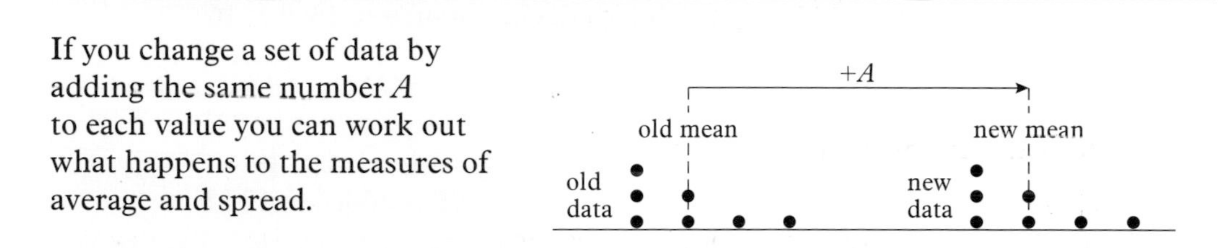

Adding A to the data translates all the data values $+A$ along the number line. The pattern of the data stays the same so the measures of spread will be the same. The mean value of the new data will be A more than the mean of the old data. This will also be true for the median and the mode.

$$\text{new mean} = \text{old mean} + A$$

$$\text{new mode} = \text{old mode} + A$$

$$\text{new median} = \text{old median} + A$$

$$\text{new range} = \text{old range}$$

$$\text{new standard deviation} = \text{old standard deviation}$$

---

6 **a** Find the mean and standard deviation of: 16, 18, 14, 10, 13, 15, 17.
b Write down the mean and standard deviation of: 26, 28, 24, 20, 23, 25, 27.
c Write down the mean and standard deviation of: 19, 21, 17, 13, 16, 18, 20.
d Write down the mean and standard deviation of: 14, 16, 12, 8, 11, 13, 15.

7 **a** Find the mean, median and standard deviation of the data: 4, 6, 11, 9, 16, 3.
b Multiply each item of data by 3. Find the mean, median and standard deviation of this new set of data.
c Use your answers to parts **a** and **b** to explain the effect on the mean, median and standard deviation of multiplying each item of data by 3.

If you change a set of data by multiplying each value by the same number k, you can work out what happens to the measures of average and spread.

$$\text{new mean} = \text{old mean} \times k$$
$$\text{new mode} = \text{old mode} \times k$$
$$\text{new median} = \text{old median} \times k$$
$$\text{new range} = \text{old range} \times k$$
$$\text{new standard deviation} = \text{old standard deviation} \times k$$

---

8 **a** Find the mean and standard deviation of: 1.5, 2.6, 1.9, 3.2
b Write down the mean and standard deviation of: 3.0, 5.2, 3.8, 6.4
c Write down the mean and standard deviation of: 15, 26, 19, 32
d Write down the mean and standard deviation of: 7.5, 13, 9.5, 16

9 Ken is a gardener. Ken has listed the charges before VAT is added, of his last seven jobs.
The mean charge is £75.45 and the standard deviation is £37.19
Find the mean and standard deviation when VAT is added at 17.5%.

You can find the mean and standard deviation from a frequency table.
To do this you multiply the columns for x and x^2 by the frequency values.

Example Find the mean and standard deviation of the data in the table.

x	Frequency, f
5	2
6	8
7	5

You need to add some extra columns

x	Frequency, f	$f \times x$	x^2	$f \times x^2$
5	2	10	25	50
6	8	48	36	288
7	5	35	49	245
	$\Sigma f = 15$	$\Sigma fx = 93$		$\Sigma fx^2 = 583$

The number of values of data, n, is now replaced by Σf

The formula $s = \sqrt{\dfrac{\Sigma x^2}{n} - \left(\dfrac{\Sigma x}{n}\right)^2}$ becomes $\sqrt{\dfrac{\Sigma fx^2}{\Sigma f} - \left(\dfrac{\Sigma fx}{\Sigma f}\right)^2}$

$\overline{x} = \dfrac{93}{15} = 6.2 \qquad s = \sqrt{\dfrac{583}{15} - \left(\dfrac{93}{15}\right)^2} = 0.65319\ldots$

The mean is 6.2 and the standard deviation is 0.653 to 3 s.f.

Exercise 4:7

1 **a** Copy the table. Fill in the missing values.
b Find the mean and standard deviation of the data.

x	Frequency, f	$f \times x$	x^2	$f \times x^2$
3	24			
7	15			
8	21			

2 Find the mean and standard deviation of the data in each table.

a

x	Frequency, f
10	14
11	19
12	11
13	8

b

x	Frequency, f
2.5	12
3.1	26
4.5	9
5.5	7

3 **a** Set your calculator in statistics mode.
b Enter this data: 14.2, 16.9, 16.3, 15, 12.6, 17.2, 14.6, 16, 13.7, 12.9
c Find these values: (1) Σx (2) $\overline{x}$ (3) Σx^2 (4) s.
d Enter this extra data: 15.6, 13.8, 17.2
e Re-do part **c** with the extra data included.

4 Use your calculator to find the mean and standard deviation of the data in each table. Remember to clear your calculator memories first.

a

x	Frequency, f
2.5	14
2.9	26
3.3	21

b

x	Frequency, f
21.6	31
23.6	47
25.6	39

You can estimate the mean and standard deviation of grouped data. To do this you use the middle value of each group as a representative value.

x	Mid-value	Frequency, f
$10 \leqslant x < 15$	12.5	8
$15 \leqslant x < 20$	17.5	11
$20 \leqslant x < 25$	22.5	15
$25 \leqslant x < 30$	27.5	10

From the calculator $\bar{x} = 20.568\ldots, s = 5.1363\ldots$
These values need to be rounded as they are only estimates.
Rounding to 1 d.p. the mean is 20.6 and the standard deviation is 5.1

---

5 **a** Copy the table.
b Fill in the missing values.
c Estimate the mean.
d Estimate the standard deviation.

x	Mid-value	Frequency, f
$5 < x \leqslant 10$		6
$10 < x \leqslant 15$		9
$15 < x \leqslant 20$		16
$20 < x \leqslant 25$		8
$25 < x \leqslant 30$		7

6 The prices paid for new cars in a garage for one week are shown in the table.

Price	£5000–	£10 000–	£15 000–	£20 000–	£25 000–£40 000
Number sold	14	29	35	21	15

a Calculate an estimate for the mean and standard deviation of the prices paid.
b The mean and standard deviation for prices at a second garage are £13 412 and £4981 respectively. Compare the prices at the two garages.

7 The bar-chart shows the number of houses sold in Tanley in 1999.

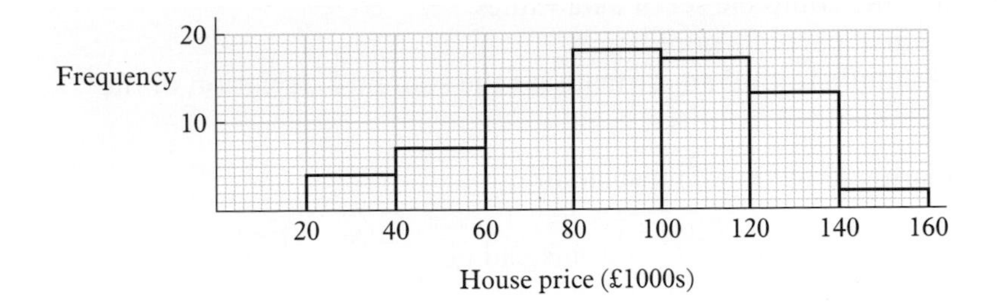

a How many houses were sold in 1999 in Tanley?

b Calculate an estimate of the mean and standard deviation of the house prices.

c The table gives estimates of the mean and standard deviation of house prices in 1999 for two other towns.

	Cordworth	Lornton
Mean	£85 104	£70 497
Standard deviation	£10 971	£22 048

Compare the prices in the three towns.

8 The graph shows the length of calls made from a village telephone box.

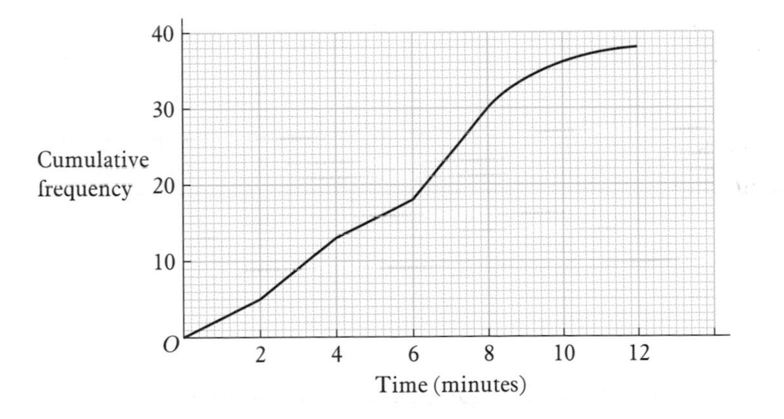

a Use the curve to complete the frequency table.

Length of call (minutes)	0–	2–	4–	6–	8–	10–12
Frequency						

b Calculate an estimate for the mean and standard deviation of the length of calls.

A population is the complete set of data values.
A sample is a small part of a population.
You can use a sample to find an estimate for the standard deviation of the population.

To do this you divide by $(n - 1)$ instead of n.

So the formula changes to $\sqrt{\dfrac{\Sigma(x - \bar{x})^2}{n - 1}}$ instead of $\sqrt{\dfrac{\Sigma(x - \bar{x})^2}{n}}$

Example The sample values 39, 46, 51 and 36 are taken from a population. Find an estimate of the population standard deviation.

Use a table to show your working.

x	$x - \bar{x}$	$(x - \bar{x})^2$
39	−4	16
46	3	9
51	8	64
36	−7	49
$\Sigma x = 172$		$\Sigma(x - \bar{x})^2 = 138$

$\bar{x} = 172 \div 4 = 43$

An estimate of the population standard deviation $= \sqrt{\dfrac{\Sigma(x - \bar{x})^2}{n - 1}}$

$= \sqrt{\dfrac{138}{3}}$

$= 6.78$ (3 sf)

9 **a** These samples are taken from the same population.
Work out an estimate for the population standard deviation for each sample.

(1) 14 18 13 17 13
(2) 15 16 16 14 19
(3) 16 17 11 17 14

b Explain why the answers to part **a** are different.

1 The table below shows the distribution of wages of the 11 workers in a small business.

Director	£54 000
Manager	£40 000
Salesman	£22 000
Foreman	£18 000
Welder	£17 000
Patternmaker	£15 000
1st Metalworker	£12 000
2nd Metalworker	£12 000
3rd Metalworker	£12 000
4th Metalworker	£12 000
Apprentice	£6 000

a For the distribution of wages find the
- **i** median *(1)*
- **ii** lower quartile *(1)*
- **iii** upper quartile *(1)*
- **iv** interquartile range *(1)*
- **v** mean. *(1)*

b The director was given a wage rise of £22 000. Find the **new** values for the
- **i** median *(1)*
- **ii** lower quartile *(1)*
- **iii** upper quartile *(1)*
- **iv** interquartile range *(1)*
- **v** mean. *(1)*

NEAB, 1998, Paper 2

2 In a certain engineering factory the numbers of faulty items produced each day over a 100 day period were recorded as follows.

No. of faulty items	0	1	2	3	4	5	6	7 or more
No. of days	14	28	25	18	9	4	2	0
Cumulative frequency	14							

a Copy the table above. Complete the cumulative frequency row. *(2)*

b On graph paper, draw the cumulative frequency step polygon for these data. *(2)*

c Identify from this polygon:
i the median *(1)*
ii the inter-quartile range *(2)*
iii the 60th percentile. *(1)*

d What is the probability of three randomly selected days from this sample each recording a daily total of four faulty items? *(3)*

NEAB, 1997, Paper 3

3 In an attempt to devise an aptitude test for applicants seeking work on a factory's assembly line, it was proposed to use a simple construction puzzle. The times taken to complete the task by a random sample of 90 employees were observed with the following results.

Times to complete the puzzle (seconds), x	Number of employees	Cumulative frequency
$10 \leqslant x < 20$	5	5
$20 \leqslant x < 30$	11	
$30 \leqslant x < 40$	16	
$40 \leqslant x < 45$	19	
$45 \leqslant x < 50$	14	
$50 \leqslant x < 60$	12	
$60 \leqslant x < 70$	9	
$70 \leqslant x < 80$	4	

a Copy the table. Complete the cumulative frequency column. *(2)*

b Construct, on graph paper, a cumulative frequency polygon for this data. *(3)*

c Identify from this polygon:
i the median (Q_2) *(1)*
ii the upper quartile (Q_3) *(1)*
iii the lower quartile (Q_1). *(1)*

d It is decided to grade the applicants on the basis of their times taken, as good, average or poor.
The percentages of applicants in these grades are to be approximately 15%, 70% and 15%, respectively.
Estimate, from your cumulative frequency polygon, the grade limits. *(4)*

NEAB, 1998, Paper 3 (pt.)

4 The frequency table records the number of books borrowed by the first twenty people to enter a library last Wednesday.

Number of books	Frequency
1	5
2	7
3	4
4	2
5	2

a Calculate the mean number of books per person borrowed by these people. *(2)*

b Calculate the standard deviation of the data. *(2)*

On Saturday the mean number of books per person borrowed by the first twenty people to enter the library was 3.5 and the standard deviation was 2.

c i Give **one** reason which could explain the difference in the standard deviations on these two days. *(1)*

ii Give **one** reason which could explain the difference in the value of the means on these two days. *(1)*

The modal number of books borrowed was 2.

d Give **one** reason why this might be more useful than the mean as a measure of average for the data. *(1)*

SEG, 1996, Paper 2

5 The diagram represents the response times to the ringing of the telephone in an office

a Copy the table below. Use the diagram to construct the cumulative frequency distribution for these response times. *(2)*

Response time (seconds)	Cumulative frequency

b Draw the cumulative frequency curve on graph paper. *(2)*
c Obtain an estimate for the interquartile range of these response times. *(2)*

Six months later a similar survey of response times gave an interquartile range of 10 seconds.

d What does this new value suggest about the change in the distribution of response times at this office? *(2)*

SEG, 1998, Paper 1

6 The table shows the weight, measured to the nearest kilogram, of 48 adult foxes.

Weight (kg)	Frequency
9	8
10	13
11	16
12	8
13	3

a Calculate the mean and the standard deviation of the weight of the foxes. *(3)*
b What is the maximum range in the weight of the foxes? *(2)*

Another fox was weighed and its weight was the same as the mean of the other 48 foxes.

c If the weight of this fox is added to the previous data, how will the following change?
i The mean of the weight of the foxes. *(1)*
ii The standard deviation of the weight of the foxes. *(1)*
iii The range of the weight of the foxes. *(1)*

SEG, 1999, Paper 3

7 The number of traffic accidents recorded per day over a 200-day period at a busy road junction in the town of Fernlea were as follows.

Number of accidents per day, x	Number of days
0	24
1	88
2	61
3	20
4	5
5	2
	200

a Calculate for this data:

i the mean $(\bar{x})$.

ii the standard deviation (s), to 2 decimal places.

(You may use the following formulae

$$\text{Mean} = \frac{\Sigma fx}{\Sigma f} \quad \text{Standard deviation} = \sqrt{\frac{\Sigma fx^2}{\Sigma f} - \bar{x}^2} \text{ or } \sqrt{\frac{\Sigma f(x - \bar{x})^2}{\Sigma f}}$$

or any suitable alternative, or the statistical functions on your calculator.) *(6)*

An earlier study conducted over a similar time period at the same road junction produced the following summarised results.

Daily road accidents	
Mean $(\bar{x})$	Standard deviation (s)
3.2	1.15

A comparison of the relative variation present in each of the two sets of data is to be undertaken using the following measure:

$$\frac{s}{\bar{x}}$$

b Calculate, to 2 decimal places, this measure for **each** data set. *(2)*

c Comment on the results which you obtained in parts **a** and **b**. *(2)*

NEAB, 1998, Paper 3

8 The table shows the annual wage of women working for a company.

Wage (£ per annum)	Frequency
<8 000	0
8000 ⩽ wage < 10 000	14
10 000 ⩽ wage < 12 000	33
12 000 ⩽ wage < 15 000	38
15 000 ⩽ wage < 20 000	30
20 000 ⩽ wage < 25 000	17
25 000 ⩽ wage < 35 000	11
35 000 ⩽ wage < 60 000	7

a On graph paper, draw a cumulative frequency graph for these data. *(4)*

b Estimate the proportion of women with an annual wage of more than £31 000. *(2)*

c Use your graph to estimate the median annual wage. *(1)*

d **i** Use your graph to estimate the range between the 1st and 9th decile. *(3)*

The range between the 1st and 9th decile for the **male** workers at this company was £24 000.

ii What does this tell you about the wages of men and women working for this company? *(1)*

iii Give **one** advantage of using an interdecile range. *(1)*

At Christmas all the women receive the same bonus of £300.

e What effect will this bonus have on:

i the median; *(1)*

ii the range between the 1st and 9th decile? *(1)*

SEG, 1999, Paper 3

Lower quartile The **lower quartile** is the value one-quarter of the way through the data.

Upper quartile The **upper quartile** is the value three-quarters of the way through the data.

Interquartile range The **interquartile range** = upper quartile − lower quartile

Semi-interquartile range The **semi-interquartile range** = $\dfrac{\text{interquartile range}}{2}$

Cumulative frequency diagram A **cumulative frequency** diagram shows how the frequency changes as the data values increase. The data is shown on a continuous scale on the horizontal axis. The cumulative frequency is shown on the vertical axis.

You plot the upper ends of each group against the cumulative frequency. You then join the points with straight lines or a curve. If you use straight lines the diagram is called a cumulative frequency polygon. If you use a curve then the diagram is called a cumulative frequency curve.

If the distribution is for discrete data you draw a step polygon.

Standard deviation The **standard deviation,** s, of a set of data is given by the formula

$$s = \sqrt{\frac{\sum(x-\bar{x})^2}{n}}$$

The bigger the standard deviation the more spread out the data is.

Alternative formulas for the standard deviation are:

$$s = \sqrt{\frac{\sum x^2}{n} - \left(\frac{\sum x}{n}\right)^2} \quad \text{and} \quad s = \sqrt{\frac{\sum fx^2}{\sum f} - \left(\frac{\sum fx}{\sum f}\right)^2}$$

If you change a set of data by adding A to each item of data:

New mean = old mean + A

New standard deviation = old standard deviation

If you change a set of data by multiplying each item of data by the same number k:

New mean = old mean × k

New standard deviation = old standard deviation × k

Working with *Excel*

Drawing a cumulative frequency polygon and a cumulative frequency curve

Eighty car owners were selected at random and asked the age of their car.
These are the results:

Age, a (years)	Frequency
$0 \leq a < 3$	15
$3 \leq a < 6$	36
$6 \leq a < 9$	14
$9 \leq a < 12$	8
$12 \leq a < 15$	5
$15 \leq a < 18$	2

Use *Excel* to: **a** draw a **cumulative frequency polygon** for this data.
b draw a **cumulative frequency curve** for this data.

Open a new *Excel* document

Enter the information on car ages

In Cell A1, type **Age in years**. In Cell A2, type **0 up to 3**, in Cell A3, type **3 up to 6** and so on.

In Cell B1, type **Frequency**. In Cell B2, type **15**, in Cell B3, type **36** and so on.

Calculate the cumulative frequency

In Cell D1, type **Age in years**. In Cell D2, type **Up to 3**, in Cell D3, type **Up to 6** and so on.

In Cell E1, type **Cumulative Frequency**. In Cell E2, insert the formula: **=B2**. Remember to press the **Enter** key.

In Cell E3, insert the formula: **=E2+B3**. Remember to press the **Enter** key.

Select Cell E3.

Place the cursor over the bottom right-hand corner of Cell E3 and drag down to Cell E7.

This copies the formula into each cell. Cells E2 to E7 display the cumulative frequency.

Select Cells D1 to E7

Cells A1 to C7 are not selected as these columns are not used to draw the cumulative frequency polygon or curve. The upper limits are used to draw these.

Use the Chart Wizard button

Drawing the cumulative frequency diagram

From **Chart type**, choose **XY (Scatter)**.

From **Chart sub-type**, choose top left diagram.

Labelling the cumulative frequency diagram
Click the Next> button until you get to the Chart Options window.
In the Chart title box, type Age of Cars in Years.
In the Value (X) axis box, type Age in Years.
In the Value (Y) axis box, type Cumulative Frequency. Click Finish.

Drawing the cumulative frequency polygon
Place the cursor on one of the points on the diagram.
Right-click and choose Format Data Series …
In the Format Data Series box, click the Patterns tab.
Under Line, choose Automatic.
Click OK.
You can now print the cumulative frequency polygon.
The lines joining the points are straight.
Delete the diagram by clicking on a blank area and pressing the Delete key.

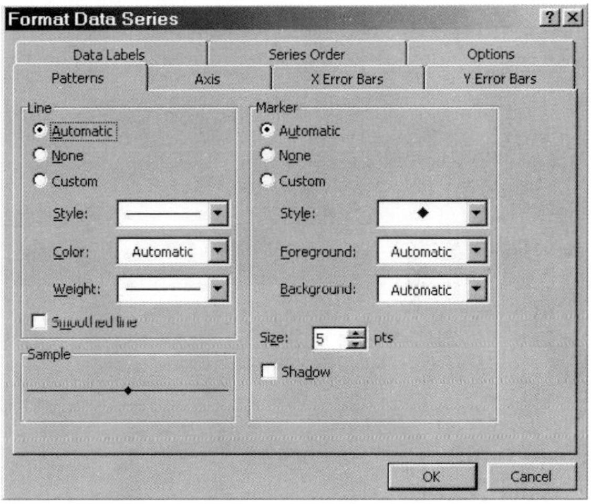

Drawing the cumulative frequency curve
Follow all the instructions from **Select Cells D1 to E7**, but at the end choose Smoothed line before clicking OK.
The lines joining the points are curved.

Exercise 4:8

1 Yuri has done a survey on the age of people using his local bus.
He carried out his survey at 11:00 a.m. and 4:00 p.m.
These are his results.

Age of passengers (years)	Frequency (11:00 a.m.)	Frequency (4:00 p.m.)
$0 \leqslant \text{age} < 10$	3	9
$10 \leqslant \text{age} < 20$	8	21
$20 \leqslant \text{age} < 30$	15	18
$30 \leqslant \text{age} < 40$	23	7
$40 \leqslant \text{age} < 50$	14	5
$50 \leqslant \text{age} < 60$	9	4
$60 \leqslant \text{age} < 70$	9	2

a Use *Excel* to draw a cumulative frequency curve for the 11:00 a.m. data.
b Use *Excel* to draw a cumulative frequency polygon for the 4:00 p.m. data.

Calculating a standard deviation

These are the scores of two players over six rounds of golf.

Martin 87, 71, 90, 82, 84, 82
Jane 77, 91, 85, 90, 70, 67

Use *Excel* to calculate the **standard deviation** of this data.

Open a new *Excel* document

Enter the information on golf scores
In Cell A1, type **Martin's Scores**. In Cell B1, type **Jane's Scores**.
In Cell A2, type **87**, in Cell A3, type **71** and so on.
In Cell B2, type **77**, in Cell B3, type **91** and so on.

Calculate the standard deviation
In Cell A8, type **Martin's Standard Deviation**. In Cell B8, type **Jane's Standard Deviation**.
In Cell A9, insert the formula: **=STDEVP(A2:A7)**
In Cell B9, insert the formula: **=STDEVP(B2:B7)**
Remember to press the **Enter** key.
Cells A9 and B9 display the standard deviations of the sets of data.

Calculating the mean and standard deviation of related data

a Use *Excel* to calculate the **mean and standard deviation** of the data:
35, 41, 32, 33, 32, 37, 36
b Sam adds 4 to each number.
Use *Excel* to find the **mean and standard deviation** of these new numbers.
c Jason multiplies each of the original numbers by 5.
Use *Excel* to find the **mean and standard deviation** of these new numbers.
d Compare the means and standard deviations calculated in parts **a**, **b** and **c**.

Enter the information for part a in a new *Excel* document
In Cell A1, type **Original Numbers**. In Cell A2, type **35**, in Cell A3, type **41** and so on.

Calculate the mean and standard deviation
In Cell A9, type **Mean**. In Cell A10, insert the formula: **=AVERAGE(A2:A8)**
In Cell A11, type **Standard Deviation**. In Cell A12, insert the formula: **=STDEVP(A2:A8)**
Cell A10 displays the mean. Cell A12 displays the standard deviation.

Enter the information for part b

In Cell B1, type **Numbers +4**.

In Cell B2, insert the formula: **=A2+4**

Select Cell B2.

Place cursor over the bottom right-hand corner of Cell B2 and drag down to Cell B8.

This copies the formula into each cell.

Calculate the mean and standard deviation

Repeat the previous method to calculate the mean and standard deviation of these numbers.

Enter the information for part c

In Cell C1, type **Numbers × 5**.

In Cell C2, insert the formula: **=A2*5**

Copy the formula into each cell.

Calculate the mean and standard deviation

Repeat the previous method to calculate the mean and standard deviation of these numbers.

	A	B	C	D
1	Original Numbers	Numbers + 4	Numbers x 5	
2	35	39	175	
3	41	45	205	
4	32	36	160	
5	33	37	165	
6	32	36	160	
7	37	41	185	
8	36	40	180	
9	Mean	Mean	Mean	
10	35.14285714	39.14285714	175.7142857	
11	Standard Deviation	Standard Deviation	Standard Deviation	
12	2.996596709	2.996596709	14.98298355	
13				
14				

In part **b**, the mean is increased by 4. The standard deviation is not changed.
In part **c**, the mean is multiplied by 5. The standard deviation is multiplied by 5.

Exercise 4:9

1 These are the marks of 10 students in their History and Science exams. The History marks are out of 50 and the Science marks are out of 100.

History	31	11	46	39	40	28	23	41	17	25
Science	67	45	60	73	75	51	63	69	50	48

a Use *Excel* to calculate the mean and standard deviation for each subject.

b The History marks are out of 50. The marks are multiplied by 2 to get a mark out of 100. Write down the new mean and standard deviation for the History marks.

c Use the means and standard deviations to compare the marks in the two subjects.

1 Find: **a** the interquartile range **b** the semi-interquartile range of:
41 65 39 50 55 62 74 49 44 76 23 58.

2

Value	17	18	19	20	21	22	23	24
Frequency	5	11	16	17	22	15	12	2

a Find (1) the 74th percentile (2) the 23rd percentile.
b Find (1) the 3rd decile (2) the 9th decile.

3 A vet has carried out a survey into the weights of dogs.
He has weighed 100 dogs. The table shows his results.

Weight, w (kg)	Frequency
$0 \leqslant w < 10$	15
$10 \leqslant w < 20$	22
$20 \leqslant w < 30$	30
$30 \leqslant w < 40$	14
$40 \leqslant w < 50$	10
$50 \leqslant w < 60$	9

a Draw a cumulative frequency polygon for the data.
b Use your polygon to estimate how many dogs weighed less than 35 kg.
c Find the median weight of the dogs.
d Find the interquartile range of the weights.
e Find the 6th decile of the weights.

4 Find the variance and the standard deviation of: 56, 72, 39, 46, 77.

5 Fran has worked out these values for a set of 40 numbers:

$\Sigma x = 387 \qquad \Sigma x^2 = 6381$

Find the standard deviation of her numbers.

6 **a** Estimate the mean and standard deviation of these times that patients waited to see Dr Job.

Time, t (minutes)	Frequency
$0 \leqslant t < 5$	7
$5 \leqslant t < 10$	11
$10 \leqslant t < 15$	9
$15 \leqslant t < 20$	4

b Another set of times for Dr Thompson has a mean of 8.49 min and a standard deviation of 6.45 min. Compare the two sets of times.

3 Averages

1 Choosing a representative figure
Calculating the mode, median and mean for a frequency distribution
Calculating the mode, median and mean for a grouped frequency distribution of discrete data
Calculating the mode, median and mean for a grouped frequency distribution of continuous data

2 Calculations involving the mean
Combining two sets of data
Calculating a weighted mean
Assumed mean
Geometric mean

3 Estimation
Estimation of the population mean
Estimation of the population proportion

4 Control charts
Drawing control charts
Interpreting control charts

EXAM QUESTIONS

SUMMARY

ICT IN STATISTICS

TEST YOURSELF

1 Choosing a representative figure

We often talk about the average person. This picture shows how difficult it can be to choose a person that represents the rest.

There are three different types of average in statistics.

Mode

The **mode** is the most common or most popular data value. It is sometimes called the **modal value**.

The mode is most useful when one value appears much more often than any other. If the data values are too varied, then the mode should not be used. It is the only average that can be used for qualitative data.

Median

To find the **median** of a set of data, put the values in order of size. The **median** is the middle value.

For n data values, $\dfrac{n+1}{2}$ gives the position of the median.

If there are 10 data values the median is the $\dfrac{10+1}{2}$ th value.
So the median is the $5\frac{1}{2}$th value

This is halfway between the 5th and 6th values.
The median is not affected by extreme values.

Mean

To find the **mean** of a set of data:

1 find the total of all the data values

2 divide the total by the number of data values.

The mean is a very important average, as it considers every piece of data. However, it can be affected by extreme values.

5 Presenting data

1 Two-way tables
Defining two-way tables
Designing and completing a two-way table

2 Stem and leaf diagrams
Drawing a stem and leaf diagram
Changing the scales
Small numbers of stems
Using the stem and leaf diagram
Comparing two sets of data using back to back stem and leaf diagrams

3 Population pyramids
Drawing a population pyramid

EXAM QUESTIONS

SUMMARY

TEST YOURSELF

1 Two-way tables

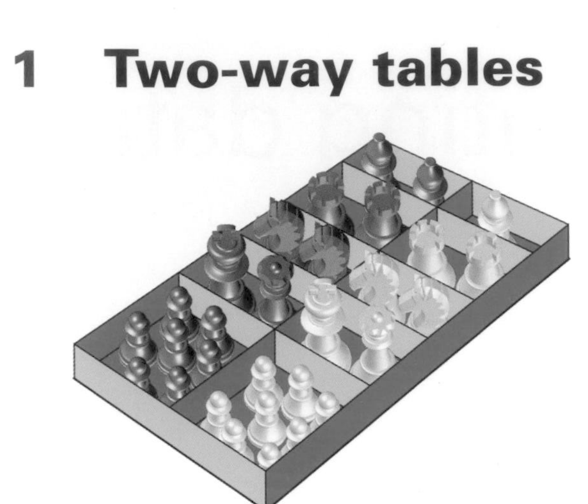

There are two pieces missing from this chess set. You could not tell which ones were missing by looking at a pile of pieces. Once the pieces are placed in the box it is easy to see which are missing. This is because they are arranged by both colour and shape.

Two-way tables

Two-way tables are used to show two pieces of information.
This type of table has both rows and columns.
The rows show one piece of information.
The columns show the second piece of information.

This two-way table shows the favourite science, from physics, chemistry and biology, of a group of boys and girls.

	Physics	Chemistry	Biology	Total
Boys	55	35	20	110
Girls	10	28	52	90
Total	65	63	72	200

The **red row** gives information about **boys**.
Number of boys = 55 + 35 + 20 = 110

The green row gives information about girls.
Number of girls = 10 + 28 + 52 = 90

	Physics	Chemistry	Biology	Total
Boys	55	35	20	110
Girls	10	28	52	90
Total	65	63	72	200

The **blue column** shows the number of pupils who prefer **physics**.
The **purple column** shows the number of pupils who prefer **chemistry**.
The **brown** column shows the number of pupils who prefer **biology**.

By using both the **red row** and **blue column** it is possible to find the number of **boys** who prefer **physics**.

	Physics	Chemistry	Biology	Total
Boys	55	35	20	110
Girls	10	28	52	90
Total	65	63	72	200

So 55 **boys** prefer **physics**.

Exercise 5:1

1 Debbie counts how many videos and DVDs she has. She also places them into three categories of Film, Comedy and Sport. She records the information in a table.

	Videos	DVDs	Total
Film	43	16	59
Comedy	12	5	17
Sport	21	3	24
Total	76	24	100

Use the table to find out the number of Debbie's:

a sports videos
b comedy DVDs
c videos
d films.

2 Jamie investigated the relationship between eye and hair colour as part of a project. He recorded his results in a two-way table.

	Fair	Dark	Total
Blue	8	5	13
Other	7	10	17
Total	15	15	30

Use the table to find the number of:

a dark-haired, blue-eyed pupils
b fair-haired pupils.

3 Fifty children are asked about the destination and accommodation of their summer holiday. Here are the results.

	Villa	Hotel	Caravan	Tent	Total
Europe	3	2	10	7	22
America	4	5	5	3	17
Japan	5	6	0	0	11
Total	12	13	15	10	50

Use this table to find the number of children who are staying in:

a a hotel in Japan

b a caravan in Europe

c a tent in America.

You can put information in a table and then find missing values.

A group of 60 Year 7 and Year 8 pupils are asked whether they prefer individual or team sports. Twenty four from a total of 32 Year 7 pupils prefer team sports. In total, 21 pupils prefer individual sports.

You can use a two-way table to show this information.
First, draw the table and put in the given information.

	Year 7	Year 8	Total
Individual sport			21
Team sport	24		
Total	32		60

Now, work out each of the missing numbers in turn.
Look for a column or a row with just one missing value.
The first column has just one value missing.

	Year 7	Year 8	Total
Individual sport			21
Team sport	24		
Total	32		60

The number of Year 7 pupils who prefer individual sports = Total − Team sport
= 32 − 24 = 8

Put 8 in the table.

The first row now has just one value missing.

	Year 7	Year 8	Total
Individual sport	8		21
Team sport	24		
Total	32		60

The number of Year 8 pupils who prefer individual sports

$= \text{Total} - \text{Year 7}$

$= 21 - 8$

$= 13$ Put 13 in the table.

Carry on like this until the table is complete.
Use the third row to calculate the Year 8 total.
Use the Year 8 total to calculate the Year 8 team sport.
Complete the table by adding Year 7 team sport and Year 8 team sport.

	Year 7	Year 8	Total
Individual sport	8	13	21
Team sport	24	$28 - 13 = 15$	$15 + 24 = 39$
Total	32	$60 - 32 = 28$	60

4 A theme park kiosk sells four products. There is a special offer on certain pairs of these products. The manager needs to know which combinations are popular. He records the information in a two-way table. Copy the manager's table and complete it.

	Ice cream	Drinks	Total
Chocolates		207	365
Toffees	99		
Total			507

5 A tour operator recorded bookings made last Saturday. Some of the information is given in the table. Copy the table and complete it.

	France	Spain	Germany	Total
Car/ferry	15	8		28
Plane				12
Total	18	14		40

6 Carol loves cars. She stands outside her house, and records the make and colour of the cars which pass by. The table shows her data.

	Volvo	Renault	Ford	Total
Grey		9	14	31
Red	4	11		23
Blue				
Total	12		25	60

Copy and complete the table.

7 There are 260 people at a meeting in London. They all complete a questionnaire. In the questionnaire, they are asked if they are a **student**, a **tourist** or a **retailer**. Another question asks if they have travelled to the meeting by **car**, **coach** or **train**.
From the 102 students, 37 travelled by train and 20 travelled by car.
Out of the total 150 who travelled by train, 53 were retailers.
Only 62 people at the meeting came by car and 12 of these were tourists.
In total, only 73 tourists attended the meeting.

a Design a two-way table to show this information.

b Use the table to calculate how many more tourists travelled by car than by coach.

8 Three hundred visitors to a Rock and Pop Museum are asked their age and what type of television they prefer. Out of the 300, 90 preferred Satellite and 60 preferred Terrestrial. From the 97 under 21s, 48 preferred Satellite and 28 preferred Terrestrial. From the 116 over 45s, 19 preferred Satellite and 86 preferred Cable.

a Copy and complete the table, by entering the number of visitors in each category.

	Under 21s	21–45	Over 45s	Total
Satellite				
Terrestrial				
Cable				
Total				

Using the table, write down:

b the total number of visitors who prefer Cable

c the number of visitors who are in the age group 21–45 and prefer either Satellite or Terrestrial.

2 Stem and leaf diagrams

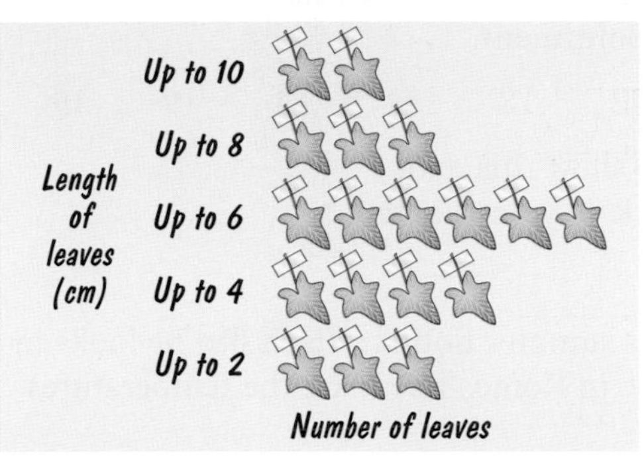

Dawn has produced a chart showing the lengths of leaves on an ivy plant. She has used ivy leaves of about the same size to show the distribution of lengths.

Stem and leaf diagram

A **stem and leaf diagram** shows the shape of a distribution.
It is like a bar-chart, with the numbers themselves forming the bars.
It is only suitable for small amounts of quantitative data.

These are the times, in seconds, taken by nine children to solve a logic puzzle.

28 11 22 34 43 39 28 25 23

These times can be shown in a stem and leaf diagram.

Step 1 Divide each number into two parts, the tens and the units.
The **tens digits** form the **stem**.
The **units digits** form the **leaves**.

28 11 22 34 43
39 28 25 23

Step 2 Draw a vertical line.
Place the **tens (stems)** in numerical order on the left of this line (**1, 2, 3, 4**).
Place the **units (leaves)** on the right of this line.

Stem	Leaf
1	1
2	8 2 8 5 3
3	9 4
4	3

Step 3 Put the **units (leaves)** in numerical order.
Add a title and a key to your final diagram.

Times taken to solve a puzzle

Stem	Leaf
1	1
2	2 3 5 8 8
3	4 9
4	3

where 2 | 8 means 28 seconds.

Exercise 5:2

1 Jacky visits her doctor frequently. These are the times, in minutes, that she has spent waiting for her appointment.

11 24 28 36 12 18 33 14 19 19

Draw a stem and leaf diagram to show this data.
Remember to add a key and a title to your diagram.

2 Roberto is looking forward to his summer holiday. Each day he looks in the newspaper at the temperature in Rome. These are the temperatures, in °C, that he recorded.

32 18 30 26 28 19
28 25 24 27 26 33

Draw a stem and leaf diagram to show these temperatures.

You can use different values for the stems.

Graham enjoys watching the Grand Prix. He is interested in the length of time it takes a team of mechanics to change the tyres at a pit stop. These are the times, in seconds, that he records.

8.2 7.6 5.7 6.3 4.9 6.6
7.8 6.2 5.3 7.4 5.1 6.5

This time you use the units as the stems.

Step 1 Divide each number into two parts, the units and the tenths.
The **units** form the **stem**. **8.2 7.6 5.7 6.3 4.9 6.6**
The **tenths** form the **leaves**. **7.8 6.2 5.3 7.4 5.1 6.5**

Step 2 Draw the first diagram.
The **units** are the **stems**.
The **tenths** are the **leaves**.

Stem	Leaf
4	9
5	7 3 1
6	3 6 2 5
7	6 8 4
8	2

Step 3 Put the leaves in numerical order.
Add a title and a key to your final diagram.

Times taken to change tyres

Stem	Leaf
4	9
5	1 3 7
6	2 3 5 6
7	4 6 8
8	2

where 4 | 9 means 4.9 seconds.

---

3 James cycles to school each day. He records the time the journey takes. These are his times to the nearest tenth of a minute.

7.2	6.1	8.3	8.8	6.3	7.7
9.9	7.8	8.4	6.8	7.4	7.9

Draw a stem and leaf diagram to show this data.

4 Anne measures 21 sunflower leaves as part of her biology project.
These are the lengths, to the nearest tenth of a centimetre.

17.2	14.7	16.6	17.1	14.7
15.1	15.2	18.9	15.3	14.7
14.8	15.5	15.8	16.7	18.9
15.9	16.5	16.4	16.6	17.8
18.8				

Draw a stem and leaf diagram to show these measurements.
Use 14, 15, 16, … as the stems.

5 Mr Jones has 20 piano students. They all recently took examinations and passed. These are the marks they scored.

139	101	119	113	102	124	118	140	125	124
104	124	131	103	122	136	112	118	119	108

Use this key to draw a stem and leaf diagram for the marks.
Key: 12 | 4 means 124 marks.

6 Gemma enjoys taking part in swimming competitions. When training, she records the length of time it takes her to complete one length of the pool. These are her times to the nearest tenth of a second.

22.0	20.6	19.8	20.1	21.3	19.2	19.4	20.5
19.6	21.2	19.5	20.6	19.7	20.3	22.1	20.7

Draw a stem and leaf diagram to show these times.

7 Nathan is the leader of a local Scout group. He is interested in the different heights of the boys in his pack. He measures and records these heights. These are his measurements to the nearest centimetre.

174	157	143	173	164	152	168	154	168	177
183	169	156	169	156	145	168	165	175	182

Draw a stem and leaf diagram to show this data.

Some stem and leaf diagrams only have a small number of stems.

These are drawn in a different way.

Each number from the stem is put in two separate rows.

The first row contains the leaves 0, 1, 2, 3 and 4.

The **second row** contains the leaves **5, 6, 7, 8** and **9**.

Example Matthew is a computer enthusiast and enjoys browsing the Internet. He records the length of time he spends surfing the web. These are his times to the nearest minute.

24	11	20	10	25	24	36
37	23	18	29	32	18	23

Step 1 There are just three tens digits: 1, 2 and 3. These will form the stems. Because there are only a small number of stems you use two rows for each one.

Stem	Leaf
1	1 0
1	**8 8**
2	4 0 4 3 3
2	**5 9**
3	2
3	**6 7**

The first row contains the leaves 0, 1, 2, 3 and 4.
The **second row** contains the leaves **5, 6, 7, 8** and **9**.

Step 2 Now put the leaves in numerical order.
Add a title and a key.

Lengths of time spent on the Internet

Stem	Leaf
1	0 1
1	**8 8**
2	0 3 3 4 4
2	**5 9**
3	2
3	**6 7**

where 1 | 8 means 18 minutes.

Exercise 5:3

1 Charlotte has a daily newspaper round. She keeps a record of the length of time this takes. These are her times to the nearest minute.

39	30	41	32	48	54	36	34
40	38	31	37	54	55	33	46

Draw a stem and leaf diagram using two lines for each tens digit.

2 Martin is the captain of the school cricket team. He has to keep a record of all the runs. These are the scores he recorded for one season.

136	138	120	131	138	126	132	128
137	139	130	121	139	137	127	134

Draw a stem and leaf diagram to show these scores. Use two lines for each different stem value.

3 Louise enjoys listening to popular music. On one CD, she has noticed that the length of the tracks varies greatly. These are the times in minutes, to the nearest tenth of a minute, of each track.

4.6	3.8	4.0	3.1	4.2	3.2	3.5	4.7
4.7	3.3	4.7	4.1	3.6	3.4	4.6	5.9

Draw a stem and leaf diagram to show these times. Use two lines for each different stem value.

4 Julian is considering whether or not to change telephone companies. To help him decide, he writes down the lengths of all the local calls made from his house. These are the times in minutes to the nearest tenth of a minute.

9.7	7.6	7.9	7.1	8.1	9.5	8.6	7.3	7.9	9.7
9.8	9.6	9.2	7.5	7.8	7.2	7.8	8.4	9.4	9.7

Using two lines for each different stem value, draw a stem and leaf diagram to show these times.

5 A group of students compare how much they earn each week in their part-time jobs. These are the amounts to the nearest pound.

£38	£28	£41	£43	£36	£34
£48	£32	£37	£34	£29	£35

Using two lines for each different stem value, draw a stem and leaf diagram to show these amounts.

A **stem and leaf diagram** displays the data in order of size.
It can be used to find the mode and the median.

This stem and leaf diagram shows the length of time it takes Ruth to travel to work on the train.

Stem	Leaf
2	5 6 7 8
3	2 2 4 6 8 8
4	3 3 3 4 6 7
5	0 6

where 2 | 5 means 25 minutes.

43 is the time that appears the most. The **mode** is **43** minutes.

There are 18 values: 4 with stem 2, 6 with stem 3, 6 with stem 4 and 2 with stem 5.

The position of the median is $\frac{18 + 1}{2} = 9\frac{1}{2}$th value.

Counting through the leaves the 9th and 10th values are 38 and 38.

So the **median** is $\frac{38 + 38}{2} = 38$ minutes.

A stem and leaf diagram can also be used to find the lower and upper quartiles and the interquartile range.

There are 9 values in the lower half of the data.
The lower quartile is the 5th value.
The **lower quartile** is 32 minutes.

The upper quartile is the 9th + 5th value = 14th value.
The 14th value is 44.
The **upper quartile** is 44.

$$\begin{aligned}\text{The interquartile range} &= \text{upper quartile} - \text{lower quartile}\\ &= 44 - 32\\ &= 12 \text{ minutes}\end{aligned}$$

You can work out the semi-interquartile range by halving the interquartile range.

The semi-interquartile range = 6 minutes.

Exercise 5:4

1 Christina grows tomatoes in her greenhouse. She records the highest temperature of her greenhouse each day. The stem and leaf diagram shows the temperatures, in °C, for 23 days in July.

Stem	Leaf
1	3 3 4 4
1	5 5 6 6 7 8 8 9 9
2	3 3 3 3 4
2	5 8 8 9
3	2

where 2 | 3 means 23 °C.

Find:

a the modal temperature

b the median temperature.

2 Mrs Parker is a teacher. She records the Biology mock examination marks for her Year 11 class. The stem and leaf diagram shows these results.

Stem	Leaf
3	2 5 8 9
4	1 2 3 8 8 9 9
5	2 4 6 7 7 7 8 9
6	2 5 6
7	1 8
8	3

where 5 | 2 means 52 marks.

Use this diagram to find:

a the modal mark

b the interquartile range.

3 Sam is interested in the distance her friends travel to school. The stem and leaf diagram shows these distances in miles to the nearest tenth of a mile.

Stem	Leaf
1	8 9
2	3 4 5 5
3	6 7 9
4	2

where 2 | 3 means 2.3 miles.

Use this diagram to find:

a the median distance

b the semi-interquartile range.

Back to back stem and leaf diagrams

Two stem and leaf diagrams can be drawn using the same stem. These are called **back to back stem and leaf diagrams**. The leaves of one set of data are put on the right of the stem. The leaves of the other set of data are put on the left.

John and Kira compared the length of time they spent each evening on their homework. Their times are shown in the back to back stem and leaf diagram.

				John		Kira			
6	5	5	3	2	2				
		8	6	5	3	6	7		
			3	2	4	4	6	6	
				1	5	2	3	4	5
					6	4	8		

where 4 | 6 means 46 minutes.

Back to back stem and leaf diagrams can be used to compare two sets of data. By looking at the diagram it is easy to see that most of John's times are between 22 and 38 minutes while most of Kira's times are between 44 and 68 minutes.

Kira spends more time on homework than John.
You could also find the medians to compare the two sets of data.

For John the middle value is **35**. His median time is **35 minutes.**
For Kira the middle value is 52. Her median time is **52 minutes.**

This also supports the idea that Kira spends more time on her homework than John because her median time is greater than John's.

Exercise 5:5

1 Last summer, Janet spent two weeks on holiday in France and two weeks in Scotland. Each day, she recorded the temperature at 3 p.m. These temperatures are shown in the back to back stem and leaf diagram.

					France		Scotland				
						1					
					9	1	6	7	8	8	
		4	3	3	3	2	2	2	2	2	4
9	8	8	6	6	5	2	5	5	6	7	8
		4	2	2	1	3	1				

where 2 | 5 means 25 °C.

a Find the median and modal temperature in France.
b Find the median and modal temperature in Scotland.
c Does the median or the mode give the better value to represent the data? Explain your answer.

2 Alan has just passed his driving test. He wants to find out if male and female students needed different numbers of lessons before they passed. He asks his friend at college how many lessons they each had and records this data in a back to back stem and leaf diagram.

				Male		Female				
			9	8	1	9	9			
8	8	5	3	2	2	3	4	5	7	7
	8	4	4	4	3	5	5	5	7	
			3	2	4	3	4			

where 2 | 3 means 23 lessons.

a For the female students, find the mode and median.
b For the male students, find the mode and median.
c Write down any conclusions which Alan can make.

3 Beth and Josie enjoy visiting the local zoo. They want to compare the length of time that each one of them spends there. These are the times, to the nearest tenth of an hour, of 10 visits each.

Beth	2.3	5.5	2.5	4.3	5.3	5.7	2.4	4.5	3.2	5.8
Josie	3.6	4.3	4.1	3.8	4.6	5.2	5.5	3.8	4.5	5.4

a Draw a back to back stem and leaf diagram to compare their visits.
b For each girl, find the median time and the semi-interquartile range. Show your working.
c Write down any conclusions which can be made.

4 The table gives the average hours, usually worked per week by full-time employees, by gender, EU comparison 1999.

a Draw a back to back stem and leaf diagram to compare the hours of males and females. Use 36, 37, 38 … as the stems.
b Find the median time and the interquartile range for (1) the males (2) the females.
c Write down any conclusions that you can make.

	Hours	
	Males	Females
United Kingdom	45.2	40.7
Greece	41.7	39.3
Portugal	41.5	39.4
Irish Republic	41.3	38.0
Spain	41.1	39.6
Germany	40.5	39.4
Luxembourg	40.5	38.0
Austria	40.3	39.9
Sweden	40.2	39.9
France	40.2	38.6
Finland	40.1	38.3
Italy	39.7	36.3
Denmark	39.6	37.9
Netherlands	39.2	38.3
Belgium	39.1	36.9

Source: Labour Force Surveys, Eurostat

3 Population pyramids

Tahar Davis supported the heaviest ever human pyramid on 17th December 1979. There were 12 people in the pyramid in three levels. The total weight was 771 kg.

Population pyramid

A **population pyramid** compares two sets of data. It uses two horizontal bar-charts, drawn back to back. The vertical axis represents the age groups of the population. The horizontal axis represents the percentage of people in each age group for each of the two sets of data.

Juliet has researched the age group and gender of the residents of a new housing estate. She has worked out the percentage of males and females in each age group. This is her data.

Age group (years)	0–9	10–19	20–29	30–39	40–49	50–59	60–69
Males (%)	10	7	6	13	8	7	3
Females (%)	7	5	4	12	9	5	4

To draw a population pyramid, start by drawing a horizontal axis. Label the centre 0 and label the percentages to the left and right. Put the age groups on the left-hand side.

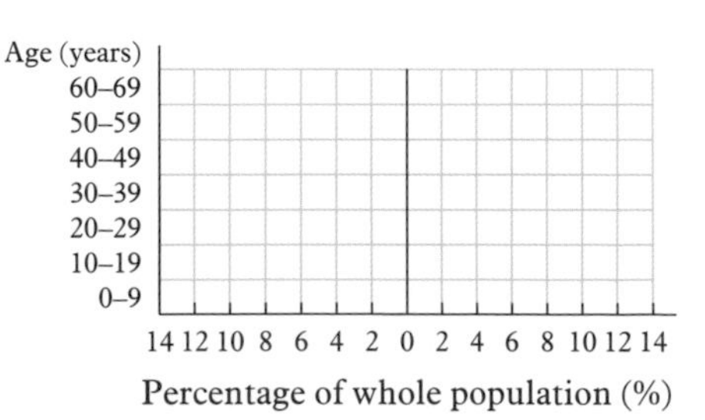

Draw a vertical line at 0.
Now draw bars for each age group.
Put the males on the left and the females on the right.
Label the left side Males and the right side Females.

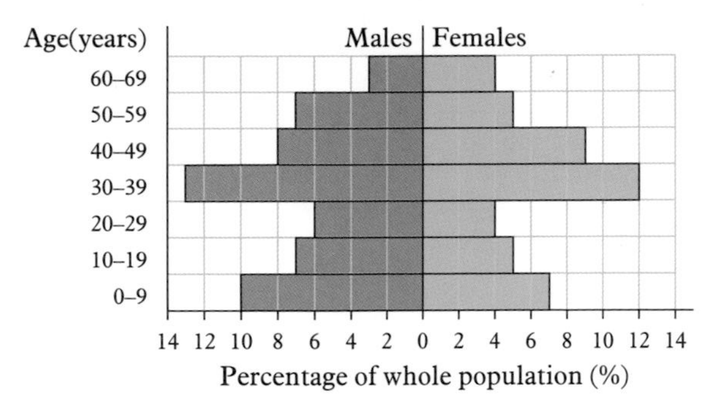

You can use the population pyramid to obtain information. For example:

Percentage of males under 40 years $= 10 + 7 + 6 + 13$
$= 36\%$

Percentage of the population in the age group 30–39 years $= 13 + 12$
$= 25\%$

Percentage of population at least 50 years old $= 7 + 3 + 5 + 4$
$= 19\%$

Exercise 5:6

1 Jennifer is the leader of a village youth group. To help her plan activities, she records the age and sex of the members. The diagram shows her data.

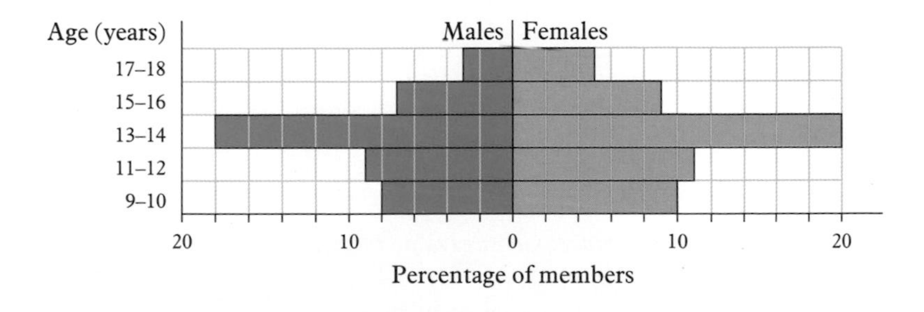

a Find the percentage of members who are male and are at least 13 years old.

b Find the age group which has a total membership of 38%.

c Find the percentage of members who are female and are at most 16 years old.

2 South Lea is a seaside resort and Winchford is a small rural community. The table gives the percentage of the population in different age groups for each village.

Age group (years)	Population of South Lea (%)	Population of Winchford (%)
0–9	7	12
10–19	6	8
20–29	7	11
30–39	12	13
40–49	11	16
50–59	14	13
60–69	18	11
70–79	13	9
80–89	6	6
90–99	2	1

Use this table to draw a population pyramid. Copy the axes from the Example on page 136. You will need to include extra age groups.

3 The population of rural and urban areas of India, who are under 50 years of age in 1991, is given in a table.

Age group (years)	Population of rural areas (%)	Population of urban areas (%)
0–4	12.7	10.7
5–9	13.7	11.9
10–14	11.9	11.5
15–19	9.2	10.2
20–24	8.5	9.9
25–29	8.0	9.1
30–34	6.7	7.6
35–39	6.0	6.9
40–44	5.0	5.4
45–49	4.3	4.4

(Source: Census Data Online)

a Use this table to draw a population pyramid.

b Find the percentage of people under 30 who live in a rural area.

c Find the percentage of people who are between 20 and 39 and who live in urban areas.

d What is the most popular area for those between the ages of 15 and 19?

1 Lisa carried out a survey to find out people's views on music. These are the two questions that she asked:

1. Do you prefer pop music or classical music?
2. Are you aged under 25?

Design a two-way table to summarise her results. *(3)*

NEAB, 1996, Paper 2

2 Zena and Charles played nine rounds of crazy golf on their summer holidays. Their scores are shown on the back to back stem and leaf diagram.

Zena		Charles
	3	0 0 2
1	4	1 1 1 2
9 3 1 0 0	5	2
6 5 4	6	8

Charles' lowest score was 30.

a What was Zena's lowest score? *(1)*
b What was Charles' modal score? *(1)*
c What was Zena's median score? *(1)*

In crazy golf the player with the lowest score wins.
Charles actually made the highest score that summer but was still chosen as the better player.

d Give a reason for this choice. *(1)*

SEG, 1996, Paper 1

3 The table below shows the results of a survey about the three basic skills. A total of 2875 people were asked about any difficulties they had experienced with basic skills since leaving school.

Age	All	72-74	62-64	52-54	42-44	32-34	22-24
No. of people	2,875	437	456	444	498	551	489
Reading (%)	4	4	3	4	4	4	6
Writing/Spelling (%)	13	6	2	15	12	16	15
Numberwork (%)	4	3	3	5	5	5	5
Any difficulty (%)	17	10	14	18	16	19	20

Source: ALBSU 1994

a What percentage of 22–24 year olds found reading difficult? *(1)*
b 16% of which age group had difficulty with writing/spelling? *(1)*
c Some people had difficulty with more than one basic skill. How can you tell this from the table? *(2)*
d There is a mistake in the 62–64 column. How can you tell? *(2)*

NEAB, 1996, Paper 2

4 The stem and leaf diagram below shows the ages, in years, of 26 people who wished to enter a 10-mile walking competition.

Stem	Leaf
0	3 4
1	4 4 6 9
2	1 3 6 6 6 8
3	3 5 5 7 9
4	0 2 3 3 7 7
5	1 2 4

KEY
1 | 4 means 14 years old

a How many people were less than 20 years old? *(1)*

b Write down the modal age. *(1)*

c Exactly half the people entering the walk are more than a certain age. What is that age? *(2)*

d Shaun says that two people will not be allowed to enter. Using the information in the stem and leaf diagram, suggest a reason for this. *(1)*

NEAB, 1998, Paper 1

5 Ten men and ten women were asked how much television they watched last weekend. Their times, in minutes, were as follows:

Men 40, 41, 42, 52, 52, 52, 64, 65, 65, 71
Women 40, 41, 51, 62, 63, 75, 87, 88, 93, 95

a Copy and complete the stem and leaf diagram which would represent these two sets of data. *(2)*

Men		Women
	4	
	5	
	6	
	7	
	8	
	9	

b What is the modal time for the men's data? *(1)*

c Calculate the median of each set of data. *(2)*

d Calculate the mean and the standard deviation for the times of the ten men.
You must use the statistical functions on your calculator or the following formula. You must show all your working.

Standard deviation $= \sqrt{\dfrac{\Sigma fx^2}{\Sigma f} - \left(\dfrac{\Sigma fx}{\Sigma f}\right)^2}$ *(3)*

The data were checked and it was found that all the times of the men should have been increased by 10 minutes.

e **i** What is the mean of the new times of the men? *(1)*

ii What is the standard deviation of the new times of the men? *(1)*

SEG, 1997, Paper 1

6 The diagrams below show the percentage of the whole population of the United Kingdom by age and sex for the years 1901 and 1991.

a **i** In 1901 what percentage of the population was male and under 15 years? *(1)*

ii In 1991 what percentage of the population was female and in the age group 20 to 40 years? *(1)*

b **i** Comment on the differences in the age structure of the population between 1901 and 1991. *(2)*

ii Give a reason for these differences. *(1)*

c Compare the percentages of males and females aged 60–90 years in 1991. *(1)*

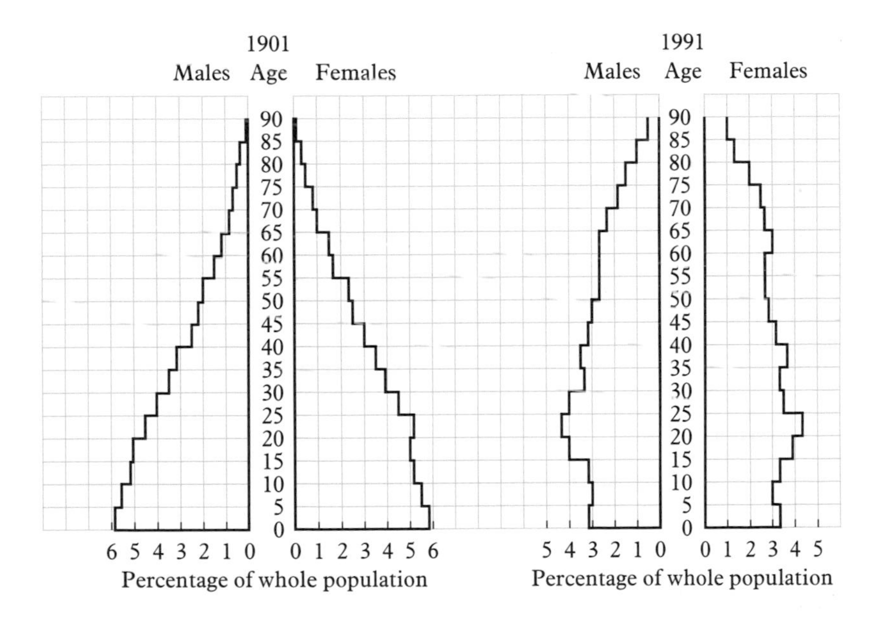

Source: Adapted from *Annual Abstract of Statistics 1992*

NEAB, 1997, Paper 3

7 A population pyramid is to be drawn showing the percentage of the population in different age groups. The data for females have already been drawn.

a Use the following table to copy and complete the population pyramid. *(2)*

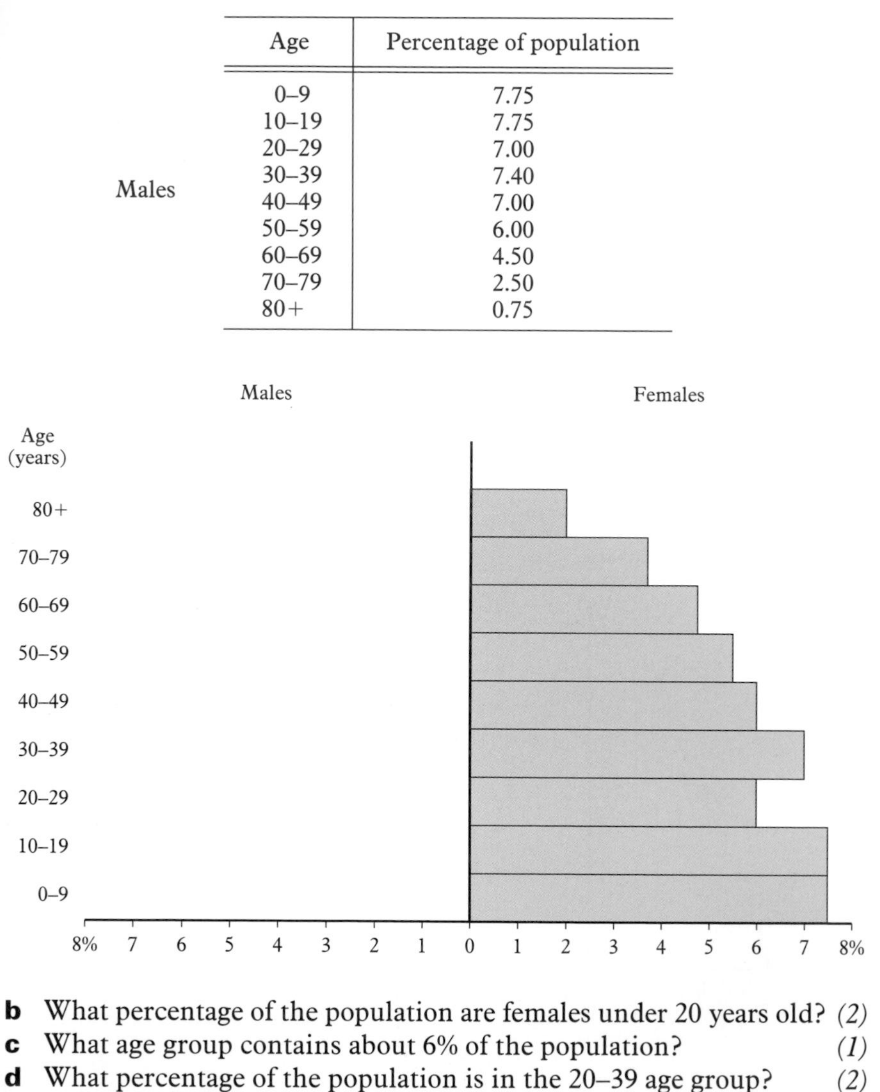

Males

Age	Percentage of population
0–9	7.75
10–19	7.75
20–29	7.00
30–39	7.40
40–49	7.00
50–59	6.00
60–69	4.50
70–79	2.50
80+	0.75

b What percentage of the population are females under 20 years old? *(2)*

c What age group contains about 6% of the population? *(1)*

d What percentage of the population is in the 20–39 age group? *(2)*

e Using the diagram, make one comment about the population who are over 60 years old. *(1)*

SEG, 1998, Paper 2

Two-way tables

Two-way tables show two facts at one time.
This table shows the number of boys and girls who choose to study either French or German in Year 9.

	French	German	Total
Boys	61	31	92
Girls	57	53	110
Total	118	84	202

31 boys study German. **118** pupils study French.

Stem and leaf diagrams

A **stem and leaf diagram** shows the shape of a distribution. It is like a bar-chart with the numbers themselves forming the horizontal bars.

Times in minutes spent on the phone

Stem	Leaf
2	2 2 5
3	4 6 8 9 9

where 3 | 4 means 34 minutes.

Times in minutes spent on the phone

Internet		Friends
2	2	3
3 1	3	6 7 8
4 3	4	1 2 3 4
9 8 7 6 2	5	2 3

where 4 | 1 means 41 minutes.

Two stem and leaf diagrams can be drawn back to back. These can be used to compare data. They are called **back to back stem and leaf diagrams**.

You can find the mode, median and both of the quartiles from a stem and leaf diagram.

Population pyramids

Population pyramids can be used to compare two sets of data. The two sets of horizontal bars are drawn as two bar-charts placed back to back.

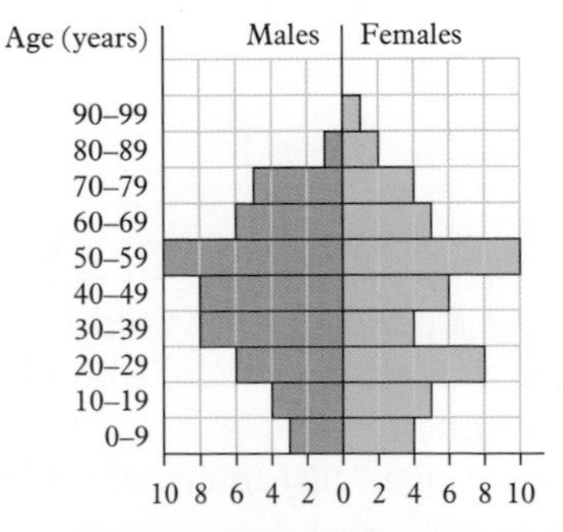

1 Joni asks the pupils in her class for their favourite primary colour. She records the results in a two-way table.

	Red	Yellow	Blue	Total
Boys	10	3	4	17
Girls	5	2	9	16
Total	15	5	13	33

Use the table to find how many:

a boys preferred red

b girls preferred either red or yellow

c boys Joni asked.

2 A local bookshop carried out a survey of 120 customers. As part of this, each person was asked if they preferred their books to be hardback or paperback. The results from this survey are recorded in a two-way table.

	Men	Women	Total
Hardback	26		38
Paperback			82
Total	57	63	120

Copy the table and complete it.

3 These are Andrew's scores in the school darts competition.

78 44 88 63 86 52 77 89
75 63 88 52 89 76 61 84

Draw a stem and leaf diagram to show these scores.

4 Fli-Bi-Nite are tour operators. They record how their customers who visit coastal resorts are distributed by age. They do the same for those visiting historical towns. This information is recorded in the table.

Age group (years)	Coastal resorts (%)	Historical towns (%)
21–30	22	2
31–40	36	15
41–50	20	42
51–60	13	26
61–70	9	15

a Use this table to draw a population pyramid.

b Find the percentage of their customers who visit coastal resorts and are aged 40 years or under.

c Find the percentage of their customers who visit historical towns and are aged 51 and over.

d Compare the two sets of data.

6 Probability

1 It all adds up to one
Probability scales
Equally likely events
Probability of an event
Probabilities add up to one
Discrete uniform distribution
Sample space diagrams

2 Conditional probability
Finding conditional probabilities

3 Relative frequency
Finding relative frequencies
Using graphs for relative frequencies
Expected number
Finding probabilities
Simulation

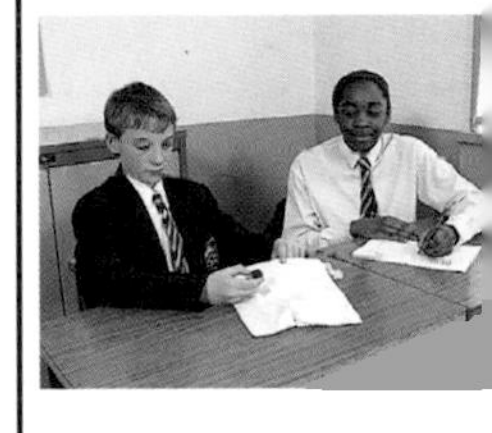

EXAM QUESTIONS

SUMMARY

ICT IN STATISTICS

TEST YOURSELF

1 It all adds up to one

Moses and Whitney are watching the weather forecast.
The presenter says that the probability of a hurricane is 40%.
Moses is worried.
Whitney is more cheerful.
She says the probability of the hurricane missing them is 60%.

Probabilities can also be written as fractions or decimals.

Probability

Probability tells you how likely something is to happen.

Probabilities are often shown on a scale with 'impossible' at one end and 'certain' at the other.

This is a probability scale.

b			c			a
impossible	very unlikely	unlikely	even chance	likely	highly likely	certain

These probabilities are shown on the scale.

a The sun will rise tomorrow
b Ice will be found on the sun
c The sun will shine on the first day of spring.

Exercise 6:1

1 Draw a probability scale.
Mark points **a**, **b** and **c** to show how likely you think each one is.
a A £1 ticket will win the jackpot in the National Lottery.
b A £1 ticket will win £10 in the National Lottery.
c A ticket will not win anything in the National Lottery.

2 Write down two things which

a are highly likely

b are impossible

c have an even chance of happening

d are certain to happen.

A probability scale often has numbers on it instead of words.
Something which is impossible has a probability of 0.
Something that is certain has a probability of 1.

These probabilities are shown on the scale.

a The Pacific ocean will freeze over tomorrow.

b If I toss a coin it will land tail up.

c Louise will drink some fluid tomorrow.

a b c
0 0.5 1

The value of a probability can only be between 0 and 1.

---

3 Draw a probability scale with numbers.
Mark points **a**, **b** and **c** to show how likely you think each of these is.

a A car aged 2 will break down in the next year.

b A car aged 15 will break down in the next year.

c A car aged 7 will break down in the next year.

4 Write down two things which have a probability of

a 1 **b** 0 **c** 0.5 **d** 0.1 **e** 0.8 **f** 0.25

5 John says that the probability he will fail his maths exam is 1.5
Explain what is wrong with this.

6 Lucy says that the probability of her getting a car for her birthday is −0.8
Explain what is wrong with this.

Equally likely

Two events are **equally likely** if they have the same chance of happening.

The chance of getting a red with this spinner is the same as the chance of getting a blue. Getting a red and getting a blue are equally likely.

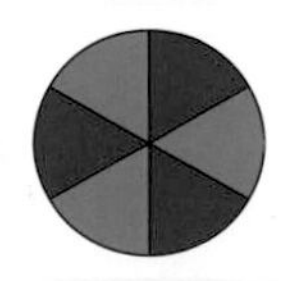

You are more likely to get a red with this spinner than a blue. Getting a red and getting a blue are not equally likely with this spinner.

7 Look at these spinners
For each spinner which colour is **a** most likely **b** least likely

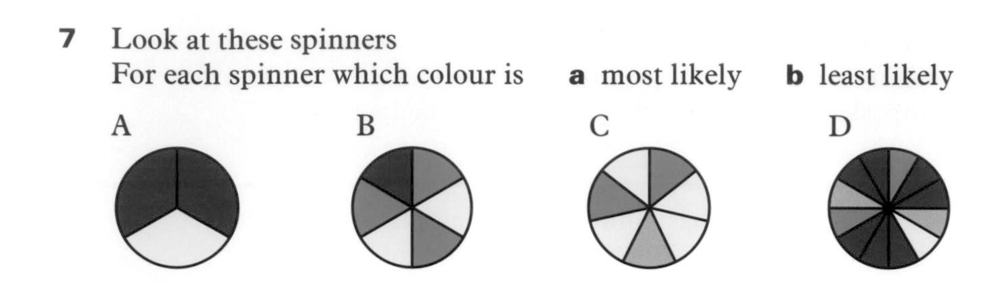

8 Which of these spinners give equally likely events?

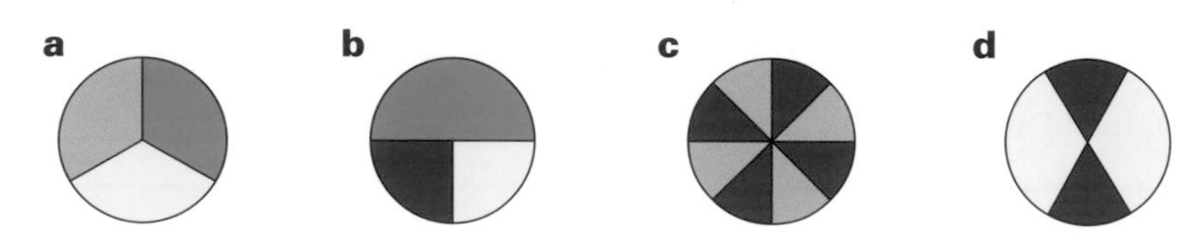

9 Tom picks a chocolate at random from a box of plain and milk chocolates.
He is equally likely to choose plain or milk.
What does this tell you about the number of plain and milk chocolates in the box?

You can use probability to decide whether something is fair or not.

Example

Julia and Peter are deciding who is going to eat the last Rolo.

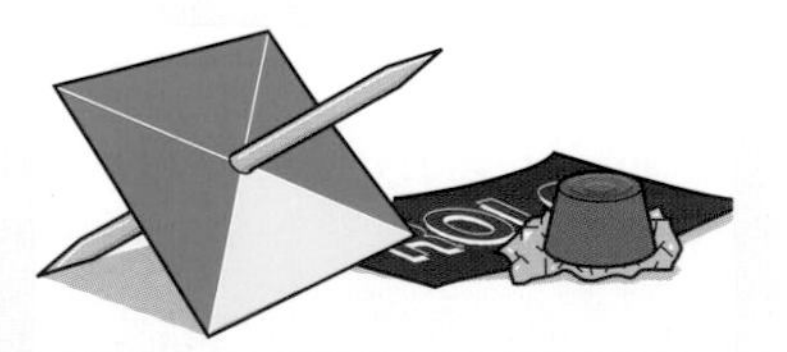

They are going to use this spinner.
Julia will eat it if the spinner lands on blue.
Peter will eat it if the spinner lands on yellow.
Why is this unfair?

There are more blue sections on the spinner than yellow.
There is a greater chance it will land on a blue section.
Julia has a better chance of getting the last Rolo!

---

10 Sue and Liam have only one ticket for a pop concert.
They decide to roll a dice to see who gets the ticket.
Sue gets the ticket if the dice shows a 1 or 2.
Liam gets the ticket if the dice shows a 3, 4, 5 or 6.
Is this fair? Explain your answer.

11 Jenny and Lisa are playing a game with a fair dice.
They take turns to roll the dice.
If the dice shows an even number Jenny writes it down.
If the dice shows an odd number Lisa writes it down.
The winner is the first person whose numbers add up to 24.

a Explain what is meant by a fair dice.

b Explain why the game is unfair.

You can use equally likely events to work out probabilities.
Paula uses this spinner.
There are 5 equal sections.

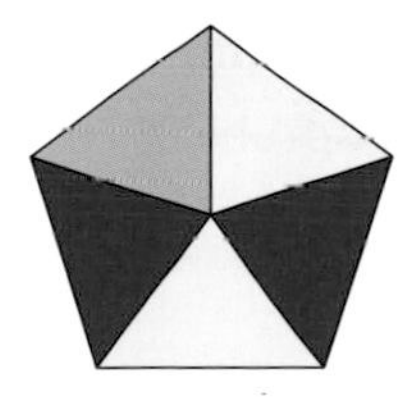

Probability of an event A

The **probability of an event A** is:

$$\frac{\text{the number of ways that the event A can happen}}{\text{the total number of things that can happen}}$$

The probability of getting a green $= \frac{1}{5}$ only 1 of the 5 sections is green.

The probability of getting a red $= \frac{2}{5}$ 2 of the 5 sections are red.

Exercise 6:2

1 Alan has three black pens and five blue pens in his pencil case.
He picks out a pen at random.
Write down the probability that he chooses

a a black pen **b** a blue pen.

2 Ria throws a dice.
Write down the probability that she gets

a 1

b 5 or 6

c an odd number

d a prime number.

3 Joggers crisps have a special offer. Eight bags in every 100 contain a voucher for a free packet of crisps. Prodeep buys a packet of crisps. Write down the probability that his packet contains a voucher. Give your answer as:

a a fraction **b** a decimal **c** a percentage.

4 Janet picks a card at random from a pack of 52 playing cards. Write down the probability that the card is:

a a three

b a heart

c the five of spades

d a queen or a king.

Probabilities always add up to one.
A dice is thrown.

The probability of getting a 6 is $\frac{1}{6}$.

Not getting a 6 means that you get a 1, 2, 3, 4 or 5.

The probability of not getting a 6 is $\frac{5}{6}$. $1 - \frac{1}{6} = \frac{5}{6}$

Example The probability that Jane is late is $\frac{1}{4}$.
Find the probability that Jane is not late.
The probability that Jane is not late is $1 - \frac{1}{4} = \frac{3}{4}$

Example Tim picks a cube at random from a box of cubes. The box contains red, blue and yellow cubes.
The probability of picking out a red cube is $\frac{5}{17}$.
The probability of picking out a blue cube is $\frac{9}{17}$.
Find the probability of picking out a yellow cube.

The probability of picking a red or a blue cube $= \frac{5}{17} + \frac{9}{17}$
$= \frac{14}{17}$

The probability of picking out a yellow cube $= 1 - \frac{14}{17}$
$= \frac{3}{17}$

---

5 The probability that a phone is engaged is $\frac{1}{8}$.
Find the probability that the phone is not engaged.

6 The probability that a dentist runs late for his appointment is 40%.
Find the probability that he doesn't run late.

7 Debbie passes through one set of traffic lights on her way to college.
The probability that the lights are showing green is $\frac{5}{8}$.
The probability that the lights are showing amber is $\frac{1}{8}$.
Find the probability that the lights are showing red.
Give your answer as:
a a fraction **b** a decimal **c** a percentage.

8 Branston Rovers can win, lose or draw their hockey match.
The probability that they will lose is $\frac{4}{15}$.
The probability that they will draw is $\frac{2}{15}$.
Find the probability that they will win.

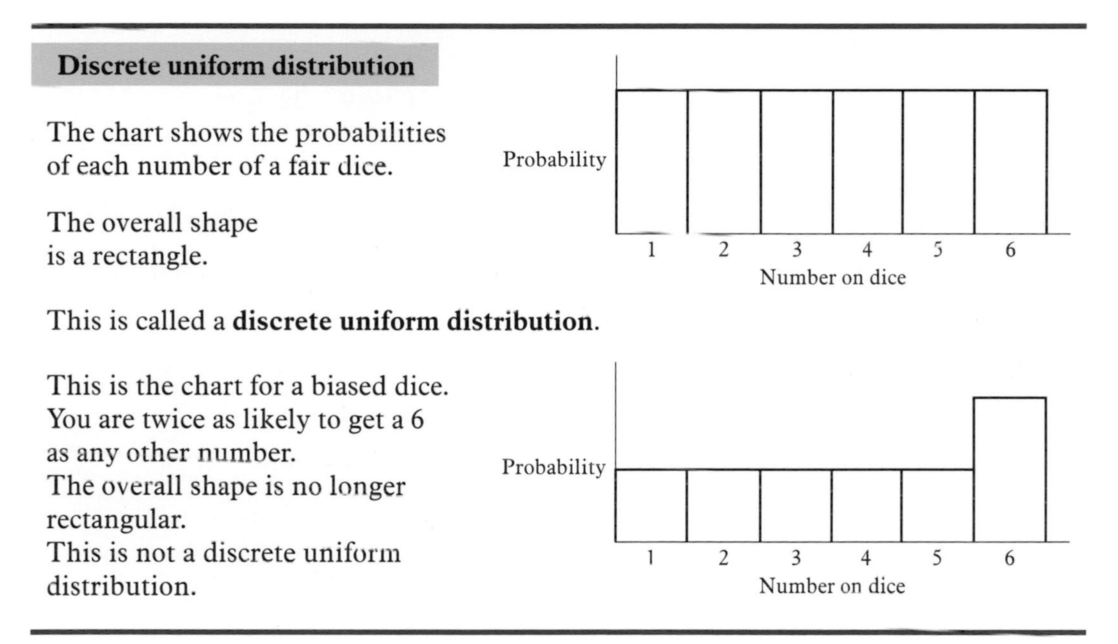

Discrete uniform distribution

The chart shows the probabilities of each number of a fair dice.

The overall shape is a rectangle.

This is called a **discrete uniform distribution**.

This is the chart for a biased dice.
You are twice as likely to get a 6 as any other number.
The overall shape is no longer rectangular.
This is not a discrete uniform distribution.

9 Which of these spinners gives a discrete uniform distribution?

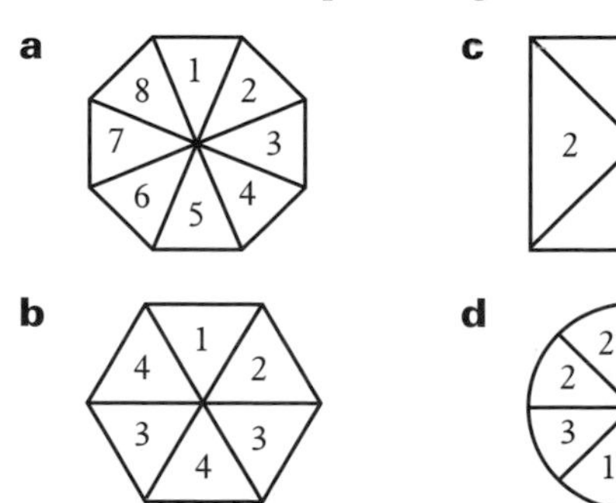

You can get data from tables to find probabilities.

Example The table shows the type of person that took part in a survey.

	Male	Female
Child	18	21
Adult	19	32

A person is chosen at random from the survey.
Find the probability that the person is:

a a male adult **b** a child.

a 19 male adults took part in the survey.
The total number of people in the survey $= 18 + 21 + 19 + 32 = 90$.
The probability that the person is a male adult is $\frac{19}{90}$.

b There are $18 + 21$ children in the survey.
The probability that the person is a child $= \frac{39}{90}$.

10 The table gives the membership of a club.

	Male	Female
Child	24	26
Adult	32	38

A member is chosen at random.
Find the probability that this member is:

a a child
b a girl
c male
d an adult
e a female adult
f female.

11 The table gives the GCSE grades A* to C and grades D to G for the Year 11 pupils in Maths, English and Science.

	A* to C	D to G
Maths	110	90
English	122	78
Science	105	95

A pupil is chosen at random. Find the probability that the pupil

a achieved A* to C in Maths
b achieved D to G in English
c did not achieve D to G in Science.

12 A shop sells sweatshirts in three sizes: small, medium and large.
The sweatshirts come in three different colours.
The table shows how many of each size and colour they have in stock.

	Small	Medium	Large
White	14	23	12
Red	19	14	8
Black	5	15	20

a How many sweatshirts does the shop have in stock?

A sweatshirt is chosen at random.
Find the probability that it is:

b a medium red

c a large white

d black

e small.

Outcome Each thing that can happen in an experiment is called an **outcome**.

Sample space A **sample space** is a list of all possible outcomes.

Sample space diagram A table which shows all of the possible outcomes is called a **sample space diagram**.

Example Toby rolls a dice and spins a £1 coin.

a Draw a sample space diagram to show all the possible outcomes.

b Write down the probability that Toby gets a Head and a 5.

a Draw a table. Put all the numbers for the dice along the top.
Use H and T, for Head and Tail, down the side of the table.

		Dice					
		1	2	3	4	5	6
Coin	H	H, 1	H, 2	H, 3	H, 4	H, 5	H, 6
	T	T, 1	T, 2	T, 3	T, 4	T, 5	T, 6

b There are 12 possible outcomes. They are equally likely.
A Head and a 5 appears once.
The probability of getting a Head and a 5 is $\frac{1}{12}$.

Exercise 6:3

1 Carla uses this spinner and throws a coin.

a Copy this sample space diagram. Fill it in.

Write down the probability of getting:

b red and a Head

c blue and a Tail.

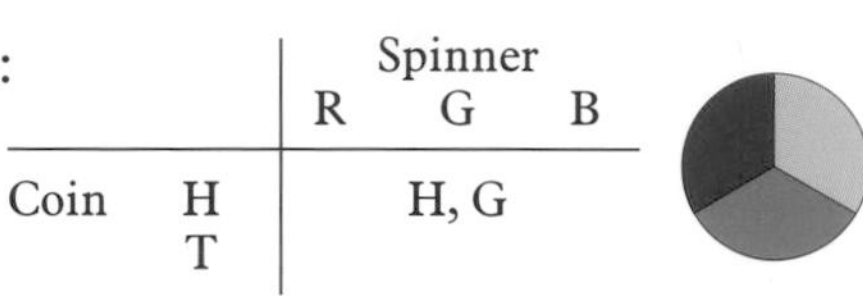

2 Sean has two boxes of counters.
Each box contains two red, one blue and one yellow counter.
Sean takes one counter at random from each box.

a Draw a sample space diagram to show all the possible outcomes.

Write down the probability of getting:

b two red counters

c two counters of the same colour

d one red and one yellow counter.

3 Rudi is using this set of axes as his sample space diagram.
It shows some of the posssible outcomes when a red dice and a blue dice are thrown together.

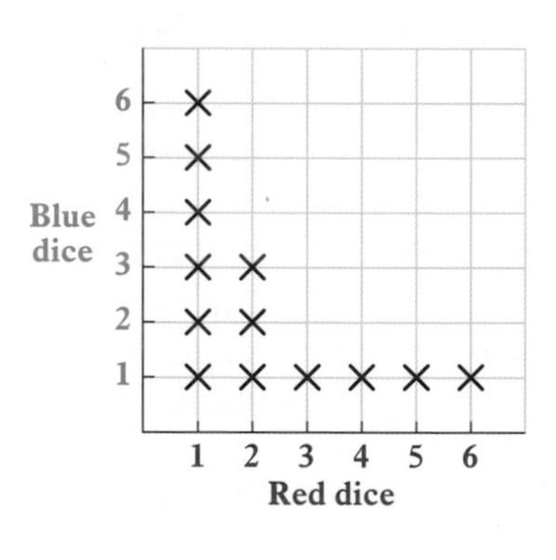

a Copy and complete Rudi's sample space diagram to show all the possible outcomes.

Write down the probability:

b of getting one even and one odd number

c of getting two even numbers

d that the sum of the two numbers is five or less

e that the sum of the two numbers is greater than ten

f that the difference between the two numbers is two.

4 Pat has two sets of four cards.
Each set contains the cards shown in the picture.
She picks one card at random from each set.

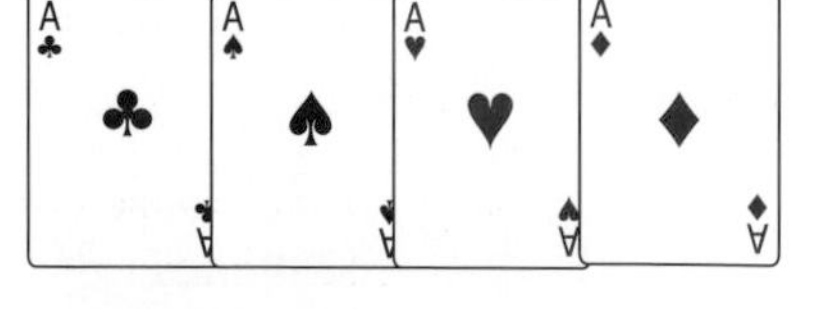

a Draw a sample space diagram to show all the possible outcomes.

Write down the probability that:

b both cards are hearts

c one card is a heart and the other is a spade

d the cards are of different suits

e the cards are both red.

2 Conditional probability

Tanya has won her tennis match. She has another match tomorrow. She thinks that she is playing well and should win that match as well.

Sometimes the value of a probability is affected by what has happened before.

Example Rita has five chocolates left. Three are plain and two are milk.
She picks out a chocolate at random and eats it.

a Write down the probability that this chocolate is milk.
b The first chocolate was milk. Rita now picks out a second chocolate at random. Write down the probability that the second chocolate is milk.

a There are five chocolates. Two are milk.
The probability that the 1st chocolate is milk $= \frac{2}{5}$.
b The choice of the 1st chocolate has an effect on the probability that the 2nd chocolate is milk.
There are now four chocolates. Only one of these is milk.
The probability the second chocolate is milk is $\frac{1}{4}$.

Conditional probability

When the outcome of a first event affects the outcome of a second event, the probability of the second event depends on what has happened. This is called **conditional probability**.

Exercise 6:4

1 Danny has to choose two players to play in a badminton match.
He puts the names of his four best players into a hat.
They are Tim, Peter, Sam and Geoff.
Danny picks out a name at random.
a Write down the probability that Tim is picked out.
b The first name picked out is Geoff. Danny doesn't put Geoff's name back into the hat. He picks another name at random.
Write down the probability that Tim's name is picked out this time.

2 Philip has seven clean shirts. Four are blue and three are white.
Philip picks a shirt at random and wears it.

a Write down the probability that this shirt is white.

b The first shirt is blue. Philip puts the dirty shirt in the laundry.
He picks out another shirt at random.
Write down the probability that this second shirt is white.

3 Anna has three raffle tickets. 100 tickets were sold.

a A ticket is picked at random for first prize.
Write down the probability that Anna wins first prize.

b Anna did not win first prize.
Another ticket is picked at random for second prize.
Write down the probability that Anna has this ticket.

c Anna won the second prize.
Another ticket is picked at random for the third prize.
Write down the probability that Anna has this third ticket.

4 A box contains five red counters and seven green counters.
Brian picks out a counter at random.
He does not put the counter back in the box.
He now picks out another counter at random.

a Write down the probability that the first counter is green.

b The first counter is green.
Write down the probability that the second counter is green.

5 Cara has a bag containing cubes. There are six white, seven red and three black cubes in the bag.
Cara takes out two cubes.
(This is the same as taking out one cube, not replacing it and then taking out another cube.)

a Write down the probability that the first cube is red.

b If the first cube is white, write down the probability that the second cube is red.

c Both the first and second cubes are white and neither are replaced.
Cara takes a third cube at random.
Write down the probability that the third cube is white.

You can find conditional probabilities using the information stored in two-way tables.

Example The Maths department enters 120 pupils for the GCSE Statistics exam. The table gives information on the entries.

	Foundation	Higher
Girls	21	35
Boys	25	39

a Write down the probability that a pupil chosen at random is entered for Foundation.

b A girl is chosen at random. Write down the probability that she has been entered for Foundation.

a The total number entered for Foundation $= 21 + 25 = 46$.
The probability that a pupil taken at random is entered for Foundation $= \frac{46}{120}$.

b You know that the pupil is a girl so you just look at the data for girls.
There are $21 + 35 = 56$ girls.
The probability that she has been entered for Foundation $= \frac{21}{56}$.

Exercise 6:5

1 Data were collected on 1000 people in work. Each person was asked if their salary was less than £15 000.
These are the results.

	Less than £15 000	£15 000 or more
Men	156	344
Women	281	219

a A person is chosen at random. Find the probability that the person is:
(1) a man earning less than £15 000 (2) a man.

b A man is chosen at random. Write down the probability that he earns £15 000 or more.

c A person is chosen at random. Given that the person earns less than £15 000, write down the probability that the person is female.

2 Data were collected on the voting habits of 600 people.
The table gives the results.

	Less than 35 years old	At least 35 years old
Did vote	135	274
Did not vote	59	132

a Find the probability that a person chosen at random:
(1) did not vote (2) is less than 35 years old.

b A person is chosen at random. This person is less than 35 years old.
Find the probability that the person did not vote.

c A person is chosen at random. This person did vote.
Find the probability that the person is at least 35 years old.

This diagram shows the sports played by the members of a club.
The circle labelled squash shows the number of people who play squash.

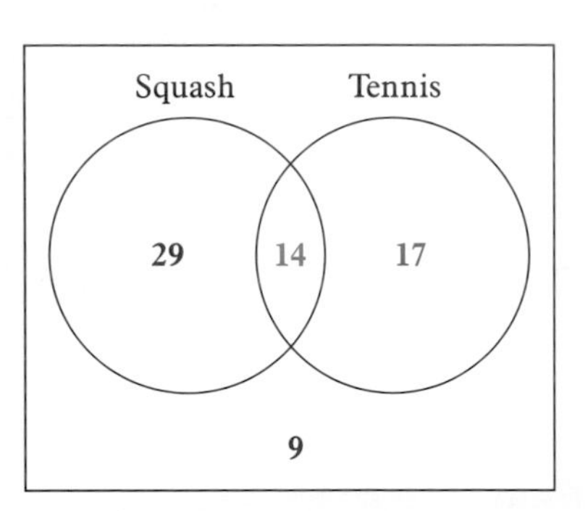

29 play squash only **14** play both squash and tennis
There are **29** + **14** = 43 people who play squash.

The circle labelled tennis shows the number of people who play tennis.
17 play tennis only **14** play both squash and tennis
There are **17** + **14** = 31 people who play tennis.

9 people play neither game.
The total number of people in the club = **29** + **14** + **17** + **9** = 69.
This type of diagram is called a **Venn Diagram**.

---

3 Use the diagram above to answer these questions.

a A member is chosen at random.
Find the probability that the member: (1) plays squash
(2) plays tennis only.

b A member is chosen at random. This member plays tennis.
Find the probability that this member plays both games.

c A member is chosen at random. This member plays squash.
Find the probability that this member plays both games.

4 The diagram shows the number of boys with blue eyes or fair hair in a youth club.

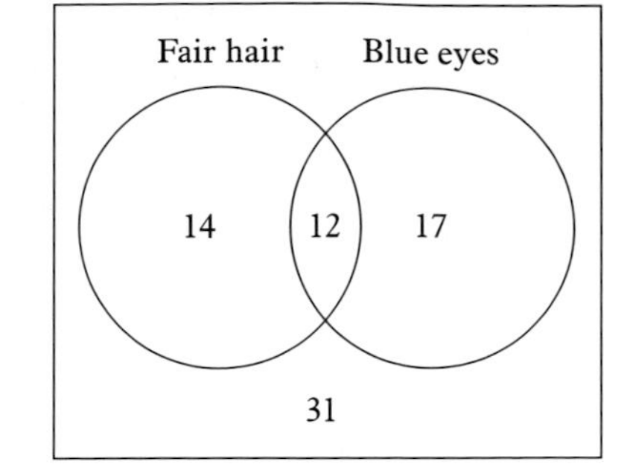

Write down how many boys there are:

a altogether
b with blue eyes
c with both blue eyes and fair hair
d with blue eyes but not fair hair
e with neither blue eyes nor fair hair.
f A boy is chosen at random.
What is the probability that he has fair hair?
g A boy is chosen at random. He has fair hair.
What is the probability that he has blue eyes?
h A boy is chosen at random. He has blue eyes.
What is the probability that he has fair hair?

5 The diagram shows the hobbies of a group of girls.

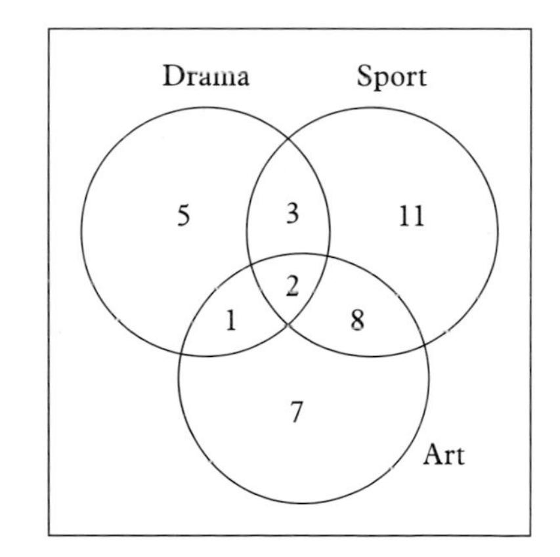

a Write down the total number of girls.
b Write down the number of girls who do:
(1) Drama
(2) Drama only
(3) Sport only
(4) Art
(5) both Drama and Sport
(6) both Sport and Art
(7) all three.
c A girl is chosen at random.
Find the probability that she does both Drama and Art.
d A girl is chosen at random. She does Drama.
Find the probability that she does Art.
e A girl is chosen at random. She does Sport.
Find the probability that she does Drama.
f A girl is chosen at random. She does both Drama and Art.
Find the probability that she does all three.

Sometimes you have to draw your own Venn diagrams.

Example Mia asks 200 sixth form students which of these types of music they listen to – pop, jazz and classical.
Her results are:

90 students listen to classical
123 students listen to pop
69 students listen to jazz
53 students listen to both classical and pop
27 students listen to both pop and jazz
34 students listen to both classical and jazz
15 students listen to all three

Draw a Venn Diagram with 3 circles overlapping.
Label the three circles.
Start where the 3 circles overlap.
Write in **15**.

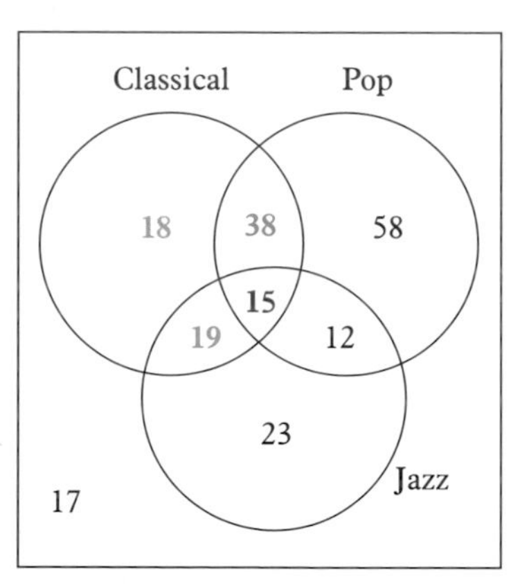

There are 53 students in the overlap of the classical and pop circles.

$53 - \mathbf{15} = 38$

So 38 is the number of students who listen to classical and pop only.
Write in 38.

Similarly the number for classical and jazz only is $34 - \mathbf{15} = 19$.

There are 90 students in the classical circle.

$90 - \mathbf{15} - 38 - 19 = \mathbf{18}$

So 18 is the number of students who listen to classical music only.
Write in **18**.

The other two circles are completed in the same way.

There are $\mathbf{18} + 38 + \mathbf{15} + 19 + 23 + 12 + 58 = 183$ students represented in the circles.
So there are $200 - 183 = 17$ students outside the circles.
These 17 students do not listen to any of the 3 types of music.

6 Tanya asks 100 members of her sports club if they listen to Tapes, CDs or MiniDiscs. These are her results.

87 listen to CDs, 59 to MiniDiscs and 57 to Tapes
53 listen to CDs and Tapes
49 listen to CDs and MiniDiscs
37 listen to Tapes and MiniDiscs
36 listen to music recorded on all three media

a Copy the Venn Diagram.

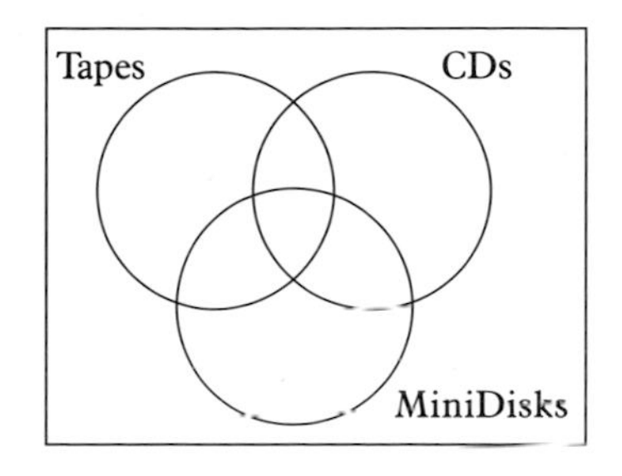

b Write in the number who listen to all three.

c The number who listen to CDs and MiniDiscs only is 49–36. Write this number in your diagram.

d Write in the number who listen to Tapes and CDs only.

e Write in the number who listen to Tapes and MiniDiscs only.

f Work out the number who listen to Tapes only.

g Complete the Venn Diagram.

7 A pizza takeaway offers up to three different extra toppings on each pizza. Pierre records the choices made by 55 of his customers.

42 chose mushrooms, 25 chose onions and 20 chose peppers
13 chose peppers and mushrooms, 11 chose peppers and onions
17 chose mushrooms and onions
5 chose all three toppings

a Draw a Venn Diagram to show this information.

b Use the diagram to work out how many of Pierre's customers have no extra topping.

8 Mr Fisher asks his class of 28 pupils which triathlon event they watched.

13 watched the swimming
8 watched the cycling and swimming
7 watched the running and swimming
2 watched the cycling only, 5 watched the running only
4 watched all three
3 did not watch any triathlon events

a Draw a Venn Diagram to show this information.

b Use your diagram to work out the number of pupils who watched the cycling and running only.

3 Relative frequency

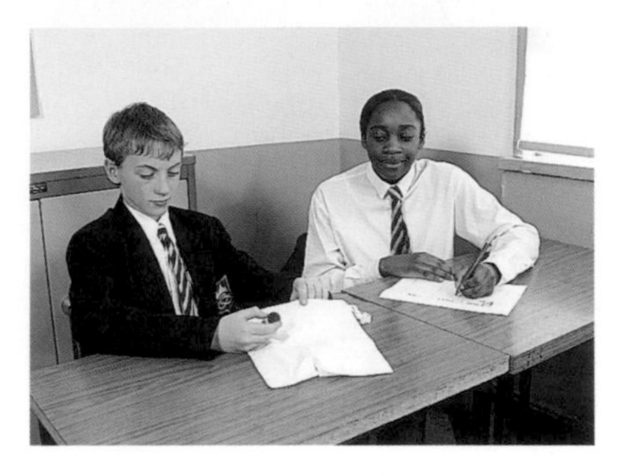

There are 10 beads in a bag. Some of the beads are red and the rest are black. David and Pavneet have to find out how many beads of each colour are in the bag without looking. They take one bead out of the bag, note its colour and replace it. They do this lots of times. How many times do they need to do this to be fairly sure about the contents of the bag?

Wendy is testing this tetrahedral dice to see if it is fair. She throws the dice 100 times and records her number each time. These are her results.

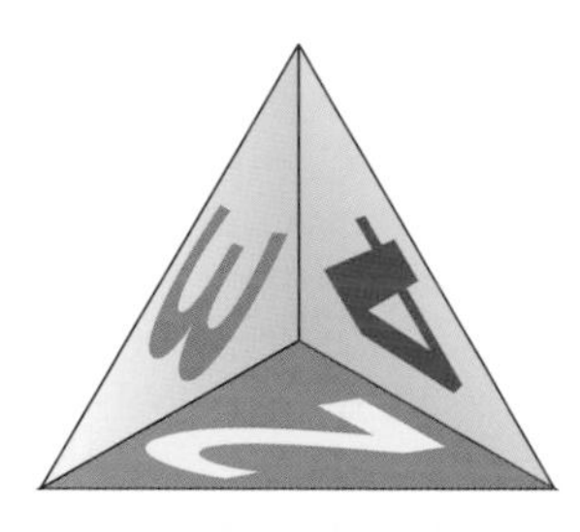

Number	Tally	Frequency
1	卌 卌 卌 卌 |||	23
2	卌 卌 卌 卌 卌		26
3	卌 卌 卌 卌 ||	22
4	卌 卌 卌 卌 卌 ||||	29

Frequency

The **frequency** of an event is the number of times that it happens.

Relative frequency

Relative frequency of an event $= \dfrac{\text{frequency of the event}}{\text{total frequency}}$

The relative frequency gives an *estimate* of the probability.

If the dice is fair the probability of getting each number is $\frac{1}{4}$ or 0.25
Wendy checks her data to see if the dice is fair.
She uses the relative frequency from her experiment.

Relative frequency of $1 = \frac{23}{100} = 0.23$
The values 0.23 and 0.25 are close.
This is also true for the numbers 2, 3 and 4 so the dice is probably fair.

Wendy can get a better value for the probability by repeating her experiment more times.

Exercise 6:6

1 Val tosses a coin 100 times.
This is her data.

Outcome	Frequency
Head	34
Tail	66

a Find the relative frequency of:
(1) head (2) tail.
b Is the coin fair? Explain your answer.

2 Jason collects data on the colours of cars passing the school gate.
His results are shown in the table.

Colour	Tally	Frequency
White	𝍸 𝍸 𝍸 𝍸 |||	23
Red	𝍸 𝍸 𝍸 𝍸 𝍸 𝍸 ||	32
Black	𝍸 𝍸 ||||	14
Blue	𝍸 𝍸 𝍸		16
Green	𝍸 𝍸		11
Other	||||	4

a How many cars did Jason include in his survey?
b What is the relative frequency of red?
Give your answer as: (1) a fraction (2) a decimal (3) a percentage.
c What is the relative frequency of green?
Give your answer as: (1) a fraction (2) a decimal (3) a percentage.
d What is the most likely colour of the next car passing the school?
e Write down an estimate for the probability that the next car will be white. Give your answer as a fraction.
f How can the estimate for the probability of white be made more reliable?

3 A machine fills boxes of cornflakes.
They should contain 500 g of cornflakes.
Chris tests the machine by weighing 100 boxes of cornflakes chosen at random. The table shows her results.

Weight of cornflakes	<500 g	Exactly 500 g	>500 g
Number of bags	6	31	63

Estimate the probability that a box chosen at random will:
a contain exactly 500 g
b not contain exactly 500 g
c be underweight
d be overweight.

An experiment can be repeated to improve the accuracy of the estimate of probability. It is useful to see what is happening to the relative frequency as the experiment is repeated.

Rosie throws a coin 20 times.
These are her results.

	Frequency	Relative frequency
Head	14	0.7
Tail	6	0.3

She throws the coin another 20 times.
These are her new results.

	Frequency	Relative frequency
Head	22	0.55
Tail	18	0.45

She throws the coin another 20 times.
These are her results.

	Frequency	Relative frequency
Head	27	0.45
Tail	32	0.55

She throws the coin another 20 times.
These are her results.

	Frequency	Relative frequency
Head	42	0.525
Tail	38	0.475

Rosie uses her results to draw a graph.

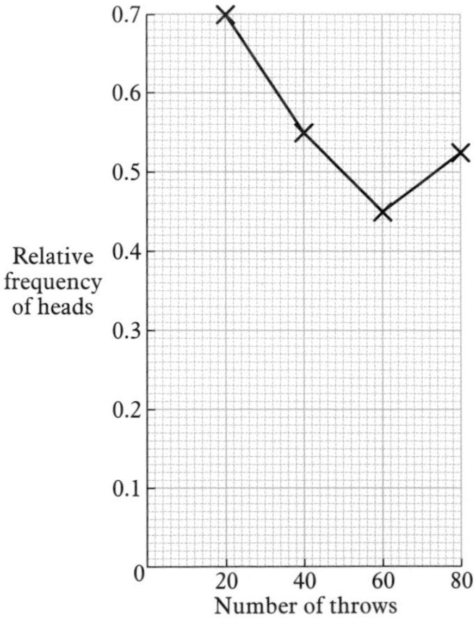

The points that she plots are (20, 0.7) (40, 0.55) (60, 0.45) and (80, 0.525).

As the experiment is repeated more times the graph should get closer to the true probability of $\frac{1}{2}$ assuming the coin is fair.

Exercise 6:7

1 Ian has thrown a dice 100 times. He worked out the relative frequency of getting a 6 after every 20 throws. These are his results.

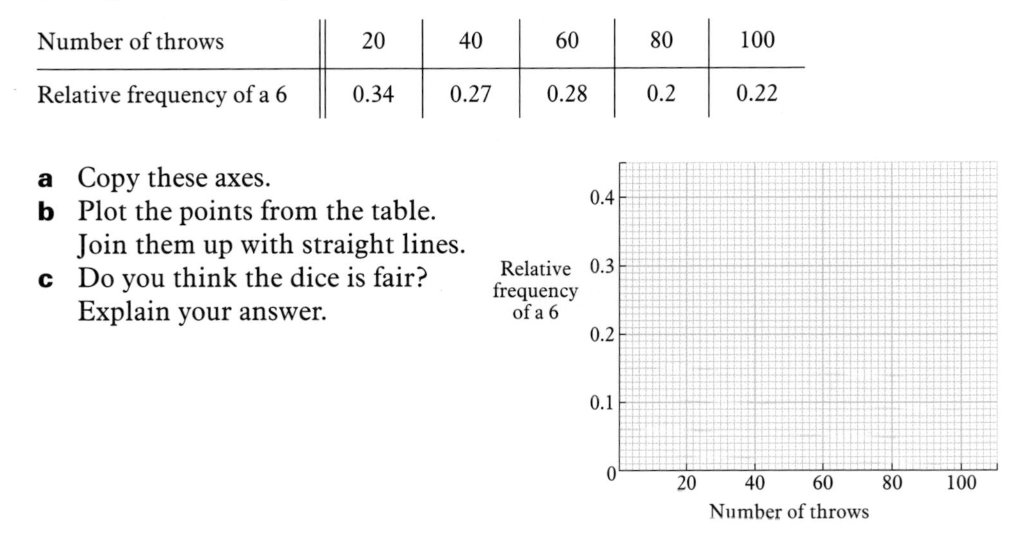

Number of throws	20	40	60	80	100
Relative frequency of a 6	0.34	0.27	0.28	0.2	0.22

a Copy these axes.
b Plot the points from the table.
Join them up with straight lines.
c Do you think the dice is fair?
Explain your answer.

2 Hacson has a bag containing 20 cubes.
He picks out a cube at random, notes the colour and then replaces it.
He does this 50 times. Every 10 times he works out the relative frequency of getting a red.
The graph shows his results.

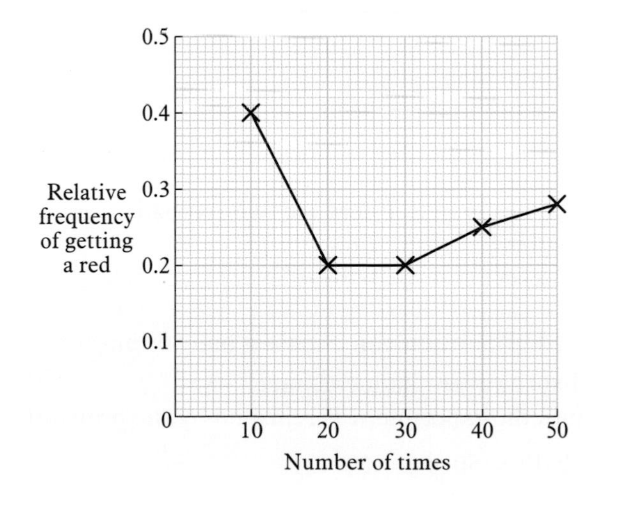

Write down the number of reds obtained in the first:

a 10 times
b 20 times
c 30 times
d 40 times
e 50 times.

3 Cheb is testing a spinner.
The graph shows the relative frequency of getting a blue.

a Write down the number of blues obtained in the first:
(1) 10 spins (2) 20 spins.

b Use your answers to part **a** to explain why one of the points must be wrong.

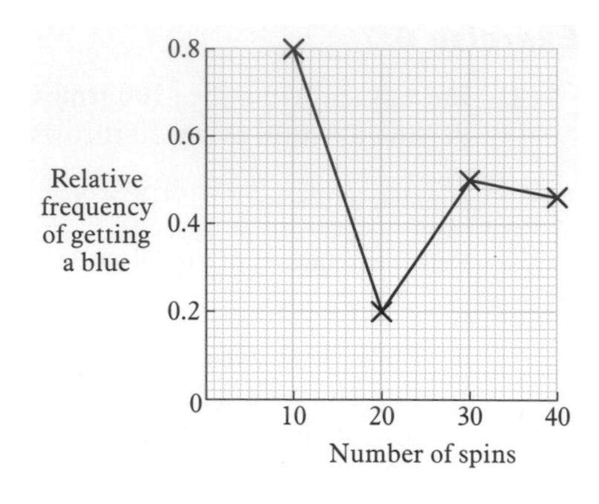

4 **a** Throw a dice 10 times.
Work out the relative frequency of getting an even number.

b Throw the dice another 10 times. Work out the new relative frequency.
Do this several times.

c Draw a graph showing your results.

d Use your graph to estimate the probability of getting an even number. Explain how you get your answer.

You can work out how many times an event is likely to happen when you repeat an experiment.

A bag contains two red balls and three blue balls.
A ball is chosen at random from the bag and replaced.
This is done 50 times.

The total number of possible outcomes is five. There are two red balls.
The probability of getting a red ball is $\frac{2}{5}$.
This means that, on average, you expect two red balls in every five chosen.

To find the expected number of red balls when you choose a ball 50 times:

Work out the probability that the event happens once.
Multiply the number of times the experiment is repeated by the probability.

Expected number of red balls $= 50 \times \frac{2}{5} = 20$

The **expected number** $=$ **number of trials** $\times$ **probability**.

Exercise 6:8

1 The probability that a first class letter is delivered within 24 hours of posting is 0.95
Diane posts 200 wedding invitations using first class stamps.
How many would you expect to be delivered within 24 hours?

2 A machine that makes tin cans has developed a fault.
Sean tests the reliability of the machine by checking a sample of 50 cans.
He finds that 12 of the cans are faulty.

a The machine produces 2400 cans in a day. How many of these would you expect to be faulty?

b The next day 589 cans are faulty. Has the fault got worse?
Explain your answer.

3 An insurance company knows that 12% of its policyholders make a claim in any one year. The company has 23 648 policyholders.

a How many claims does the company expect this year?

b Last year 3412 policyholders made a claim.
Is this more or less than expected?

4 Sally has collected data for the hospitals within 50 miles of her home.
She has worked out that the probability that a person will stay in hospital for a week or less is 0.75
How many patients, out of 2000, would you expect to stay in hospital for more than one week?

5 The probability that a player wins on a slot machine is $\frac{2}{35}$ for each go.
The cost of one go is 20p.

a How many wins would you expect for 100 goes?

b The machine pays out £1 for a win.
How much profit will the machine make on 1000 goes?

6 Janine uses a biased dice.
She expects to get 32 sixes in every 400 throws.

a What is the probability of getting a six with her dice?

b What is the probability of not getting a six with her dice?

Methods of finding probability

There are three methods of finding probability.

Method 1	Use equally likely outcomes e.g. to find the probability of getting a 2 with a dice.
Method 2	Carry out a survey or do an experiment e.g. to find the probability of getting a 2 with a biased dice.
Method 3	Sometimes you cannot do an experiment to find a probability. You need to look back at past records e.g. to find the probability that it will snow in Paris in January.

Exercise 6:9

Write down the method you would use to find the probability

1 that you will get a blue with this spinner

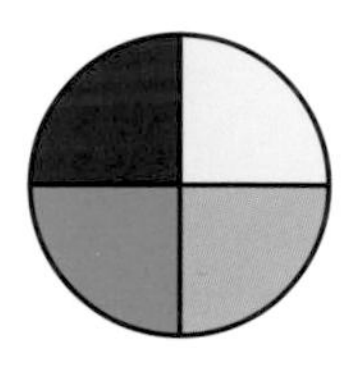

2 that there will be a measles epidemic this winter

3 that a purple counter will be drawn from a bag containing 4 white and 6 purple counters

4 that a house in the UK, chosen at random, contains 3 televisions

5 that the next person to buy a newspaper from the local shop chooses the *Mail*

6 of winning a prize in the national lottery

7 that the next car to pass the bus stop is blue

8 that the next car to leave the car park is blue

You use experiments in simulation.

Example Packets of seeds are given away with breakfast cereals.
There are 6 different packets of seeds.
How many boxes of breakfast cereal do you need to buy to collect all 6 different packets?

You can use a dice to represent each of the 6 different packets of seeds.
Roll the dice until all 6 numbers have been recorded at least once.
The results could be 3, 2, 5, 4, 4, 6, 5, 2, 3, 5, 4, 2, 1.
The dice has been rolled **13** times to get the complete set of numbers.
This suggests that **13** boxes of cereal will need to be bought.

The experiment needs to be repeated many times to get a reasonable estimate of the number of boxes of cereal.
The mode would be a sensible average to use for your estimate.
The greater the number of simulations the more accurate the estimate is likely to be.

You can use other methods of generating the numbers instead of a dice e.g. random number tables, random number function on the calculator, ICT.

Exercise 6:10

1 Photos of the 4 members of a pop group are given away in packets of sweets.

a Obtain an estimate of how many packets of sweets you need to buy to get a full set of photos. Use random number tables.

b Repeat the simulation 20 times to improve your estimate.

c How many packets should you buy to get a complete set?

2 20 different vouchers are given away in packets of crisps.
You need a full set of vouchers to get a model car.
Carry out a simulation to estimate the number of packets of crisps you need to buy to get the car.

3 Coloured toys are given away in packets of washing powder.
There are 3 different colours of toy.
Estimate the number of packets you need to buy to get all three colours.

You can also use numbers to represent proportions in simulations.

Example One third of diners in a restaurant choose fish. The rest choose meat. Use a dice to simulate the choice of the next 10 diners.

The dice has 6 numbers. $\frac{1}{3}$ of 6 numbers = 2 numbers

Allocate the numbers 1 and 2 to fish.

Allocate the numbers 3, 4, 5 and 6 to meat.

Throw the dice 10 times.

The results are 2, 5, 1, 2, 3, 1, 6, 4, 3, 4.

There are **four** 1s and 2s so **4** people choose fish.

There are **six** 3s, 4s, 5s and 6s so **6** people choose meat.

Exercise 6:11

1 20% of cars leaving a car park turn left. The rest turn right. Use random numbers to simulate the direction of turn of the next 15 cars. (*Use numbers 1 to 10 and allocate two numbers for left and 8 numbers for right.*)

2 A menu has three options – meat, fish and vegetarian. $\frac{2}{8}$ of diners choose fish, $\frac{1}{8}$ choose vegetarian and $\frac{5}{8}$ choose meat. Explain how you would simulate the choice of the next 12 diners.

3 Cars are equally likely to turn left or right out of a junction..

a (1) Use a coin to simulate the direction of turn for the next 20 cars.

(2) What percentage of the 20 cars turn left?

b (1) Use Excel to simulate the direction of turn for the next 1000 cars. See page **201**.

(2) What percentage of the 1000 cars turn left?

c Which of your two percentage answers will be closer to the true probability of a car turning left. Give a reason for your answer.

4 The two-way table shows the categories of membership of a club.

	Male	Female
Adult	37%	26%
Child	20%	17%

John wants to simulate the categories of the next 20 members.
He allocates numbers to each category in a table.

Category	Percentage	Numbers allocated
Male adult	37	00–36
Female adult	26	37–62
Male child	20	63–82
Female child	17	83–99

a Explain why the numbers 00–36 arc allocatcd to male adults.

b Explain how John can use a random numbers table to simulate the categories of the next 20 members.

c This two-way table shows the membership of a second club.

	Male	Female
Adult	21%	38%
Child	16%	25%

Allocate numbers like John to the diffferent categories and simulate the categories of the next 20 numbers.

5 Mary has analysed the results of her school hockey team.
The table shows the percentage of games won, lost or drawn.

	Won	Lost	Drawn
Percentage	58%	19%	23%

Mary wants to simulate the results of the next 4 matches.
Explain how she can do this.

1 A game is played using red, green, blue and yellow cards.
There are 20 cards of each colour.

a If one card is picked at random, what is the probability that it is red? *(2)*

b Sasha takes two of the green cards out and hides them.
All the other cards are left in.
A card is then picked at random.

i What is the probability that it is blue? *(2)*

ii What is the probability that it is green? *(2)*

NEAB, 1997, Paper 1

2 A simple board game was made for a stall at a school fete.

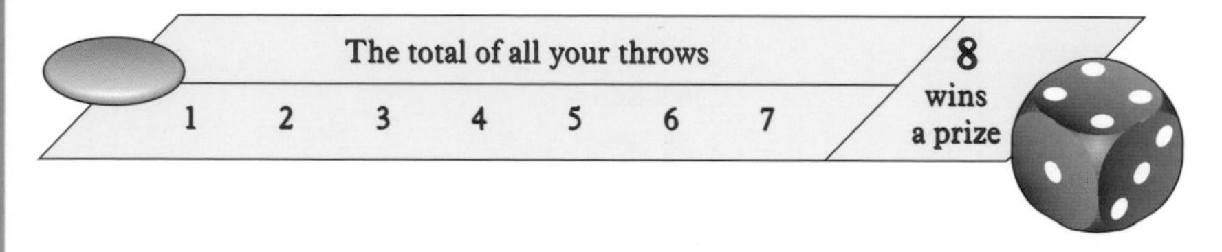

The game consists of throwing a fair dice several times and adding up all the scores.

The player can throw any number of times but to win a prize the total sum of all the throws must equal 8.

If the sum goes over 8 the game is lost.

A group of friends play the game.

a What is the probability that Louisa reaches 6 on her first throw? *(1)*

Helen throws a 5 on her first throw.

b What is the probability that she wins a prize with just one more throw? *(1)*

Ramon throws a 4 on his first throw.

c Find the maximum number of throws Ramon could now take and still win a prize. *(1)*

Reuben throws a 4 on his first throw.

d What is the probability that he loses on his second throw? *(1)*

Amelia wins a prize in two throws.

e **i** Copy the table below. Write down all the ways in which she could have thrown the dice. *(2)*

First throw			
Second throw			

ii Calculate the probability that Amelia can win in two throws. *(3)*

SEG, 1997, Paper 2

3 A survey of the age of cars was carried out at a local car park.
The cars were parked in rows.
The table below displays the data collected.

Row	Number of cars in the row	Number of cars more than 4 years old	Relative frequency of cars more than 4 years old
A	10	4	0.4
B	20	6	
C	15	3	
D	10	2	
E	10	0	
F	15	3	
G	20	4	
H	10	5	
I	20	7	
J	10	4	

a Complete the relative frequency column. *(2)*

b Calculate the best estimate of the probability of a car, selected at random from the car park, being more than 4 years old. *(3)*

SEG, 1998, Paper 4

4 A pupil threw a coin 40 times and recorded the results on the graph below.

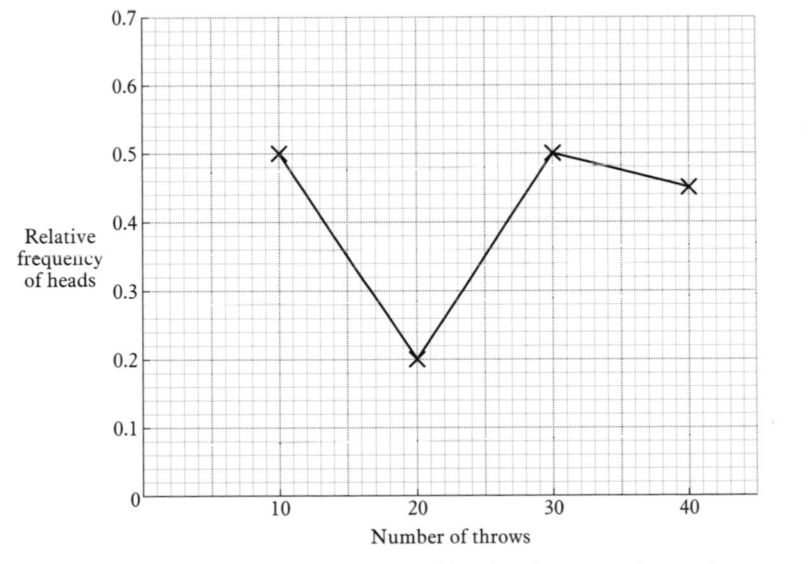

a How many Heads were obtained in the first ten throws? *(1)*

b One of the points was plotted incorrectly. Which point was plotted incorrectly? Explain your reasoning. *(2)*

c The pupil threw the coin another 10 times so that after 50 throws there were 23 heads. Copy the graph above onto graph paper. Plot the next point on the graph. *(2)*

NEAB, 1996, Paper 2

5 In a bag of 50 sweets, 10 are golden toffee, 13 are treacle toffee, 15 are milk chocolate and the remainder are plain chocolate.
Tim chooses one of the sweets at random.

a What is the probability that it is a plain chocolate? *(2)*

b What is the probability that it is a toffee? *(2)*

c Sarah chooses from an identical full bag of sweets. The sweet she takes out is a toffee.
What is the probability that it is a treacle toffee? *(2)*

NEAB, 1998, Paper 1

6 A boy threw a coin twenty times and recorded the results in the table below.

Face	Tally	Frequency
Heads	𝍸 𝍸 𝍸	
Tails	𝍸	

a Complete the frequency column. *(1)*

The diagram below shows the relative frequency of the number of Tails for particular numbers of throws throughout the experiment.

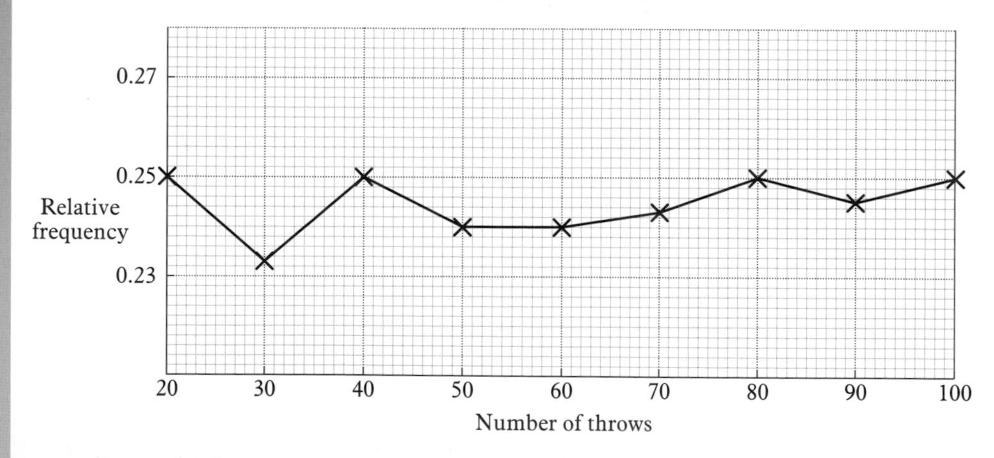

b How many Tails were obtained in the first 50 throws? *(1)*

c How many Tails were obtained in the last 50 throws? *(2)*

d Use the diagram to estimate the probability of obtaining a Tail. *(1)*

e Estimate the probability of obtaining a Head. *(1)*

f Do you think the coin was biased? Give your reasons. *(2)*

NEAB, 1998, Paper 2

7 In a board game, a counter is moved along the squares by an amount equal to the number thrown on a fair dice.

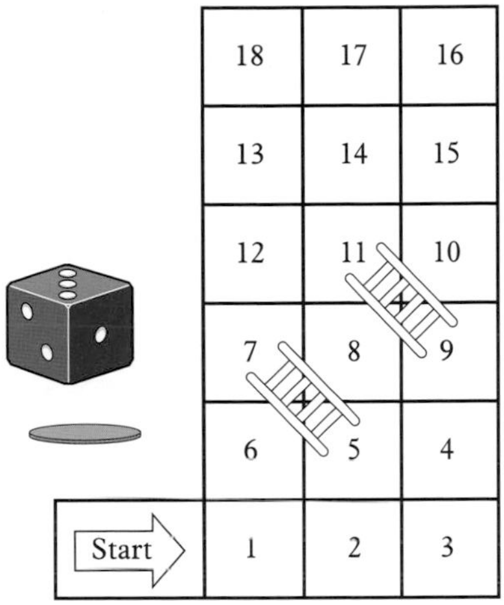

If you land on a square at the bottom of a ladder you move the counter to the square at the top of that ladder.

a What is the probability that a player reaches square 4 with one throw of the dice? *(1)*

b What is the probability that a player can reach square 7 with one throw of the dice? *(1)*

c What is the probability of taking two throws to get to square 2? *(2)*

d List the three possible ways to land on square 18 with exactly three throws of the dice.

First throw			
Second throw			
Third throw			

(3)

e Calculate the probability of landing on square 18 with exactly three throws of the dice. *(3)*

SEG, 2000, Paper 2

8 A survey of 100 households noted whether there were normally telephones in the living room, main bedroom or another room in the house.

None of the households had more than one telephone in any of these rooms, or more than three telephones in total.

Some households had no telephone.

The Venn diagram summarises **most** of the results of the survey.

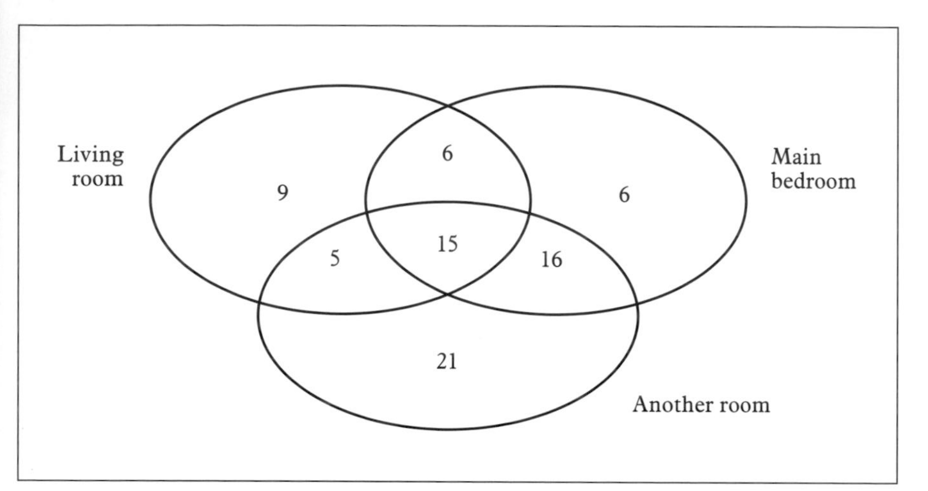

a Complete the frequency table for the number of telephones per household.

Number of telephones	Frequency
0	
1	
2	
3	

(3)

b What is the probability that a randomly selected household had

i one telephone only, *(1)*

ii no telephone in the main bedroom? *(2)*

c What proportion of households in the survey that had a telephone in the living room, also had a telephone in the main bedroom? *(2)*

9 Carla asked a sample of 50 students which of these they had in their bedroom. Video, computer, phone.

28 had a video, 37 had a computer and 12 had a phone
2 had a computer and a phone only
18 had a video and a computer
9 had a video and a phone
3 had all 3

a Draw a Venn diagram to represent these data
b How many students had none of the three items in their bedroom
c One of the students is chosen at random. Find the probability that this student
i has at least one of the three items in their bedroom
ii has only one of these items in their bedroom.

10 Bill throws a dice 60 times.
The graph shows the relative frequency of getting a 6.

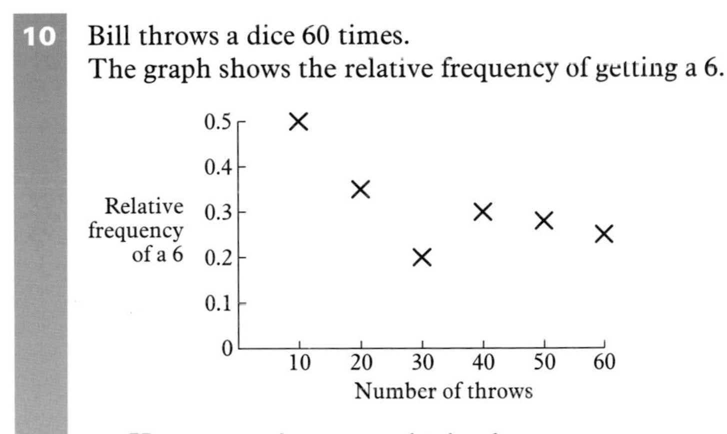

a How many sixes were obtained
(1) in the first 10 throws
(2) in the first 20 throws
(3) in the first 30 throws
b Explain why one of the first three points must be wrongly plotted.
c Do you think the dice is biased? Give your reasons.

You can use equally likely events to work out probabilities.
Paula uses this spinner.
There are five equal sections.

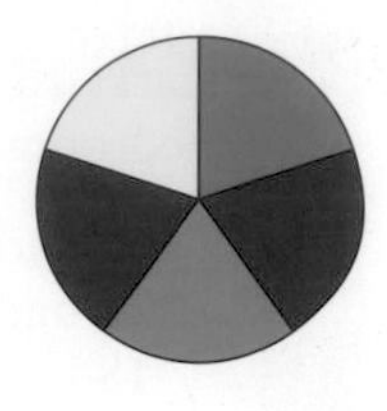

Probability of an event A The **probability of an event A** is:

$$\frac{\text{the number of ways that event A can happen}}{\text{the total number of things that can happen}}$$

The probability of getting a green $= \frac{1}{5}$ as only 1 of the 5 sections is green.
The probability of getting a red $= \frac{2}{5}$ as 2 of the 5 sections are red.

Probabilities always add up to one.

Example Tim picks a cube at random from a box of cubes. The box contains green, blue and yellow cubes.
The probability of picking out a green cube is $\frac{5}{17}$.
The probability of picking out a blue cube is $\frac{9}{17}$.
Find the probability of picking out a yellow cube.

$\frac{5}{17} + \frac{9}{17} = \frac{14}{17}$

The probability of picking out a yellow cube $= 1 - \frac{14}{17}$
$= \frac{3}{17}$

Outcome Each thing that can happen in an experiment is called an **outcome**.

Sample space A **sample space** is a list of all possible outcomes.

Sample space diagram A table which shows all of the possible outcomes is called a **sample space diagram**.

Frequency The **frequency** of an event is the number of times that it happens.

Relative frequency **Relative frequency** of an event $= \frac{\text{frequency of the event}}{\text{total frequency}}$

The relative frequency gives an *estimate* of the probability.

Expected number The **expected number** = number of trials × probability.

Working with *Excel*

Tossing a fair coin

A coin is tossed 20 times. Use *Excel* to work out the proportion of tosses that gives a head as a result.

- **Open a new *Excel* document**

- **Labelling columns**
  In Cell A1, type **Random Numbers**. In Cell B1, type **Result of Coin Toss**.
  In Cell C1, type **Total Number of Heads**. In Cell D1, type **Total Number of Tails**.
  In Cell E1, type **Total Number of Tosses**. In Cell F1, type **Proportion of Heads**.

- **Generating the first random number**
  In Cell A3, insert the formula: **=RAND()**
  The cell now displays a random number between 0 and 1.

- **Generating the first toss of a coin**
  In Cell B3, insert the formula: **=IF(A3<0.5,"Heads","Tails")**
  The cell now displays the result of one toss of a coin.
  The probability of a Head is ***0.5***
  The formula gives Heads for random numbers 0 up to 0.5 and Tails for random numbers 0.5 up to 1.

- **Generating more tosses of a coin**
  Place the cursor on Cell A3.
  Hold down the left mouse button and move the cursor to Cell B3.
  Release the mouse button.
  Cells A3 and B3 are selected.

  Place the cursor over the bottom right-hand corner of cell B3 and drag down to Cell B22.
  This copies the formulas into 19 more rows, tossing the coin 19 times more.

  Remember that the random numbers change when you change the spreadsheet.
  It is not important.

	A	B	
1	Random Numbers	Result of Coin Toss	Total Num
2			
3	0.224957821	Heads	
4	0.697085009	Tails	
5	0.283201489	Heads	
6	0.533153124	Tails	
7	0.308160105	Heads	
8	0.651719563	Tails	
9	0.749155722	Tails	
10	0.669668327	Tails	
11	0.023964318	Heads	
12	0.564898273	Tails	
13	0.077268826	Heads	
14	0.068470824	Heads	
15	[illegible]	[illegible]	

Finding the total numbers of Heads and Tails

In Cell C2, type **0**. (Use the number key for **0**.) In Cell D2, type **0**. (Use the number key for **0**.)

In Cell C3, insert the formula: **=IF(B3="Heads",C2+1,C2)**

This finds the total number of Heads so far.

In Cell D3, insert the formula: **=IF(B3="Tails",D2+1,D2)**

This finds the total number of Tails so far.

Select Cells C3 and D3.

Place the cursor over the bottom right-hand corner of Cell D3 and drag down to Cell D22.

This copies both formulas.

	C	D	E	F
Toss	Total Number of Heads	Total Number of Tails	Total Number of Tosses	Proportion of Heads
	0	0		
	1	0	1	1
	2	0	2	1
	3	0	3	1
	4	0	4	1
	5	0	5	1
	6	0	6	1
	6	1	7	0.857142857
	7	1	8	0.875
	8	1	9	0.888888889
	9	1	10	0.9
	9	2	11	0.818181818
	10	2	12	0.833333333
	11	2	13	0.846153846
	11	3	14	0.785714286
	11	4	15	0.733333333
	11	5	16	0.6875

Finding the Total Number of Tosses

In Cell E3, insert the formula: **=C3+D3**

Copy the formula into Cells E4 to E22.

Finding the proportion of Heads

In Cell F3, insert the formula: **=C3/E3**

Copy the formula into Cells F4 to F22.

Select the cells showing the Total Number of Tosses and the Proportion of Heads

Select Cells E1 to F22.

Drawing a graph showing the Number of Tosses and the Proportion of Heads

Click the **Chart Wizard** button.

From **Chart type**, choose **XY (Scatter)**.

From **Chart sub-type**, choose the bottom left diagram.

Labelling the graph

Click the **Next>** button until you get to the **Chart Options** window.

In the **Chart title** box, type **Tossing a Coin**.

In the **Value (X) axis** box, type **Number of Tosses**.

In the **Value (Y) axis** box, type **Proportion of Heads**.

Click **Finish**. You can now print this graph. It should be similar to the one shown.

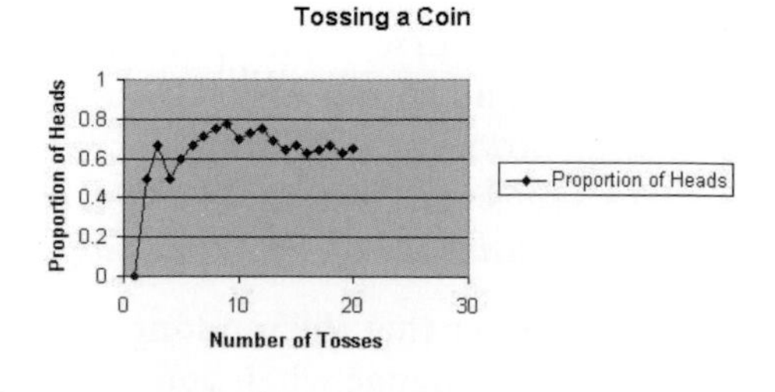

Spinning a fair pentagon

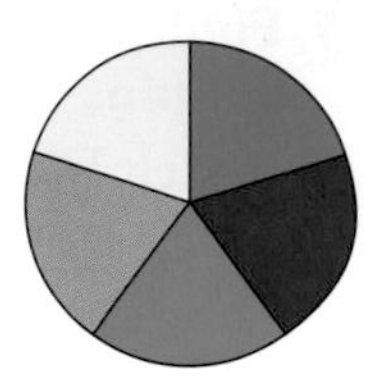

A fair spinner is formed from a pentagon. The face is divided into five triangles. Each triangle is a different colour. The spinner is spun 100 times. Use *Excel* to work out the proportion of spins that gives a yellow as the result.

❑ **Enter the information into a new *Excel* document**

In Cell A1, type **Random Numbers**. In Cell B1, type **Result of Spin**.

In Cell C1, type **Total Number of Yellows**. In Cell D1, type **Total Number of Other Colours**.

In Cell E1, type **Total Number of Spins**. In Cell F1, type **Proportion of Yellows**.

❑ **Generating the first random number**

In Cell A3, insert the formula: =**RAND**()

The cell now displays a random number between 0 and 1.

❑ **Generating the first spin of a pentagon**

In Cell B3, insert the formula:

=**IF(A3<0.2,"Yellow","Other Colour")**

The cell now displays the result of one spin of the pentagon.

The probability of a Yellow is 0.2

The formula gives Yellow for random numbers 0 up to 0.2 and Other Colours for random numbers 0.2 up to 1.

	A	B	
1	Random Numbers	Result of Spin	Total Nun
2			
3	0.644508946	Other Colour	
4	0.5373621	Other Colour	
5	0.933689199	Other Colour	
6	0.972954982	Other Colour	
7	0.602208243	Other Colour	
8	0.498482476	Other Colour	
9	0.807720025	Other Colour	
10	0.829688477	Other Colour	
11	0.581857752	Other Colour	
12	0.636307756	Other Colour	
13	0.389797548	Other Colour	
14	0.413044464	Other Colour	
15	0.721011449	Other Colour	
16	0.26476576	Other Colour	

❑ **Generating more spins of the pentagon**

Select Cells A3 and B3.

Copy the formulas into Rows 4 to 102.

This spins the pentagon 99 times more.

❑ **Finding the total numbers of Yellows and Other Colours**

In Cell C2, type **0** and in Cell D2, type **0**. (Use the number key for **0**.)

In Cell C3, insert the formula: =**IF(B3="Yellow",C2+1,C2)**

This finds the total number of Yellows so far.

In Cell D3, insert the formula: =**IF(B3="Other Colour",D2+1,D2)**

This finds the total number of Other Colours so far.

Copy the formulas in Cells C3 and D3 into Rows 4 to 102.

Finding the Total Number of Spins

In Cell E3, insert the formula:
=C3+D3

Copy the formula into Cells E4 to E102.

	C	D	E	F
of Spin	Total Number of Yellows	Total Number of Other Colours	Total Number of Spins	Proportion of Yellows
	0	0		
olour	0	1	1	0
olour	0	2	2	0
olour	0	3	3	0
olour	0	4	4	0
olour	0	5	5	0
	1	5	6	0.166666667
olour	1	6	7	0.142857143
olour	1	7	8	0.125
olour	1	8	9	0.111111111
	2	8	10	0.2
olour	2	9	11	0.181818182
olour	2	10	12	0.166666667

Finding the Proportion of Yellows

In Cell F3, insert the formula: =**C3/E3**

Copy the formula into Cells F4 to F102.

Draw a graph showing the Number of Spins and the Proportion of Yellows

From **Chart sub-type**, choose the bottom right diagram.

This omits the point markers, which would spoil the graph.

The graph should be similar to the one shown.

Spinning a Pentagon

Proportion of Yellows: 0, 0.05, 0.1, 0.15, 0.2, 0.25, 0.3, 0.35, 0.4

Number of Spins: 0, 50, 100, 150

Proportion of Yellows

Exercise 6:9

1 This tetrahedral dice has a different colour on each side.

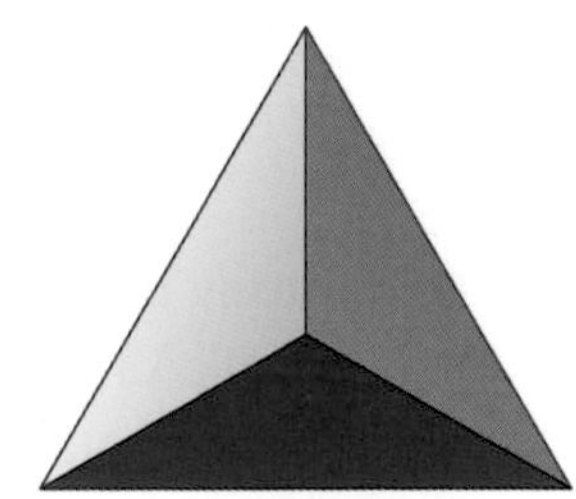

a Sian uses the dice 30 times.
Use *Excel* to work out the proportion of throws that give a purple. Print a graph showing the results.

b Chrishna uses the dice 150 times.
Use *Excel* to work out the proportion of throws that give a purple. Print a graph showing the results.

c Compare the two graphs.
Explain why they are different.

2 Rudi uses this spinner 80 times.

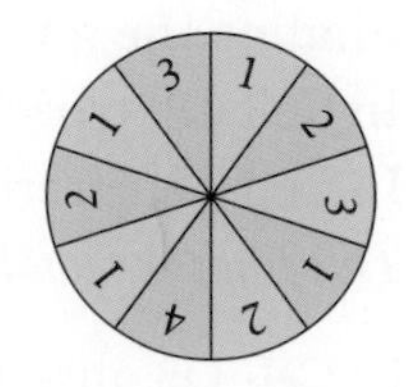

a Use *Excel* to work out the proportion of spins that give a 4.

b Use *Excel* to work out the proportion of spins that give a 2.

c Use *Excel* to work out the proportion of spins that give a 1.

d Use *Excel* to work out the proportion of spins that give a number less than 4.

1 The probability that a battery will last less than 8 hours is $\frac{2}{5}$.
The probability that it will last between 8 and 12 hours inclusive is $\frac{1}{4}$.
Find the probability that it will last more than 12 hours.

2 Edward has two bags of marbles. The first bag contains two white and three blue marbles. The second bag contains four white and two red marbles.

Edward picks a marble from each bag at random.

a Draw a sample space diagram to show all the possible outcomes.

Write down the probability of getting:

b two white balls

c two blue balls

d two balls the same colour

e only one white ball.

3 Rachel has 12 Smarties left. Four are orange, three are red and five are brown.

a She picks a Smartie at random and eats it.
Write down the probability that it is brown.

b The first Smartie was brown. Rachel now picks out a second Smartie and eats it. Write down the probability that it is orange.

c The first two Smarties Rachel picked out were red.
She now picks a third Smartie at random.
Write down the probability that this third Smartie is red.

4 The diagram shows the instruments played by 30 students.

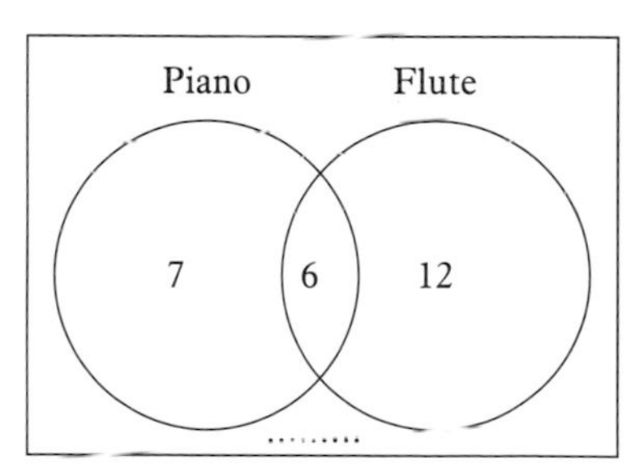

a How many students play:
(1) the piano
(2) the flute
(3) both instruments
(4) neither instrument.

b A student is picked at random.
Find the probability that this student plays the flute.

c A student is picked at random.
This student plays the piano.
Find the probability that this student plays both instruments.

5 Robert throws a coin 60 times. He writes down the number of Heads in every 10 throws. These are his results.

Number of throws	10	20	30	40	50	60
Number of Heads	2	7	13	21	27	29

a Draw a graph of the relative frequency of a Head against the number of throws.

b Is the coin fair? Explain your answer.

7 Histograms

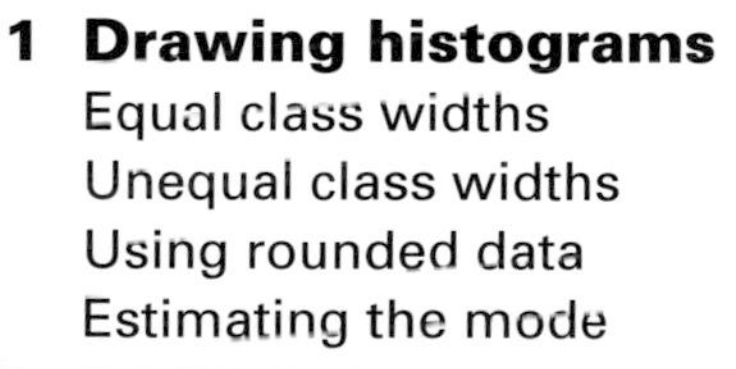

1 Drawing histograms

Equal class widths
Unequal class widths
Using rounded data
Estimating the mode

2 The shape of a distribution

Symmetrical distributions
Skewed distributions
Bimodal distributions

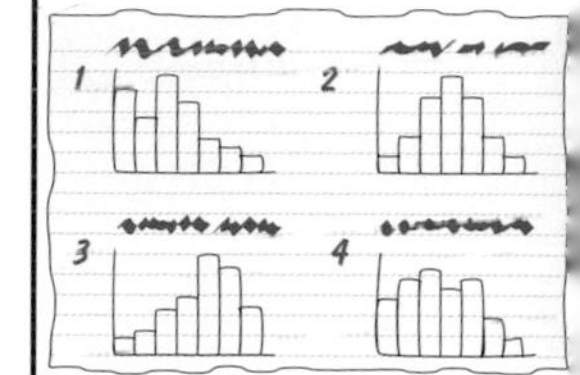

EXAM QUESTIONS

SUMMARY

TEST YOURSELF

1 Drawing histograms

Jonty finds it difficult to get up in the morning. He cycles to school as quickly as possible. Histograms are suitable diagrams for showing journey times.

Histograms

A **histogram** is drawn like a bar-chart, but there are several important differences.

1 It can **only** be used to show continuous data.
2 It can **only** be used to show numerical data.
3 The data is **always grouped**.

You can use equal width groups or unequal width groups. The groups are sometimes called classes.

Histograms with equal class widths

Sixty children are asked to solve a simple puzzle. The table records the times, in minutes, that the children took.

Time, t	$0 \leqslant t < 5$	$5 \leqslant t < 10$	$10 \leqslant t < 15$	$15 \leqslant t < 20$	$20 \leqslant t < 25$
Number of children	11	14	25	10	0

All the groups have the same width of 5 minutes.

To draw the histogram you start by drawing the two axes.

The horizontal axis must be a proper scale and not just show the groups.

The vertical axis is labelled frequency.

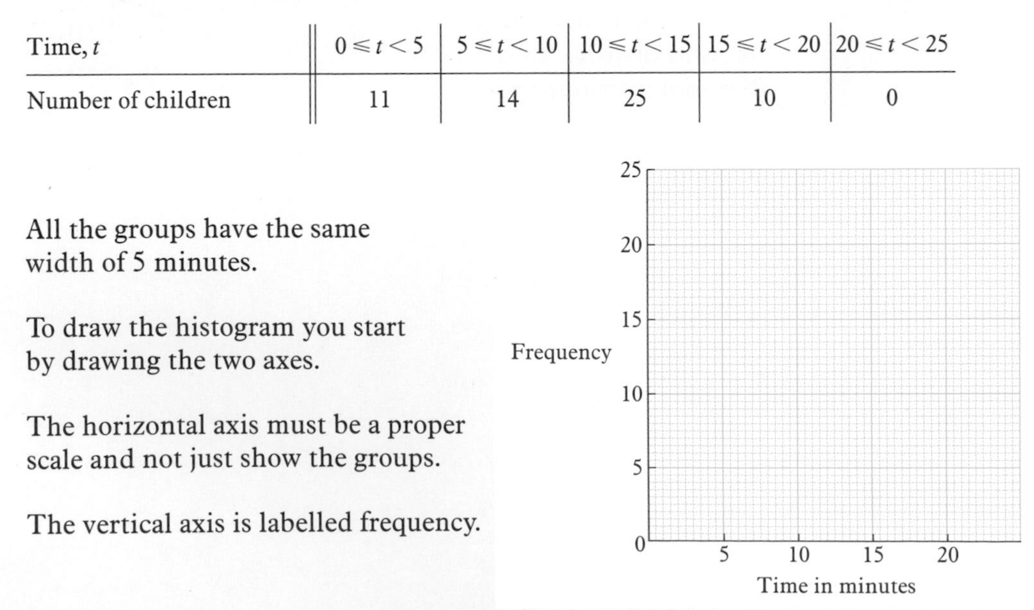

Now draw the bars for each class.
Give your histogram a title.

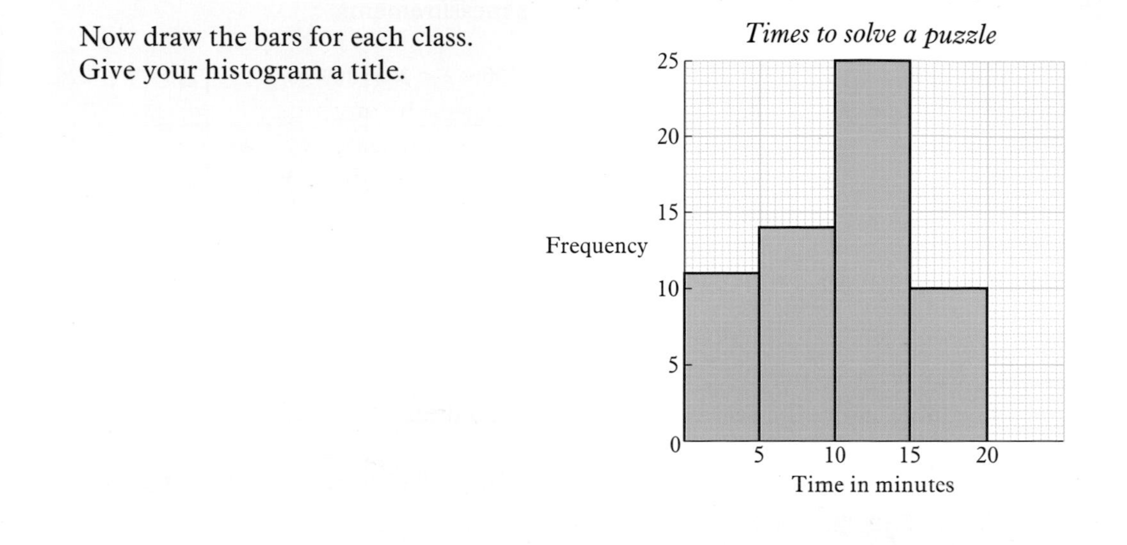

Exercise 7:1

1 A road safety officer records the speeds of cars passing a school.
The table shows the speeds that he recorded.

Speed in miles per hour (s)	$20 \leqslant s < 25$	$25 \leqslant s < 30$	$30 \leqslant s < 35$	$35 \leqslant s < 40$	$40 \leqslant s < 45$
Number of cars (frequency)	4	11	16	13	6

Draw a histogram to show the speeds of the cars.

2 Heather is looking at the weights of potatoes.
Here are the weights that she records.

Weight of potatoes in grams (w)	$50 \leqslant w < 100$	$100 \leqslant w < 150$	$150 \leqslant w < 200$	$200 \leqslant w < 250$	$250 \leqslant w < 300$
Number of potatoes (frequency)	7	8	14	7	4

Draw a histogram to show the weights of the potatoes.

3 As part of her biology investigation, Jean measures the handspan of each member of her class. The table shows her measurements.

Handspan in mm (h)	$170 \leqslant h < 180$	$180 \leqslant h < 190$	$190 \leqslant h < 200$	$200 \leqslant h < 210$	$210 \leqslant h < 220$
Number of pupils (frequency)	3	9	7	6	4

Draw a histogram to show Jean's measurements.

4 George is the captain of a swimming club.
The table shows the ages of all the junior members.

Age in years (a)	$6 \leqslant a < 8$	$8 \leqslant a < 10$	$10 \leqslant a < 12$	$12 \leqslant a < 14$	$14 \leqslant a < 16$
Number of swimmers (frequency)	8	13	20	14	10

Draw a histogram to show the ages of the junior members.

Histograms with unequal groups

When you draw a histogram with different class widths, you have to change the heights of some of the bars.
This is because it is the area of each bar that represents the frequency, not the height.

The dental manager records the waiting times of a group of patients.
The waiting times are shown in the table.

Waiting time in minutes (t)	Number of patients (frequency)
$0 \leqslant t < 2$	8
$2 \leqslant t < 6$	12
$6 \leqslant t < 8$	14
$8 \leqslant t < 14$	4

To draw a histogram of these waiting times you first look at the widths of each group.

t	Frequency	Class width
$0 \leqslant t < 2$	8	2
$2 \leqslant t < 6$	12	4
$6 \leqslant t < 8$	14	2
$8 \leqslant t < 14$	4	6

Add a third column to the table.
Write down the widths of each group.

Next add one more column to the table.
The fourth column shows **frequency** ÷ **class width**
This is called the **frequency density**.

t	Frequency	Class width	Frequency density
$0 \leqslant t < 2$	8	2	$8 \div 2 = 4$
$2 \leqslant t < 6$	12	4	$12 \div 4 = 3$
$6 \leqslant t < 8$	14	2	$14 \div 2 = 7$
$8 \leqslant t < 14$	4	6	$4 \div 6 = \frac{2}{3}$

This time the vertical axis of the histogram is labelled **frequency density**.
(Sometimes it is labelled frequency per unit interval.)
Frequency density = **frequency** ÷ **class width**

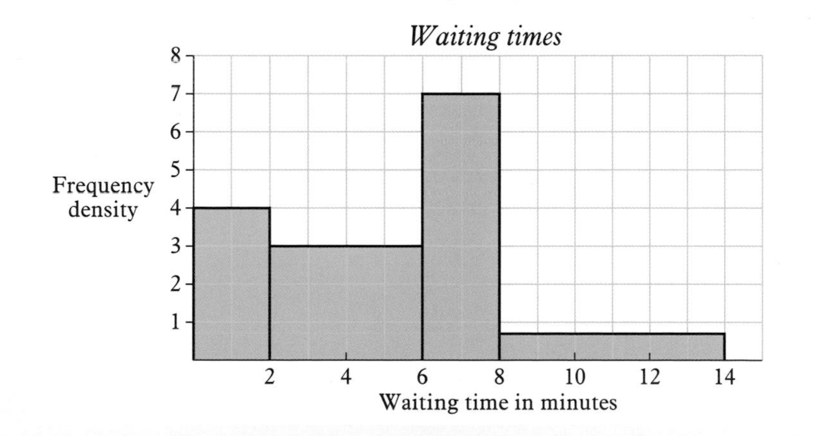

You can use the histogram to find the frequency for a group.
Frequency = **frequency density** × **class width**

For the group $2 \leqslant t < 6$ the frequency = **3** × 4 = 12

Exercise 7:2

1 Mrs Moore asks each member of her class of 33 pupils how long it takes them to travel to school. The results are shown in the table.

Time in minutes (t)	Frequency (f)	Class width	Frequency density
$0 \leqslant t < 10$	4	10	$4 \div 10 = 0.4$
$10 \leqslant t < 15$	8	5	$8 \div 5 = 1.6$
$15 \leqslant t < 20$	8		
$20 \leqslant t < 30$	10		
$30 \leqslant t < 45$	3		

a Copy the table.
Fill it in. The first two rows have been done for you.
b Draw a histogram for this data.

2 Rebecca is a member of a swimming club. She measures the height of the twenty girls in her team. Her results are shown in the table.

Height in cm (h)	Frequency (f)
$120 \leqslant h < 124$	4
$124 \leqslant h < 128$	6
$128 \leqslant h < 130$	2
$130 \leqslant h < 136$	6
$136 \leqslant h < 140$	2

a Work out the frequency density for each group.
b Draw a histogram for this data.

3 Judith works in a pet shop. She keeps a record of the weights of all the hamsters. Her results are shown in the table.

Weight in grams (w)	Frequency (f)
$50 \leqslant w < 55$	2
$55 \leqslant w < 65$	3
$65 \leqslant w < 75$	12
$75 \leqslant w < 90$	15
$90 \leqslant w < 115$	30
$115 \leqslant w < 125$	4
$125 \leqslant w < 130$	1

a Work out the frequency density for each group.
b Draw a histogram for this data.

4 Stef works in a children's nursery. He has to record the ages of all the children who attend. His results are shown in the table.

Age in months (m)	Frequency (f)
$0 \leq m < 6$	6
$6 \leq m < 12$	9
$12 \leq m < 24$	24
$24 \leq m < 36$	27
$36 \leq m < 48$	24
$48 \leq m < 54$	9
$54 \leq m < 60$	3

a Draw a histogram for this data.

b Give a reason why there are so few children in the $48 \leq m < 54$ and the $54 \leq m < 60$ age groups.

5 Jake enjoys chatting to his friends on the Internet. He asks his IT class how long they spend on the Internet each evening. The results are shown in the table.

Time in minutes (t)	Frequency (f)
$0 \leq t < 10$	1
$10 \leq t < 20$	2
$20 \leq t < 25$	5
$25 \leq t < 30$	7
$30 \leq t < 35$	10
$35 \leq t < 40$	5
$40 \leq t < 50$	1
$50 \leq t < 60$	1

a Draw a histogram for this data.

b Give a reason for the use of two different class widths in the table.

Continuous data is often rounded.

Heights can be rounded to the nearest centimetre.
Weights can be rounded to the nearest kilogram.
Times can be rounded to the nearest minute.

This affects the **drawing** and **labelling** of a histogram.
For each bar you need to look at the actual range of values before rounding.

Mr Davies times all his telephone calls. The table shows the lengths of his calls for one month. The times are given to the nearest minute.

Time (to the nearest minute)	5–9	10–14	15–19	20–24	25–29
Number of calls (frequency)	5	20	10	15	6

Look at the first class.
The times in this class are 5–9 minutes.
The lower bound is $4\frac{1}{2}$ minutes. The upper bound is $9\frac{1}{2}$ minutes.
So the class 5–9 is really $4\frac{1}{2}$ to $9\frac{1}{2}$.
So the class width $= 9\frac{1}{2} - 4\frac{1}{2}$
$= 5$ minutes

A new table is drawn to show the class widths more accurately.

Time (to the nearest minute)	5–9	10–14	15–19	20–24	25–29
Time in minutes	$4\frac{1}{2}$–$9\frac{1}{2}$	$9\frac{1}{2}$–$14\frac{1}{2}$	$14\frac{1}{2}$–$19\frac{1}{2}$	$19\frac{1}{2}$–$24\frac{1}{2}$	$24\frac{1}{2}$–$29\frac{1}{2}$
Number of calls (frequency)	5	20	10	15	6

To draw the histogram you use the values of $4\frac{1}{2}$, $9\frac{1}{2}$, $14\frac{1}{2}$, $19\frac{1}{2}$, $24\frac{1}{2}$ and $29\frac{1}{2}$ for the edges of the bars.

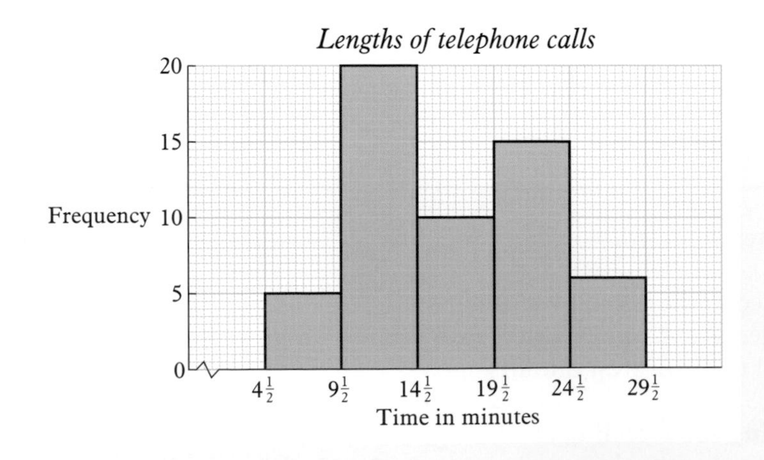

Exercise 7:3

1 Rhiannon shops each week at a local supermarket. She records the length of time that she waits in a queue at the checkout. The table shows these times, to the nearest minute.

Time (to the nearest minute)	2–4	5–7	8–10	11–13
Number of times (frequency)	6	4	2	1

a Draw a new table showing the class widths more accurately.
b Draw a histogram for the lengths of time that she waits.

2 Matthew realises that his school bag is very heavy. He compares the weight of his bag with those of his friends. He records the weights in a table.

Weight (to the nearest kg)	2–3	4–5	6–8	9–10
Number of pupils (frequency)	2	7	9	5

a Draw a new table showing the class widths more accurately.
b Draw a histogram for the weights of the bags.

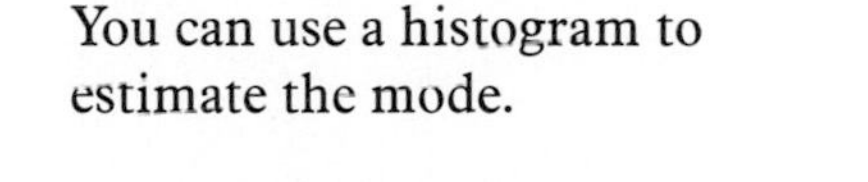

You can use a histogram to estimate the mode.

First draw the histogram. The histogram shows the distribution of lengths of the pencils used by a number of pupils.

Find the modal group. The modal group is 10 cm–15 cm.

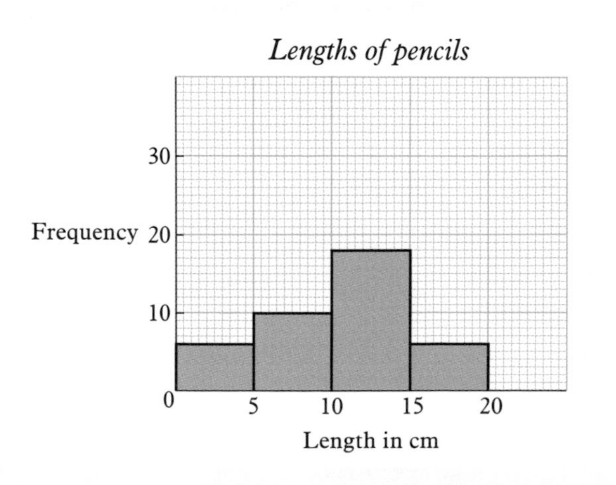

For this modal group, draw the line BD.

B is the top left corner of the modal group.

D is the top left corner of the group on the right.

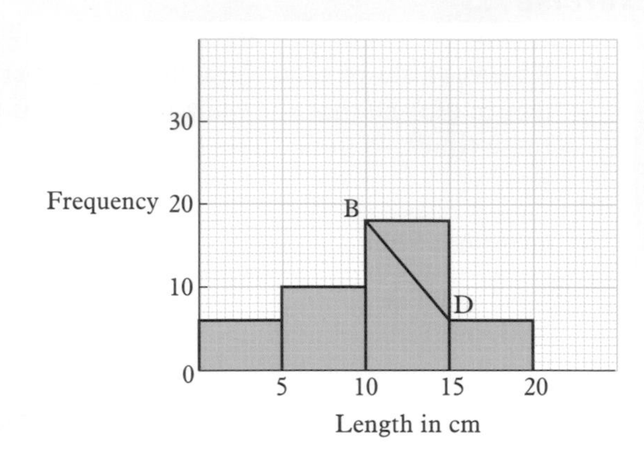

Draw the line AC.

C is the top right corner of the modal group.

A is the top right corner of the group on the left.

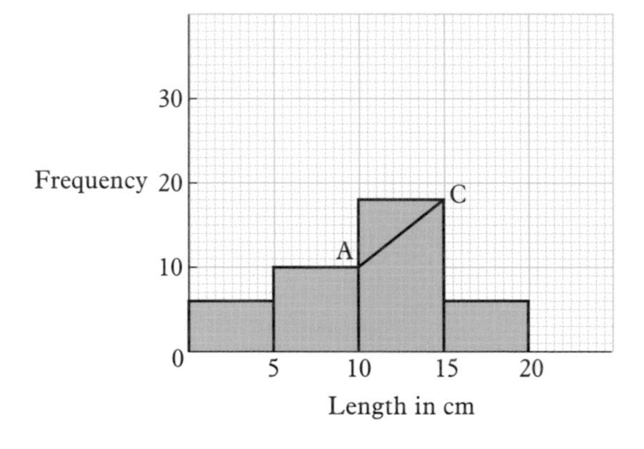

Look where BD and CA cross.

The intersection of the lines gives an estimate of the mode.
Look at the horizontal axis and read off this estimate.

An estimate of the mode = 12 cm.

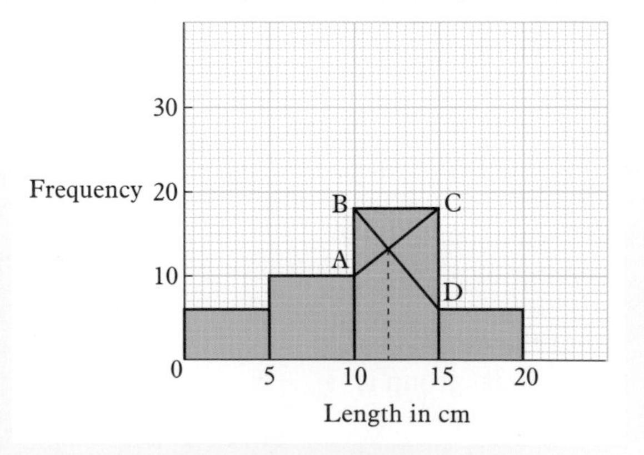

Exercise 7:4

1 Jodi regularly runs to the next village as part of her fitness campaign. She records her times in a table.

Time (to the nearest minute)	10–14	15–19	20–24	25–29
Number of evenings (frequency)	2	8	5	3

a Draw a new table showing the class widths more accurately.
b Draw a histogram of this data.
c Write down the modal group.
d Use the modal group to find an estimate of the mode.

2 Nathan is a keen cyclist. Each Sunday he tours with a local club. He keeps a record of the distances he cycles. The table shows these distances.

Distance (to the nearest km)	30–39	40–49	50–59	60–69
Number of tours (frequency)	4	8	10	4

a Draw a histogram of this data.
b Write down the modal group.
c Use the modal group to find an estimate of the mode.

3 One hundred babies are born in a hospital in a two-week period. Their weights at birth are recorded in the table.

Weight (to the nearest pound)	2–3	4–5	6–7	8–9	10–11
Number of babies (frequency)	4	29	42	20	5

a Draw a histogram of this data.
b Write down the modal group.
c Use the modal group to find an estimate of the mode.

2 The shape of a distribution

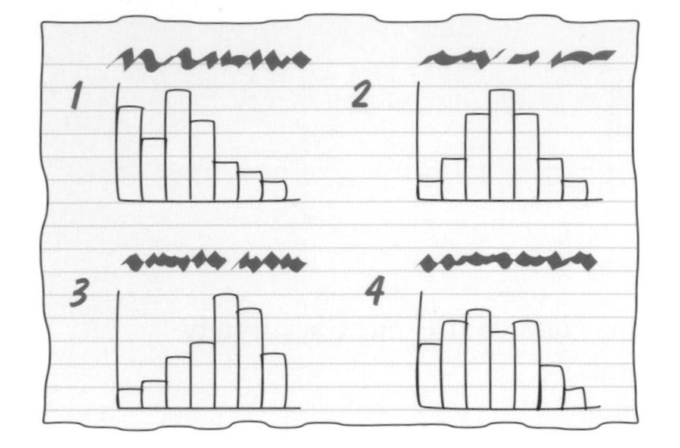

Ahmed has drawn these histograms. He notices that they have different shapes. Only the second one has an axis of symmetry.

You can use a histogram to look at the shape of a distribution.

Symmetrical distribution

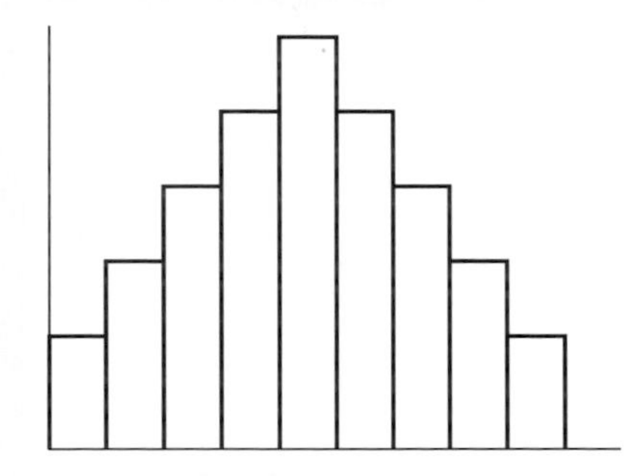

This distribution has an axis of symmetry down the middle.
It is called a **symmetrical distribution**.

For a symmetrical distribution, **mean** = **mode** = **median**.

Positive skew

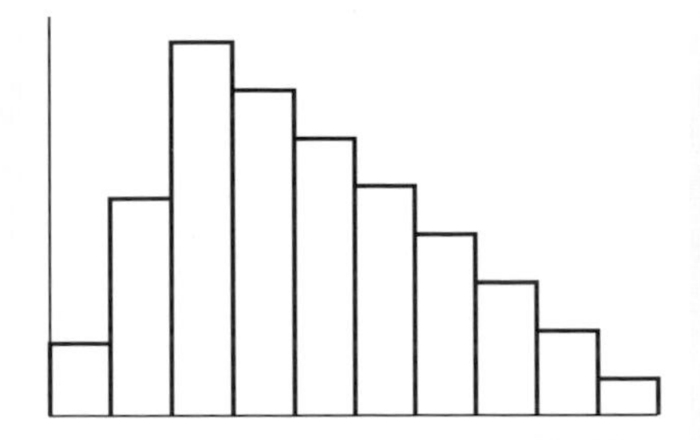

This distribution has no axis of symmetry.
It shows a lean to the left-hand side.
The distribution has a **positive skew**.

For a distribution with a positive skew, **mode** < **median** < **mean**.

Negative skew

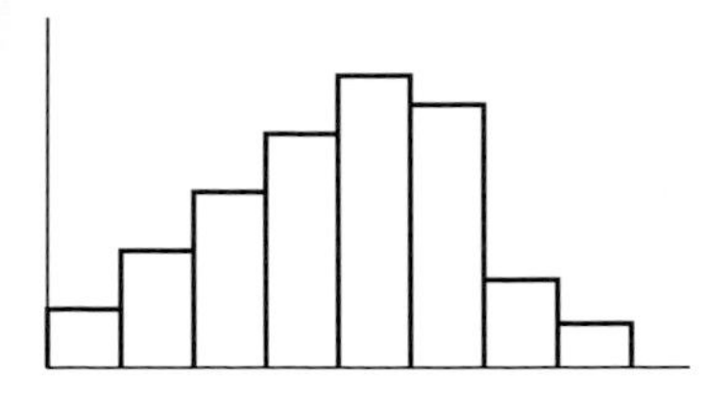

This distribution has a lean to the right-hand side.
This is called a **negative skew**.

For a negative skew, **mean < median < mode**.

The shape of a distribution has an effect on which measure of average and spread you should use.

For symmetrical distributions: The mean is a suitable average.
The standard deviation is a suitable measure of spread as all values are used.

For skewed distributions: The median is a suitable average as the mean would be affected by the skewed values.
The interquartile range is a suitable measure of spread. It only takes into account the central half of the distribution and so avoids much of the skew.

Bimodal distribution

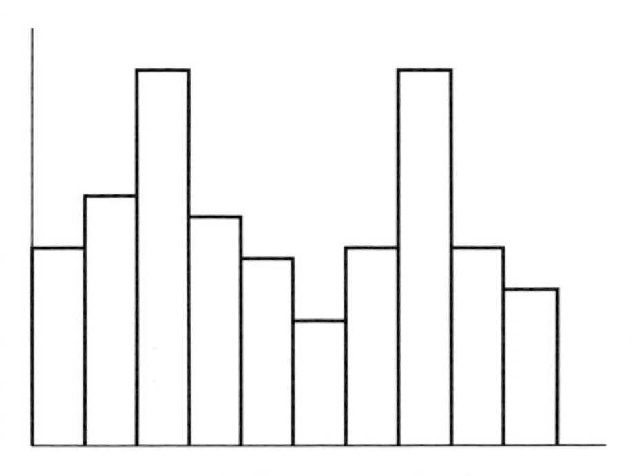

This distribution has two modes.
It is called a **bimodal distribution**.

Exercise 7:5

Look at each of the histograms.

a Write down if the histogram is symmetrical, positively skewed, negatively skewed or bimodal.

b State if the mean or the median is a suitable average.

c State if the standard deviation or the interquartile range is a suitable measure of spread.

1 **4**

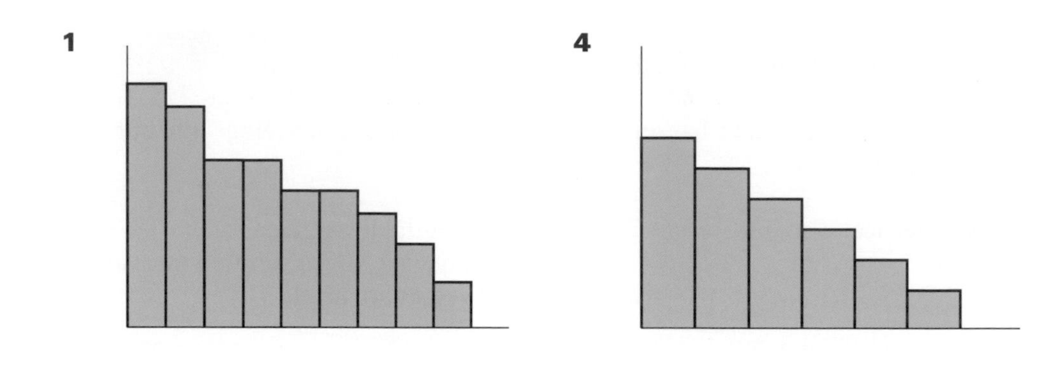

2

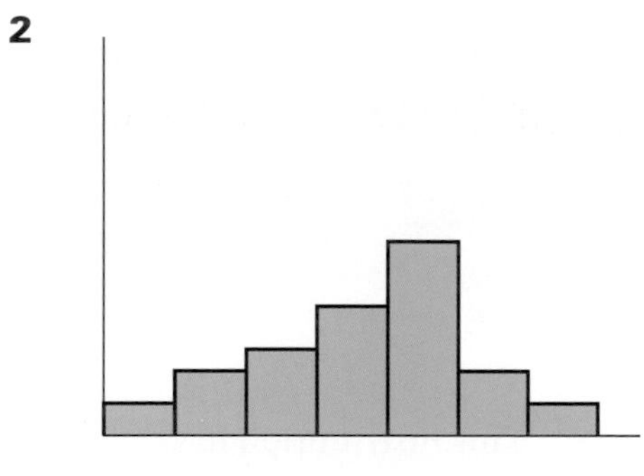

5

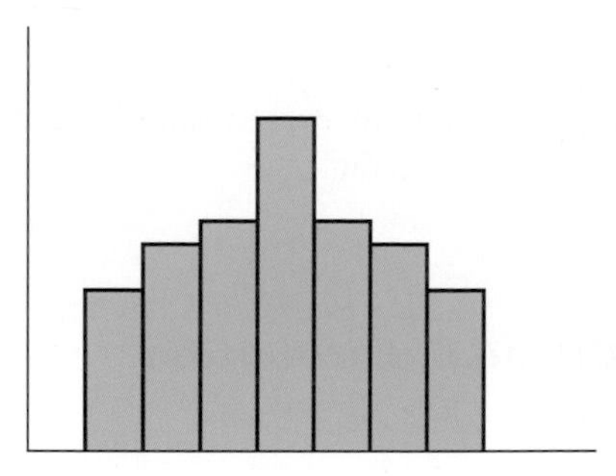

3

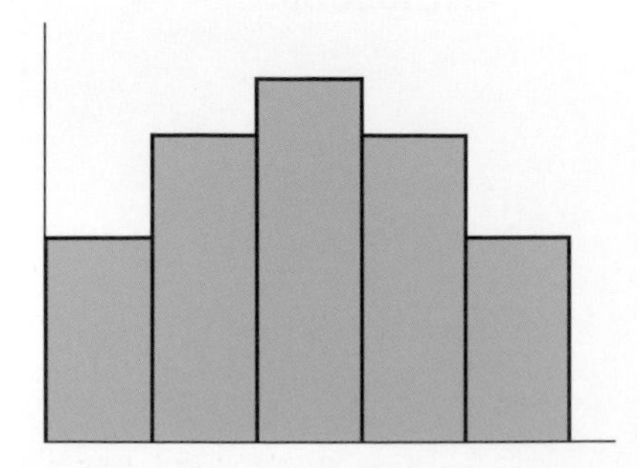

6

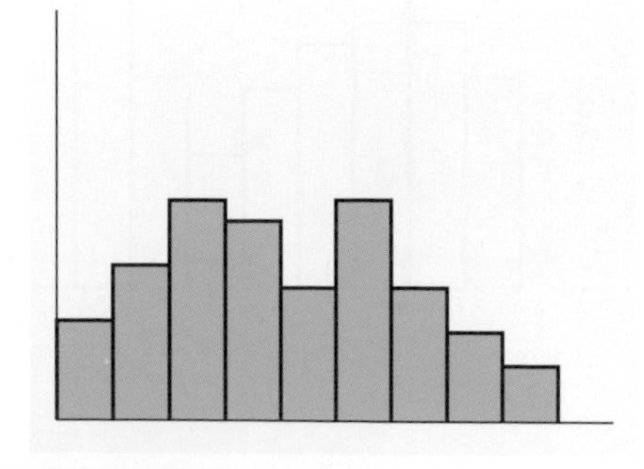

1 One hundred children were each asked to complete the jigsaw puzzle "Swan Lake".
The time, taken to the nearest minute, by each child to complete the puzzle was recorded.

Time (minutes)	0–4	5–6	7–8	9–10	11–14	15–18	19–30
Frequency	9	7	16	30	20	12	6

a On graph paper, draw a histogram to represent these data. *(6)*

b From your histogram, estimate the modal time taken to complete the puzzle. *(2)*

SEG, 1996, Paper 3

2 Louise measured the playing time, t, in seconds, of each track on several of her compact discs.
The table shows her results.

Time, t (sec)	Frequency
$180 \leqslant t < 220$	12
$220 \leqslant t < 230$	13
$230 \leqslant t < 235$	12
$235 \leqslant t < 240$	15
$240 \leqslant t < 245$	17
$245 \leqslant t < 250$	14
$250 \leqslant t < 280$	21

a On graph paper, draw a histogram to represent these data. *(6)*

b Use your histogram to find an estimate of the modal playing time of the tracks. *(2)*

SEG, 1998, Paper 4

3 The waiting time at a doctors' surgery is given in the following table.

Time (minutes)	$0 \leq t < 10$	$10 \leq t < 15$	$15 \leq t < 20$	$20 \leq t < 30$	$30 \leq t < 40$	$40 \leq t < 60$
Frequency	13	22	18	15	9	8

No patient waited for more than 60 minutes.

a On graph paper, draw a histogram to represent these data. *(6)*

b Use the histogram to obtain an estimate of the mode. *(2)*

c How many patients waited for a period of time greater than the mode? *(2)*

SEG, 1997, Paper 4

4 The histogram shows the journey time taken by 91 factory workers to get to work.

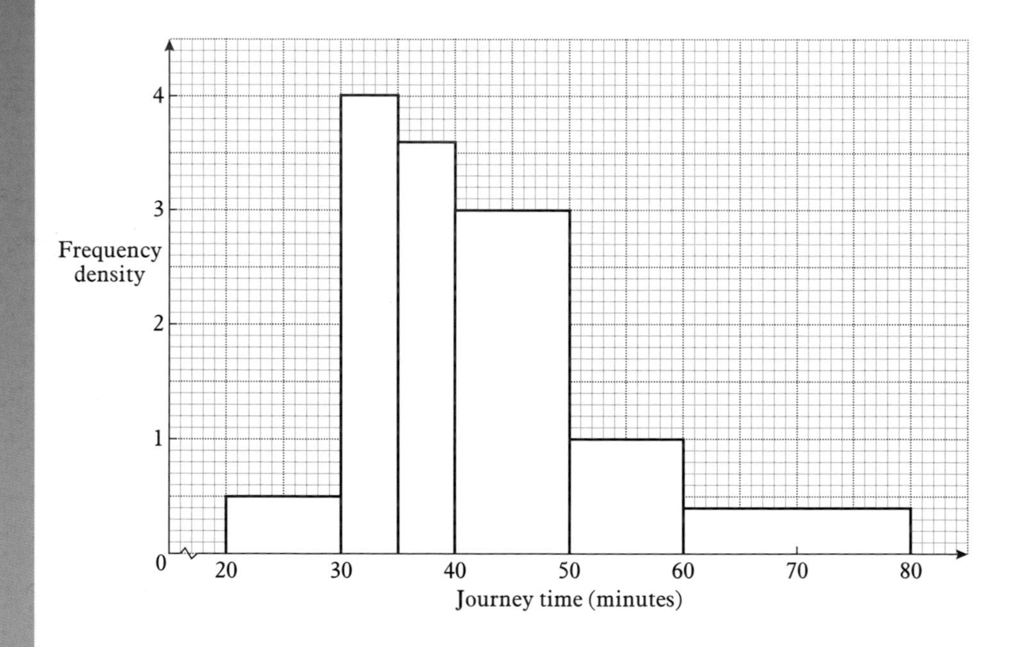

a Describe the skewness of the distribution. *(1)*

b Use the histogram to estimate the modal journey time. *(2)*

c Calculate the number of workers whose journey time was in the interval 30–35 minutes. *(2)*

d Calculate an estimate of the median journey time. *(5)*

SEG, 1999, Paper 3

5 The histogram below shows the alcohol consumption for a random sample of 100 adult **males** in the UK. The results show consumption in units of alcohol and relate to the week preceding the interview which formed the basis of the investigation.

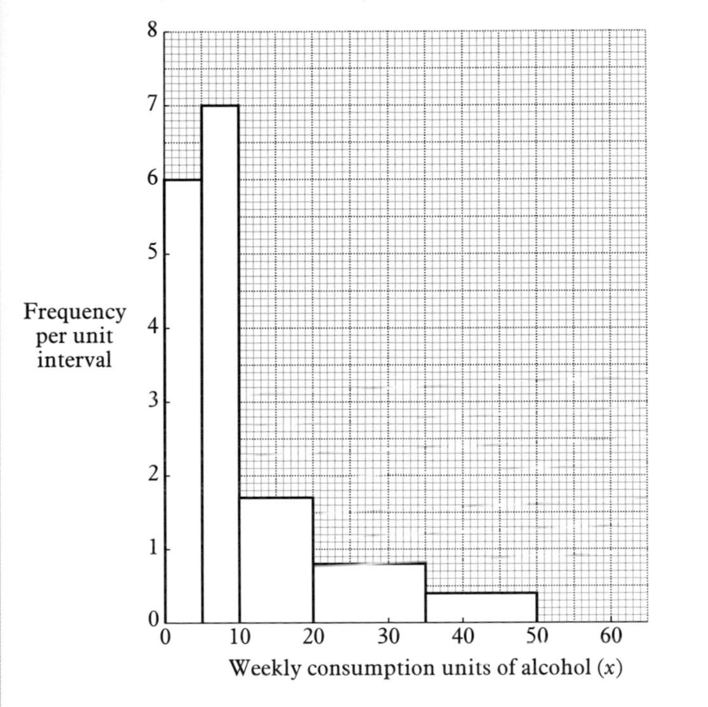

a Copy and complete the following frequency table for these data.

Consumption units of alcohol (x)	Frequency
$0 < x \leqslant 5$	
$5 < x \leqslant 10$	
$10 < x \leqslant 20$	
$20 < x \leqslant 35$	
$35 < x \leqslant 50$	

(3)

b Equivalent data for a sample of 100 **females** was:

Consumption units of alcohol (x)	Frequency
$0 < x \leqslant 5$	77
$5 < x \leqslant 10$	10
$10 < x \leqslant 20$	9
$20 < x \leqslant 35$	3
$35 < x \leqslant 50$	1

i Using the class mid-points 2.5, 7.5, 15, 27.5, 42.5 as (x), calculate, to one decimal place, estimates of the mean and standard deviation for **female** alcohol consumption per week.
(You may use the following formulas:

$$\text{Mean} = \frac{\Sigma fx}{\Sigma f} \qquad \text{Standard deviation} = \sqrt{\frac{\Sigma f(x - \bar{x})^2}{\Sigma f}}$$

or any suitable alternative, or the statistical functions on your calculator.) *(6)*

ii Hence copy and complete the following table:

	Median	Mean	Standard Deviation
Male	7.9	11.8	11.0
Female	3.3		

Comment on the key differences apparent in relation to alcohol consumption rates. *(2)*

c i Use the following formula to calculate a measure of skewness for both male and female weekly consumption of alcohol. *(3)*

$$\text{measure of skewness} = \frac{3\,(\text{mean} - \text{median})}{\text{standard deviation}}$$

ii By reference to the results obtained in part **c i** and the histogram, describe the shape of the two distributions. *(2)*

d Suggest a reason why the interviewer asked for alcohol consumption over the preceding week and not over a longer period. *(1)*

e Indicate how the survey should be continued if the results are to be representative of changes in alcohol consumption throughout the year. *(2)*

NEAB, 1997, Paper 3

6 The following table shows the annual salaries of the 100 employees of a small manufacturing company.

Salary (in £000s) x	Frequency	
$5 \leqslant x < 10$	10	
$10 \leqslant x < 12$	10	
$12 \leqslant x < 14$	22	
$14 \leqslant x < 15$	21	
$15 \leqslant x < 17$	18	
$17 \leqslant x < 20$	12	
$20 \leqslant x < 27$	7	

a Draw, on graph paper, a histogram to represent these data. *(6)*

b Use your histogram to identify the modal class for this distribution. *(1)*

c Calculate the probability that, of two randomly selected employees, both earn annual salaries in the range of £10 000–£13 999. *(3)*

NEAB, 1998, Paper 3

7 In a survey the durations, in minutes, of telephone calls were measured for a random sample of 100 local calls made at peak rate.
The results were summarised as follows:

Duration, x min	No. of calls
$0 \leqslant x < 1$	14
$1 \leqslant x < 3$	22
$3 \leqslant x < 10$	42
$10 \leqslant x < 15$	10
$15 \leqslant x < 30$	9
$30 \leqslant x < 60$	3

John was asked to draw a histogram to illustrate this distribution.
He produced the following diagram.

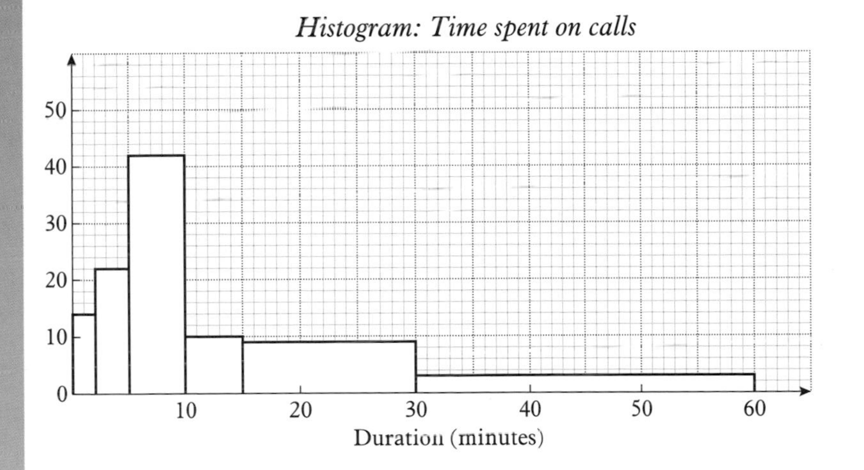

a This diagram is wrong. Give two reasons why it is wrong. *(3)*

b Draw, on graph paper, a revised histogram to represent these data and comment on the key differences displayed. *(6)*

c Use your histogram to identify the modal class for this distribution. *(1)*

NEAB, 1996, Paper 3

8 The histogram below shows the distribution of income, x, in the £1000s of a random sample of 100 factory workers.

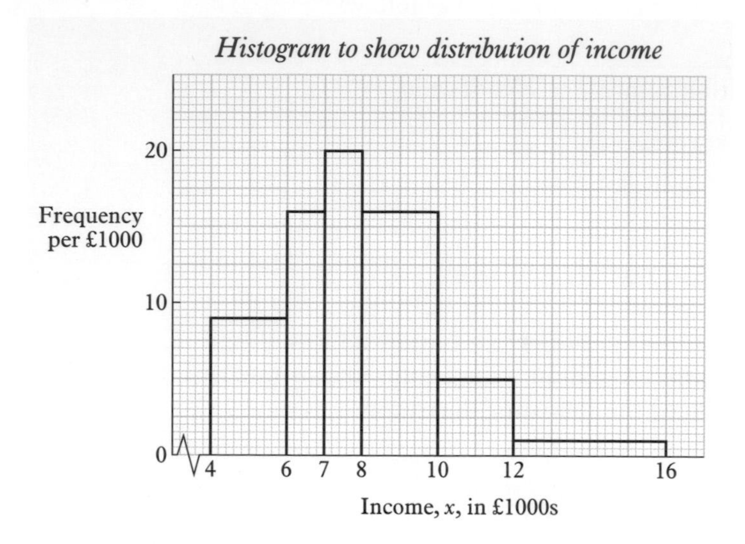

a Complete the following frequency table for this data. *(3)*

Income, x, in £1000s	Frequency	
$4 \leq x < 6$	18	
$6 \leq x < 7$		
$7 \leq x < 8$		
$8 \leq x < 10$		
$10 \leq x < 12$		
$12 \leq x < 16$		
	100	

b Describe the skewness of the distribution. *(1)*

c In which class does the median lie? *(1)*

Histogram

A **histogram** is drawn like a bar-chart, but there are several important differences.

1. It can **only** be used to show continuous data.
2. It can **only** be used to show numerical data.
3. The data is **always grouped**.

When you draw a histogram with different class widths, you have to change the heights of some of the bars.
This is because it is the area of each bar that represents the frequency, not the height.

Continuous data is often rounded.

Heights can be rounded to the nearest centimetre.
Weights can be rounded to the nearest kilogram.
Times can be rounded to the nearest minute.

This affects the **drawing** and **labelling** of a histogram.
You need to adjust the groups given for the histogram. For example, a group of 5–9 minutes becomes $4\frac{1}{2}$–$9\frac{1}{2}$ minutes.

You can use a histogram to look at the shape of a distribution.

For a symmetrical distribution, **mean** = **mode** = **median**.
For a distribution with a positive skew, **mode** < **median** < **mean**.
For a negative skew, **mean** < **median** < **mode**.

The shape of a distribution has an effect on which measure of average and spread you should use.

For symmetrical distributions:
The mean is a suitable average. The standard deviation is a suitable measure of spread as all values are used.

For skewed distributions:
The median is a suitable average as the mean would be affected by the skewed values.
The interquartile range is a suitable measure of spread because this only takes into account the central half of the distribution.

1 Mary measures the heights, to the nearest centimetre, of the 30 children in her class. She records the heights in a table.

Height (to the nearest cm)	140–144	145–149	150–154	155–159	160–164
Number of pupils (frequency)	13	10	7	6	4

a Draw a histogram to show the heights that Mary recorded.
b Write down the modal group.
c Use the modal group to find an estimate of the mode.
d Write down whether the histogram is symmetrical, positively skewed or negatively skewed.

2 Look at both of these histograms.
a Write down if the histogram is symmetrical, positively skewed or negatively skewed.
b State if the mean or the median is a suitable average.
c State if the standard deviation or the interquartile range is a suitable measure of spread.

(1)

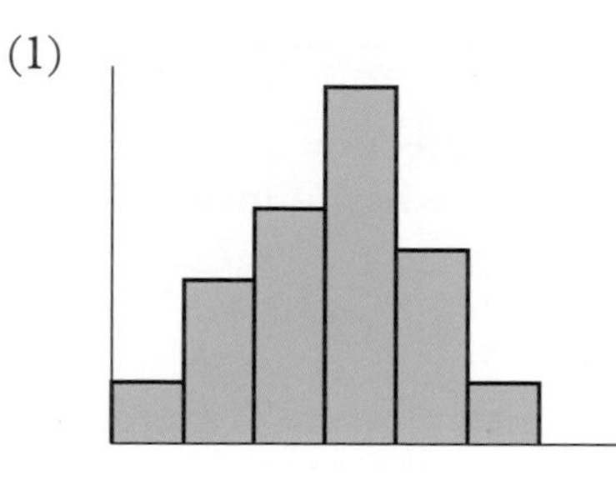

(2)

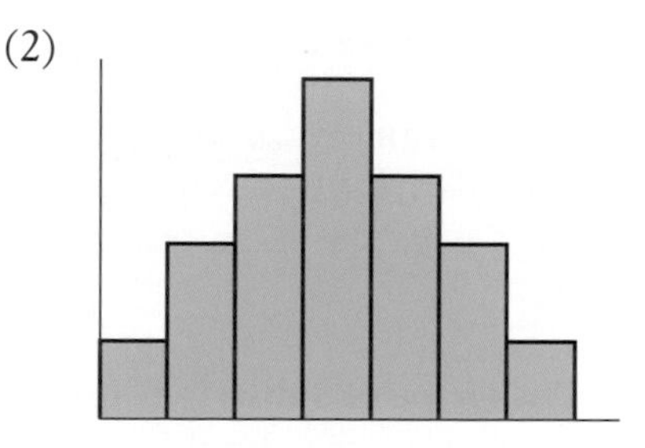

3 Janoop asks each member of his Maths class how much pocket money they receive each week. The table shows these amounts.

Amount of pocket money in £'s (m)	$0 \leq m < 3$	$3 \leq m < 4$	$4 \leq m < 5$	$5 \leq m < 6$	$6 \leq m < 10$
Number of pupils (frequency)	3	7	9	7	4

Draw a histogram to show the amounts of money his Maths class receive each week. Use a standard width of 1 unit and label the vertical axis 'frequency per unit interval'.

8 Shapes of distributions

1 Box and whisker diagrams
Drawing box and whisker diagrams
The shape of a distribution
Comparing sets of data
Identifying outliers

2 The normal curve
The shape of a normal curve
Calculating standardised scores
Using standardised scores to compare distributions

EXAM QUESTIONS

SUMMARY

TEST YOURSELF

1 Box and whisker diagrams

A cat uses its whiskers to tell how wide an opening is.

This cat is obviously confused!

Whiskers are also used in mathematics to give information about a set of data.

Box and whisker diagrams

A **box and whisker diagram** shows the distribution of data.
It can also be called a **box plot**.
It shows the median and the central half of the data.
It also shows the range of the data.

Example

John keeps a record of his journey times to school each morning. These are his times, given to the nearest minute.

29 21 16 25 21 19 18 30 21 21 12 26 19
21 20 19 30 29 16 21 18 18 27 18 20

To draw a box and whisker diagram of this data you need to find:

(1) the smallest and largest values
(2) the median and the lower and upper quartiles.

First put the data in order of size. Start with the smallest.

12 16 16 18 18 18 18 19 19 19 20 20 **21**
21 21 21 21 21 25 26 27 29 29 **30** **30**

Smallest value = **12** Largest value = **30**

The position of the median is $\frac{1}{2}(25 + 1)$th value = 13th value

So the median = **21**

Look at the data to the left of the median.

12 16 16 18 18 18 **18** 19 19 19 20 20

▲

The lower quartile is the median of this data.
The position of the lower quartile is $\frac{1}{2}(12 + 1)$th $= 6\frac{1}{2}$th value

6th value = 18 7th value = **18**

$$\text{Lower quartile} = \frac{18 + 18}{2} = 18$$

Look at the data to the right of the median.

21 21 21 21 21 25 **26** 27 29 29 30 30

▲

The upper quartile is the median of this data.
The position of the upper quartile is $\frac{1}{2}(12 + 1)$th $= 6\frac{1}{2}$th value

6th value = 25 7th value = **26**

$$\text{Upper quartile} = \frac{25 + 26}{2} = 25\frac{1}{2}$$

Now you can draw the box and whisker diagram.

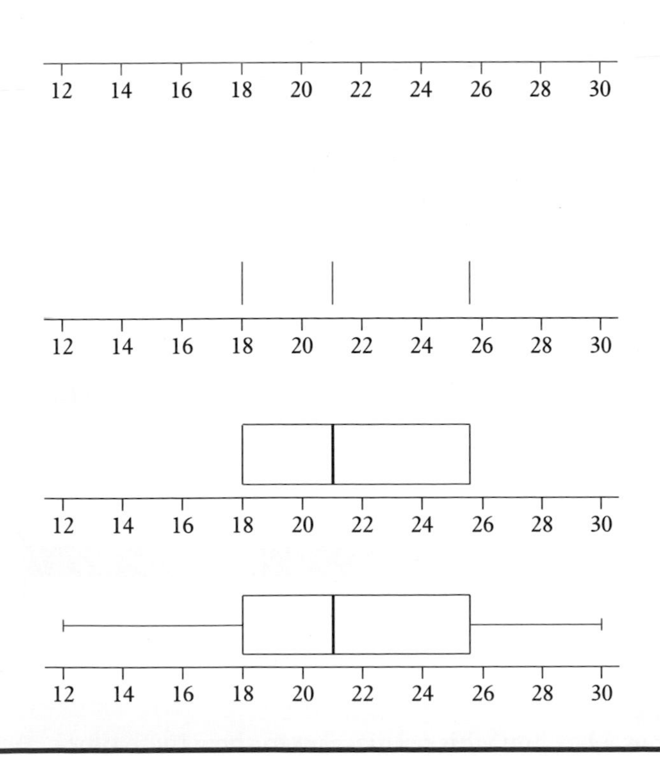

First draw a line with a scale that includes both the smallest and largest values of the data.
You can go beyond these values if you wish.

Draw vertical lines at the positions of the lower quartile, the median and the upper quartile.

Join the tops and bottoms of these lines to form the box.

Draw horizontal lines from the middle of the sides of the box to the smallest and largest values of the data.
These are the whiskers.

Exercise 8:1

1 The table gives the heights, in centimetres, of a group of pupils.

Smallest	Lower quartile	Median	Upper quartile	Largest
110	125	155	170	190

Draw a box and whisker diagram to show this data.

2 Joanna records the length of time spent by each member of her class on their last homework. These are the times recorded to the nearest minute.

85 124 55 140 120 61 95 105 118
180 55 78 130 112 70 126 60 90
115 60 142 100 105 65 100 75

a Find the median.
b Find the lower quartile.
c Find the upper quartile.
d Use these values to draw a box and whisker diagram to show Joanna's data.

3 Dale records the times taken for 11 swimmers to swim one length of a pool. These are the times, in seconds, to the nearest tenth of a second.

21.3 26.0 19.0 21.8 23.6 23.8 26.5 20.6 21.4 24.2 20.8

a Find the median.
b Find the lower quartile.
c Find the upper quartile.
d Use these values to draw a box and whisker diagram to show these times.

4 Mr Wilson travels to work by train each day. The train is frequently late. He records how late the train is for each journey. These are the times, to the nearest minute, for his last 15 journeys.

8 10 15 11 5 10 4 9
1 5 10 20 4 12 5

a Find the median.
b Find the lower quartile.
c Find the upper quartile.
d Use these values to draw a box and whisker diagram to show these times.

You can use a box and whisker diagram to look at the shape of a distribution.

Symmetrical distributions

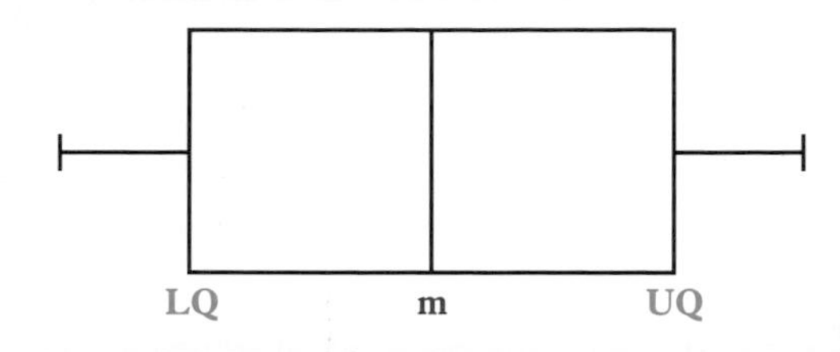

This diagram has a **symmetrical distribution**.
The median is in the centre.
The whiskers are equal lengths.

The distance between the median, **m**, and the upper quartile, **UQ**, is the same as the distance between the median and the lower quartile, **LQ**.

Upper Quartile − Median = Median − Lower Quartile
UQ − **m** = **m** − **LQ**

Positive skew

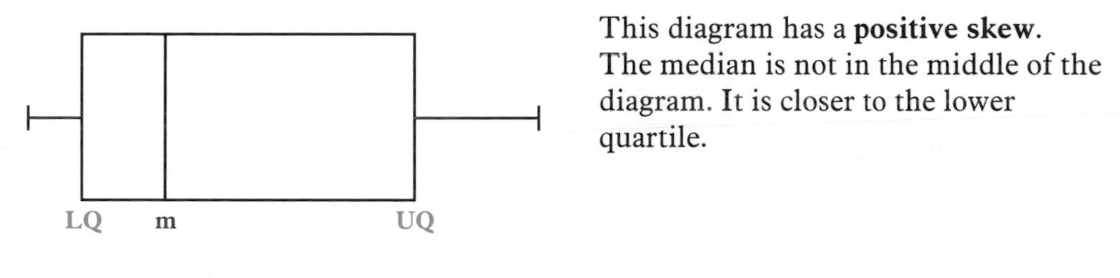

This diagram has a **positive skew**.
The median is not in the middle of the diagram. It is closer to the lower quartile.

Median − Lower Quartile < Upper Quartile − Median
m − **LQ** < **UQ** − **m**

Negative skew

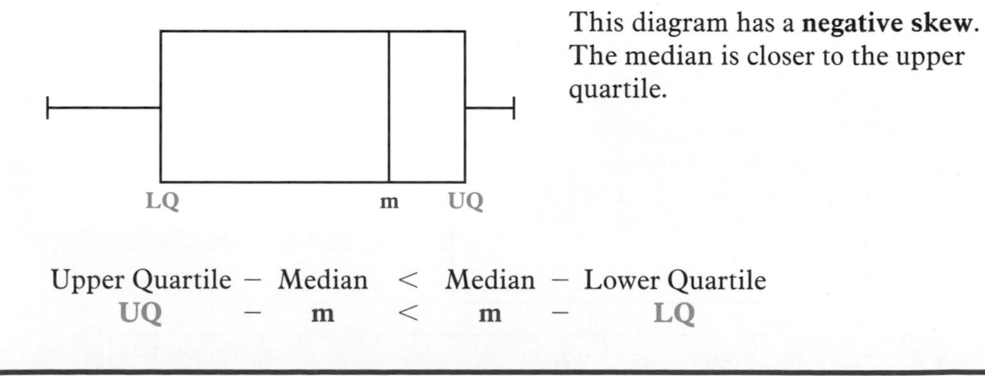

This diagram has a **negative skew**.
The median is closer to the upper quartile.

Upper Quartile − Median < Median − Lower Quartile
UQ − **m** < **m** − **LQ**

Exercise 8:2

For each box and whisker diagram:

a Write down the values of:
 (1) the lower quartile, LQ
 (2) the median, m
 (3) the upper quartile, UQ.

b Calculate m − LQ and UQ − m.

c Write down if the distribution is symmetrical, positively skewed or negatively skewed.

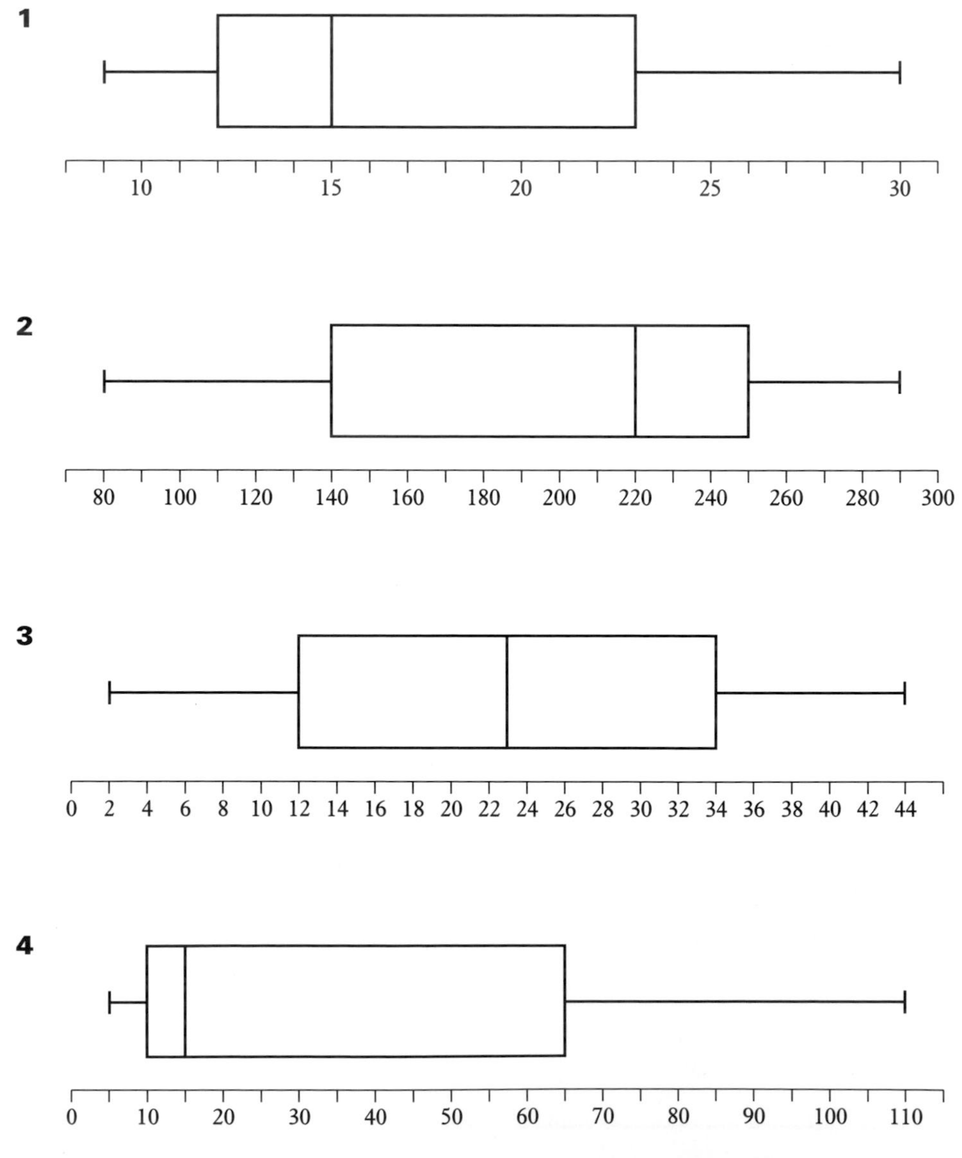

Box and whisker diagrams show the **differences** between sets of data.

The box and whisker diagrams show the results of tests in Maths and English.
Both diagrams are drawn using the same scale.

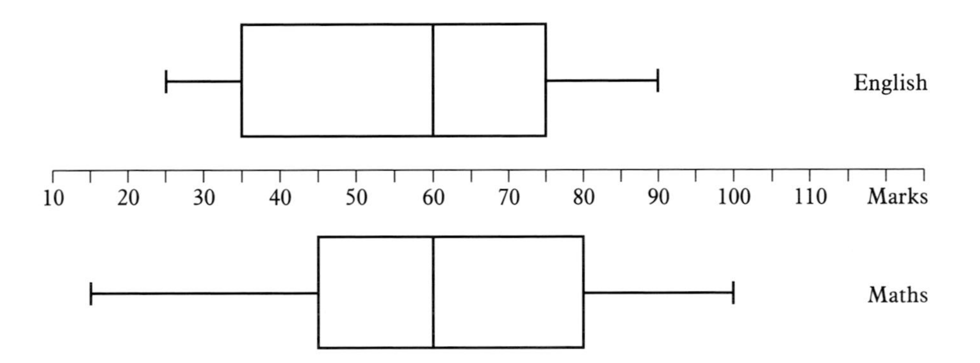

In the English tests, LQ = 35, m = 60 and UQ = 75.

$UQ - m = 75 - 60$
$= 15$

$m - LQ = 60 - 35$
$= 25$

$UQ - m < m - LQ$

The English scores are negatively skewed.

In the Maths test, LQ = 45, m = 60 and UQ = 80.

$UQ - m = 80 - 60$
$= 20$

$m - LQ = 60 - 45$
$= 15$

$UQ - m > m - LQ$

The Maths scores are positively skewed.

The medians are the same for both the English and the Maths marks.

The Maths marks have a narrower box than the English marks.
The interquartile range is smaller for Maths than English.

The whiskers extend further in Maths than English.
The range is greater for the Maths marks.

The lowest mark is in Maths.
The highest mark is in Maths.

Exercise 8:3

1 Anna and Brian record the time, to the nearest minute, that they watch TV each day for a month. The box and whisker diagrams show their data.

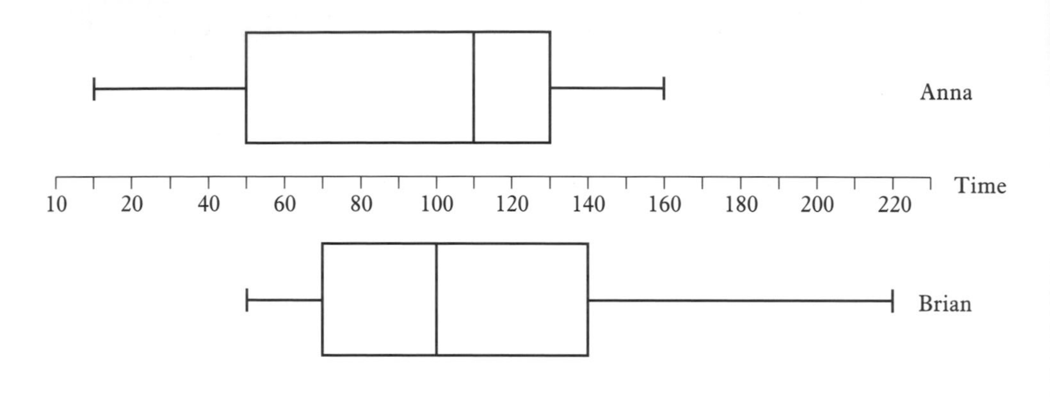

Use the diagrams to compare:

a the skewness

b the range

c the interquartile range

d the medians.

2 Mark notices that the tracks on his classical CDs seem to last longer than the tracks on his pop CDs. To investigate this, he records the length of 21 classical tracks and 21 pop tracks.
These are the track lengths, to the nearest tenth of a minute.

Lengths of time of classical tracks

1.7	8.1	1.8	3.5	3.9	10.6	11.3	4.8	1.8	2.9	1.8
5.1	11.1	12.0	9.2	11.1	6.3	2.7	7.0	11.7	11.6	

Lengths of time of pop tracks

7.0	0.4	6.9	4.2	6.1	2.5	11.3	7.2	1.3	5.0	1.0
2.5	3.0	9.2	8.0	3.0	5.8	1.2	5.5	6.3	7.8	

a For the classical CDs:
(1) find the lower quartile, the median and the upper quartile.
(2) draw a box and whisker diagram using these values.

b For the pop CDs:
(1) find the lower quartile, the median and the upper quartile.
(2) draw a box and whisker diagram using these values.

c Use the box and whisker diagrams to compare:
(1) the skewness
(2) the range
(3) the interquartile range
(4) the medians.

3 The table gives the unemployment rates for 1994 to 2000 for the North East, North West, South East and South West.

	Percentages			
	North East	North West	South East	South West
Quarterly				
1994 Q1	12.8	10.7	7.8	7.4
Q2	12.8	10.3	7.7	7.4
Q3	11.8	10.2	8.2	7.0
Q4	13.0	9.4	7.4	6.9
1995 Q1	11.8	9.1	7.8	6.9
Q2	11.6	9.1	7.8	6.5
Q3	11.2	9.2	7.4	6.5
Q4	11.2	8.5	7.1	6.2
1996 Q1	10.9	8.5	6.9	6.4
Q2	10.4	8.5	6.3	6.1
Q3	10.8	7.9	6.4	5.8
Q4	9.8	7.7	6.5	5.5
1997 Q1	9.9	7.0	5.7	5.2
Q2	9.9	7.2	5.8	5.2
Q3	8.8	7.3	5.2	4.7
Q4	8.5	6.9	5.1	4.5
1998 Q1	8.5	6.8	4.6	4.3
Q2	8.4	6.9	4.8	4.3
Q3	8.3	6.8	4.9	4.5
Q4	9.7	7.1	4.5	4.0
1999 Q1	9.7	6.7	4.9	3.9
Q2	9.6	6.3	4.5	3.9
Q3	9.7	6.3	4.4	3.8
Q4	8.4	6.0	4.2	4.1
2000 Q1	9.0	6.1	4.3	3.5
Q2	8.9	5.4	4.2	3.3
Q3	9.0	5.4	4.0	3.1

Source: Office for National Statistics

Using data for **a** the North East **b** the North West

(1) Find the lower quartile, the median and the upper quartile for the unemployment rates from 1994 Quarter 1 until 2000 Quarter 3.

(2) Draw a box and whisker diagram for each region. Use the same scale for each diagram.

c Draw similar box and whisker diagrams for the South East and the South West.

d Use your four box and whisker diagrams to make comparisons for the unemployment figures in the four regions of the country.

Outliers

Outliers are very small or very large values in a set of data. A value is called an outlier if its distance from the nearest quartile is greater than 1.5 times the interquartile range.

Outliers are often ignored because they can distort the data. An outlier could result from an error in data collection.

Pashreen asks each member of her class how many rooms they have in their house. These are her results.

11	6	16	5	8	5	7	8	9	13
9	8	15	6	5	8	5	8	8	10

To find the values of any outliers you need to find the interquartile range.

List the data in order of size. Start with the smallest.

5	5	5	5	6	6	7	8	8	8
8	8	8	9	9	10	11	13	15	16

The position of the median is $\frac{1}{2}(20 + 1)$th $= 10\frac{1}{2}$th value
The 10th value and the 11th value are both 8

$$\text{The median, m} = \frac{8 + 8}{2} = 8$$

Look at the data to the left of the median.

5 5 5 5 6 6 7 8 8 8

The position of the lower quartile is $\frac{1}{2}(10 + 1)$th $= 5\frac{1}{2}$th value
The 5th value and the 6th value are both 6

$$\text{The lower quartile, LQ} = \frac{6 + 6}{2} = 6$$

Look at the data to the right of the median.

8 8 8 9 9 10 11 13 15 16

The position of the upper quartile is $\frac{1}{2}(10 + 1)$th $= 5\frac{1}{2}$th value
The 5th value is 9 and the 6th value is 10

$$\text{The upper quartile, UQ} = \frac{9 + 10}{2} = 9\frac{1}{2}$$

$$\begin{aligned}\text{So the interquartile range} &= \text{UQ} - \text{LQ} \\ &= 9\tfrac{1}{2} - 6 \\ &= 3\tfrac{1}{2}\end{aligned}$$

Multiply the interquartile range by 1.5 $3\frac{1}{2} \times 1.5 = 5.25$

Any values more than 5.25 below the lower quartile or more than 5.25 above the upper quartile are outliers.

$LQ - 5.25 = 6 - 5.25$
$= 0.75$ Small outliers will be smaller than 0.75
There are no small outliers.

$UQ + 5.25 = 9\frac{1}{2} + 5.25$
$= 14.75$ Large outliers will be bigger than 14.75
There are two large outliers, 15 and 16.

A box and whisker diagram can be used to show the data and the outliers.

Draw the diagram as before using the median and quartiles.

The outliers are not included in the whiskers.

They are marked separately with crosses.

5 6 7 8 9 10 11 12 13 14 15 16

The whiskers end at the lowest and highest values of data that are not outliers.

Exercise 8:4

1 Rachel keeps a record of the time it takes her to prepare the meal each Sunday lunchtime. These are her times, to the nearest minute.

43	56	50	45	51	58	46	61	75	48
43	49	45	47	54	46	48	53	62	79

a For these times, find the median, the lower quartile and the upper quartile.
b Work out the interquartile range.
c Write down any small outliers.
d Write down any large outliers.
e Draw a box and whisker diagram to show this data.

2 Howard is planning a holiday in Texas. Each evening, he looks in the newspaper and records the daily temperature. These are the temperatures, in °C, that he recorded.

28	20	31	28	30	22	30	36
29	27	30	21	36	30	27	

a Find the median, the lower quartile and the upper quartile of these temperatures.
b Work out the interquartile range.
c Write down any small outliers.
d Write down any large outliers.
e Draw a box plot for this data.

3 Look at this box and whisker diagram.

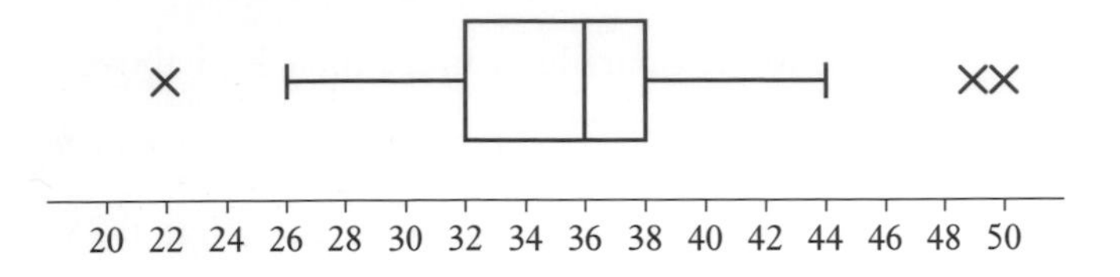

Write down the value of

a the median
b the lower quartile
c the upper quartile
d any outliers.

4 For each of these box plots

a Write down (1) the median (2) the lower quartile (3) the upper quartile (4) any outliers.

b Describe the skewness of the data

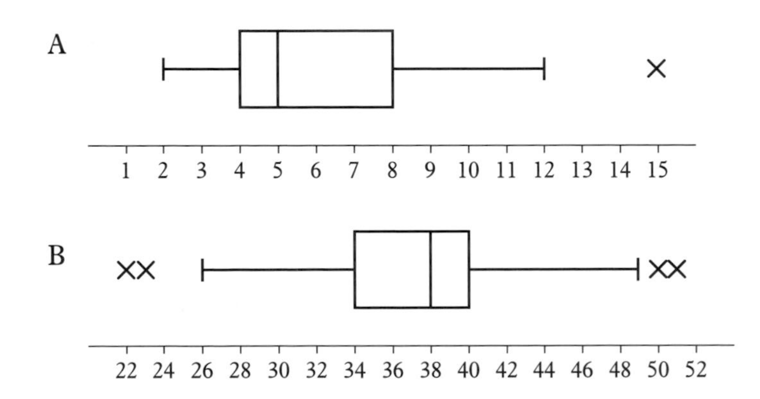

2 The normal curve

Big Ben is the name of the huge bell, which strikes the hour, from the Houses of Parliament. It is a famous landmark.

This bell shape is also very important in statistics.

The normal curve

The **normal curve** is a very special and important distribution. It is important as it occurs naturally.

The curve can look different, but some facts are always true.

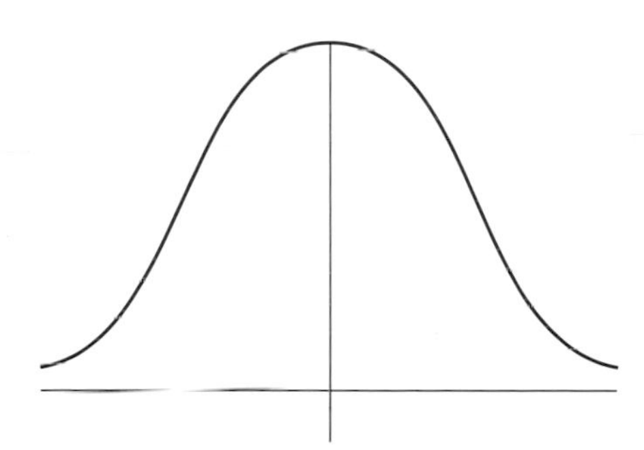

The curve is bell-shaped and symmetrical. The red line is the axis of symmetry.

The mode, median and mean are all equal. They all lie on the axis of symmetry.

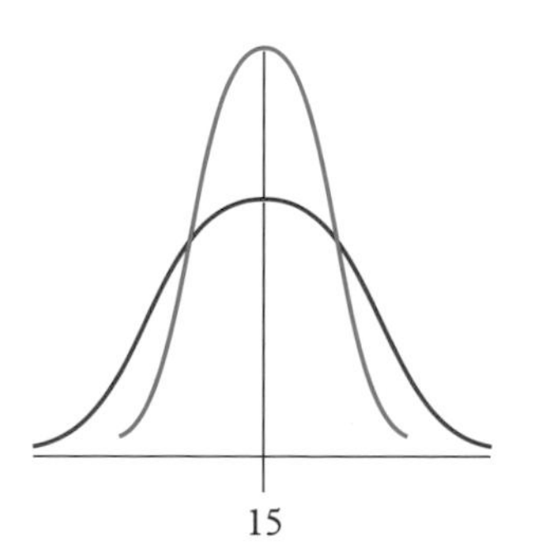

These two curves have the same mean of 15. Their axes of symmetry are in the same position.

The green curve is narrower and has the smaller spread. It has the smaller standard deviation.

The red curve is wider and has the larger spread. It has the larger standard deviation.

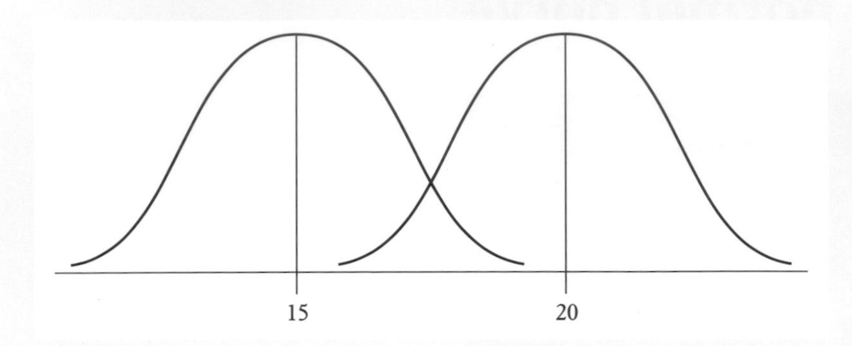

These two curves have different means.
This means they have different axes of symmetry.

The bells are the same size.
The spread is the same so the standard deviations are equal.

It is the mean and the standard deviation which change the position and shape of the bell.

Exercise 8:5

1 Here are two normal curves **A** and **B**.

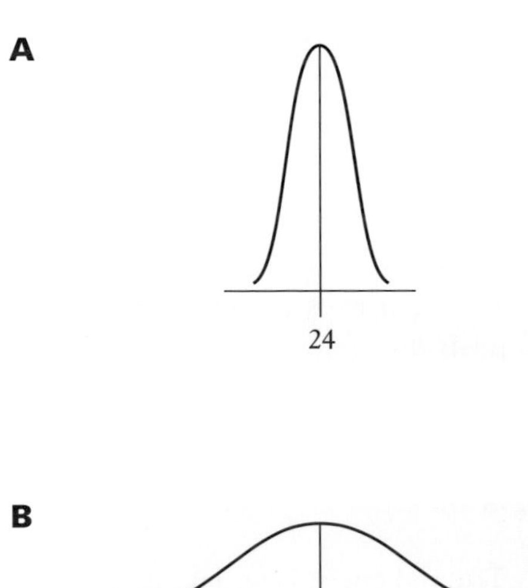

a Write down the mean for each curve.

b Write down the mode for each curve.

c Write down the median for each curve.

d Which curve has the larger spread?

e Which curve has the smaller standard deviation?

Sometimes you need to compare values from two different normal distributions. To do this you need to standardise the scores.

Standardised scores

A **standardised score** is the number of standard deviations that a value lies above or below the mean.

$$\text{The standardised score} = \frac{\text{Value} - \text{Mean}}{\text{Standard Deviation}}$$

Example

A normal curve has a mean of **60** and a standard deviation of **4**. Two values taken from the data are 68 and 56.

This is a sketch of the curve showing the mean and the two values.

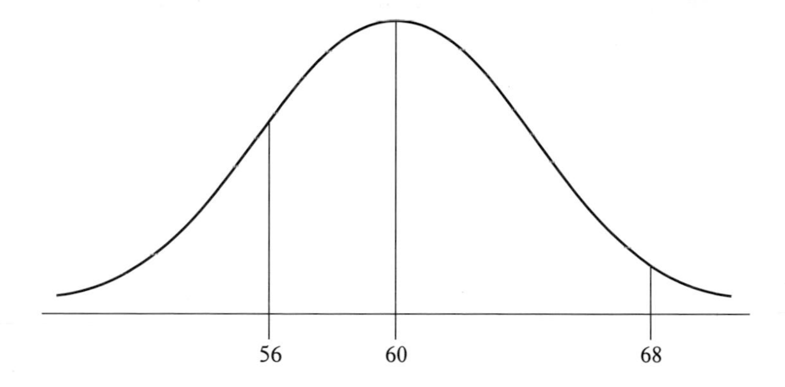

68 is to the **right** of the mean. It has a **positive standardised** score.

$$\begin{aligned}\text{The standardised score} &= \frac{\text{Value} - \text{Mean}}{\text{Standard Deviation}}\\ &= \frac{68 - 60}{4}\\ &= 2\end{aligned}$$

68 is 2 standard deviations to the right of the mean.

56 is to the **left** of the mean. It has a **negative standardised** score.

$$\begin{aligned}\text{The standardised score} &= \frac{\text{Value} - \text{Mean}}{\text{Standard Deviation}}\\ &= \frac{56 - 60}{4}\\ &= -1\end{aligned}$$

56 is 1 standard deviation to the left of the mean.

Exercise 8:6

1 A normal curve has a mean of 80 and a standard deviation of 6. Calculate the standardised scores for:
a 92 **b** 71 **c** 89 **d** 68

2 A normal curve has a mean of 68 and a standard deviation of 4. Calculate the standardised scores for:
a 72 **b** 74 **c** 62 **d** 63

3 A normal curve has a mean of 72 and a standard deviation of 3. Calculate the actual value if the standardised score is:
a 1 **b** 2 **c** -1 **d** -3

4 The marks in an English examination are normally distributed. The mean mark is 56 and the standard deviation is 6. Calculate the standardised scores for marks of:
a 59 **b** 50 **c** 65 **d** 44

5 The heights of men are normally distributed. The mean height is 1.7 m and the standard deviation is 6 cm. Calculate the standardised scores for heights of:
a 1.82 m **b** 1.79 m **c** 1.61 m **d** 1.58 m

6 The lengths of earthworms are normally distributed. The mean length is 8.2 cm and the standard deviation is 2.4 cm.
a Calculate the standardised scores for an earthworm of length:
(1) 10.4 cm (2) 6.3 cm
b Calculate the actual length of an earthworm with a standardised score of:
(1) 1.48 (2) -0.67
Give all your answers to 3 s.f.

There are some important characteristics of any normal distribution.

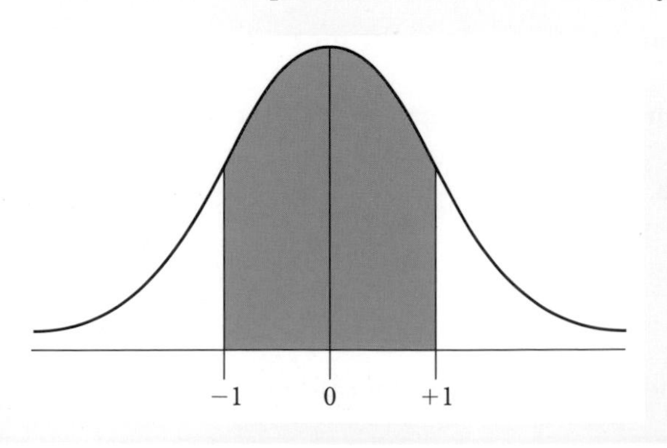

68% of all the values of any normal distribution are between the standardised scores of -1 and $+1$ i.e. ± 1 standard deviation.

Of these:

34% lie to the right of the mean

34% lie to the left of the mean.

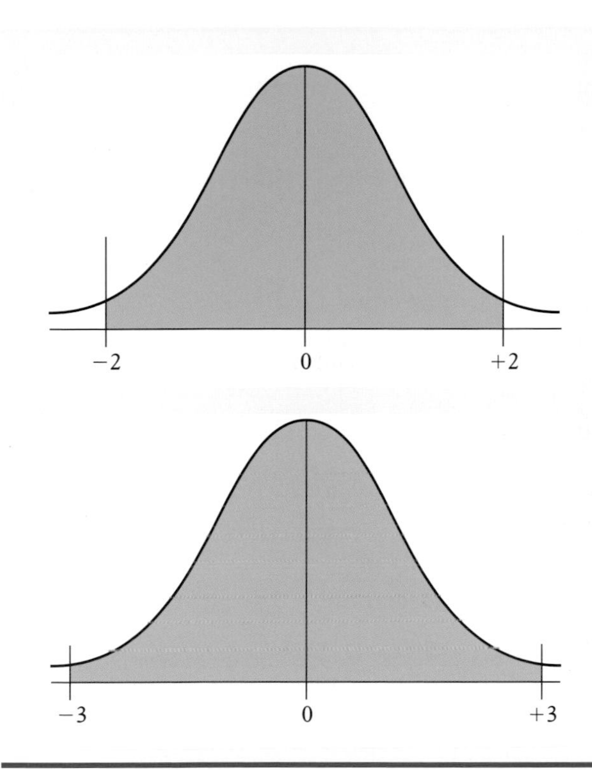

95% of all the values are between the standardised scores of −2 and +2 i.e. ±2 standard deviations.

Of these:

$47\frac{1}{2}$% lie to the right of the mean

$47\frac{1}{2}$% lie to the left of the mean.

99.7% of all the values are between the standardised scores of −3 and +3 i.e. ±1 standard deviation.

This means that almost all of the values are within the standardised scores of −3 and +3.

Exercise 8:7

Look at each of the normal curves. For each curve, find the percentage of all the values that are in the shaded area.

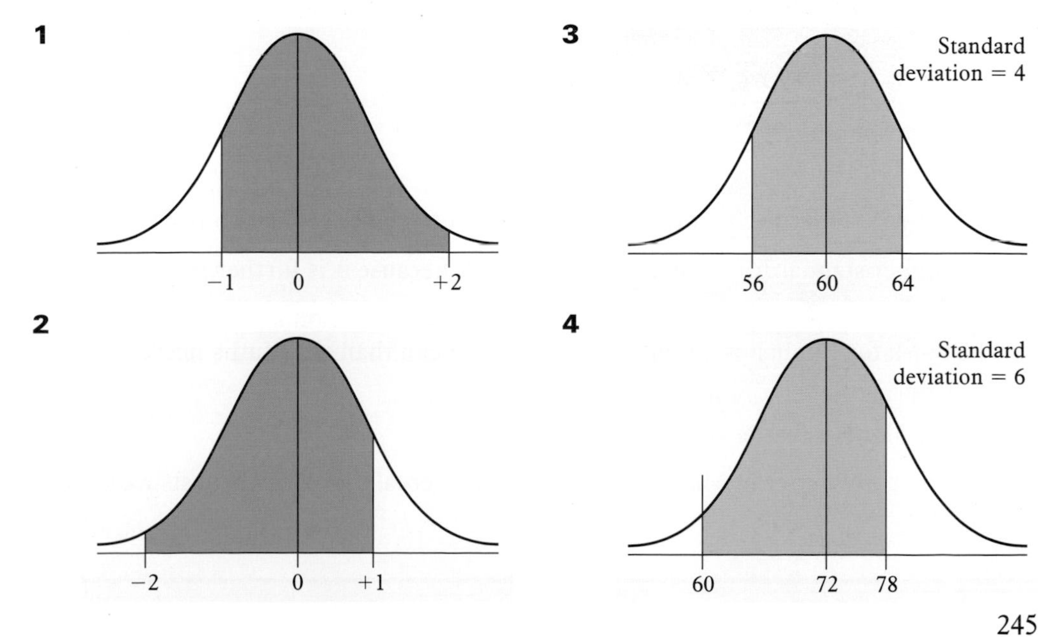

Standardised scores are used to compare values from different normal distributions.

Remya scores 72 in a Maths test and 72 in an English test. The marks in Maths and English are normally distributed.

The mean mark in Maths is **54** and the standard deviation is **13**.
The mean mark in English is **51** and the standard deviation is **20**.

To compare these two marks, work out the standardised scores in Maths and English.

$$\text{Standardised score in Maths} = \frac{\text{Value} - \text{Mean}}{\text{Standard Deviation}}$$

$$= \frac{72 - 54}{13}$$

$$= 1.38 \text{ (3 s.f.)}$$

$$\text{Standardised score in English} = \frac{\text{Value} - \text{Mean}}{\text{Standard Deviation}}$$

$$= \frac{72 - 51}{20}$$

$$= 1.05$$

These two standardised scores can now be used to compare the two marks.

The higher standardised score is a better mark because it is further from the mean.

The mark for English is closer to the English mean than the Maths mark is to its mean.

1.38 for Maths is better than 1.05 for English.

So Remya did better in Maths even though the actual scores in the tests were the same.

Exercise 8:8

1 Simon takes French and German tests. The marks for both tests are normally distributed. In French he scores 76 and in German he scores 78. The French marks have a mean of 68 and a standard deviation of 10. The German marks have a mean of 70 and a standard deviation of 12.

a Calculate Simon's standardised score in French.
b Calculate Simon's standardised score in German to 2 d.p.
c Use the standardised scores to compare his test results.

2 Evelyn studies both physics and geography. In recent exams, she scored 75 in geography and 63 in physics. Both the physics and geography marks are approximately normally distributed. The mean mark in physics is 55 and the standard deviation is 6. The mean mark in geography is 70 and the standard deviation is 5. Evelyn claims that she is better at physics than geography.
a Calculate Evelyn's standardised score in physics to 2 d.p.
b Calculate Evelyn's standardised score in geography.
c Use the standardised scores to test whether her test results support her claim. Is she right? Explain your answer.

• **3** Mr Cross has two very different trees in his garden. He knows that the lengths of the leaves on both trees are approximately normally distributed. For the first tree, the mean length is 5 cm and the standard deviation is 1 cm. For the second tree, the mean length is 8 cm and the standard deviation is 1.5 cm. He picks leaves which measure exactly 7 cm from both trees. He claims that this is more likely to happen on the second tree.
Use the standardised scores to test his claim.

• **4** As part of her geography fieldwork, Alison is measuring and comparing the lengths of pebbles on two beaches. On the first beach, the mean length is 8 mm and the standard deviation is 1.4 mm. On the second beach, the mean length is 9 mm and the standard deviation is 0.8 mm. She finds a pebble of length 10 mm on each beach. She claims that this is less likely to happen on the first beach.
Use the standardised scores to test her claim.

1 A hospital recorded the birth weights, in kilograms, of 100 girls and 100 boys. The weights are summarised in the table below.

	Median	Lower quartile	Upper quartile	Minimum	Maximum
Girls	3.1	2.5	3.7	1.3	4.5
Boys	3.3	2.4	4.0	1.0	4.8

a On graph papers, draw **two** box plots for the birth weights of girls and boys. *(4)*

b Use the information above to write a brief comparison of the birth weights of the girls and boys. *(2)*

NEAB, 1997, Paper 2

2 The box and whisker diagrams show the masses, in kilograms, of a sample of male and female rabbits.

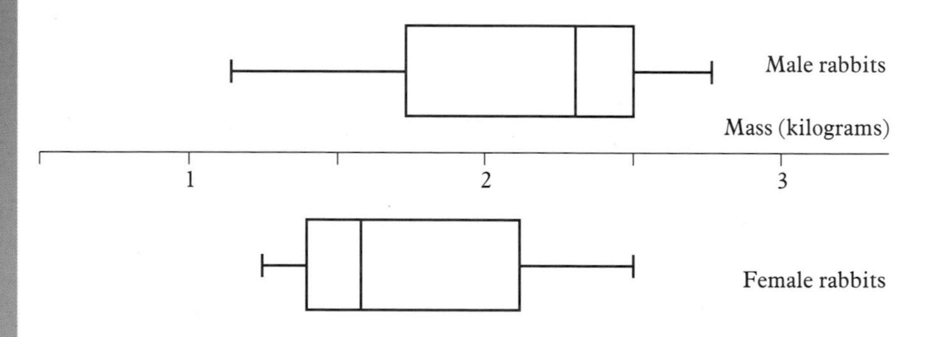

a Comment on the skewness of the distribution of the masses of the **female** rabbits. *(1)*

b Use the two box and whisker diagrams to give **two** differences, other than skewness, between the masses of male and female rabbits for this sample. *(2)*

SEG, 1998, Paper 3

3 The cumulative frequency curve represents the times taken to run 1500 m by each of the 240 members of the athletics club, Weston Harriers.

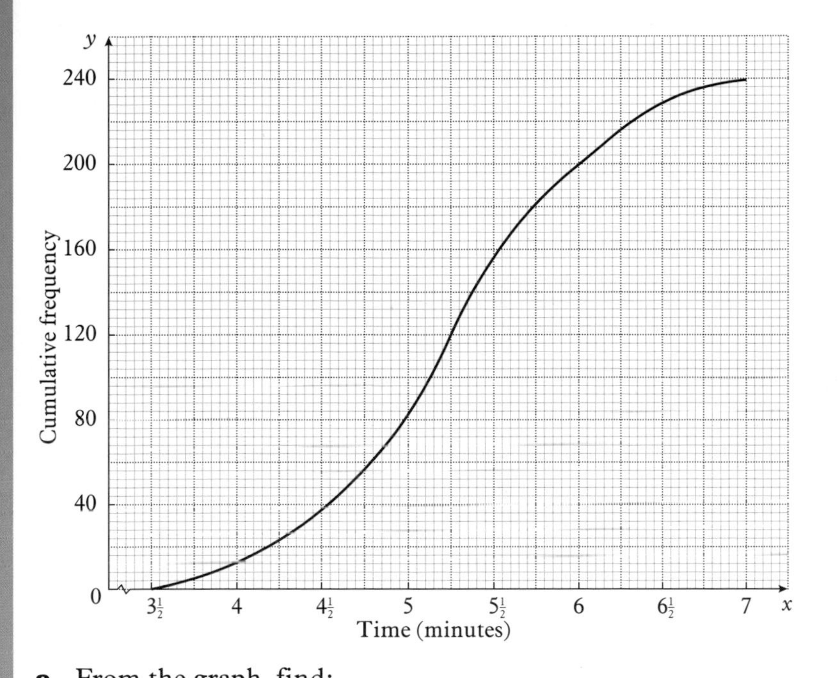

a From the graph, find:
  i the median time *(1)*
  ii the upper quartile and the lower quartile. *(2)*

b On graph paper, draw a box and whisker diagram to illustrate the data. *(2)*

c Use your box and whisker diagram to make **one** comment about the shape of a histogram for these data. *(1)*

A rival athletics club, Eastham Runners, also has 240 members. The time taken by each member to run 1500 metres is recorded and these data are shown in the following box and whisker diagram.

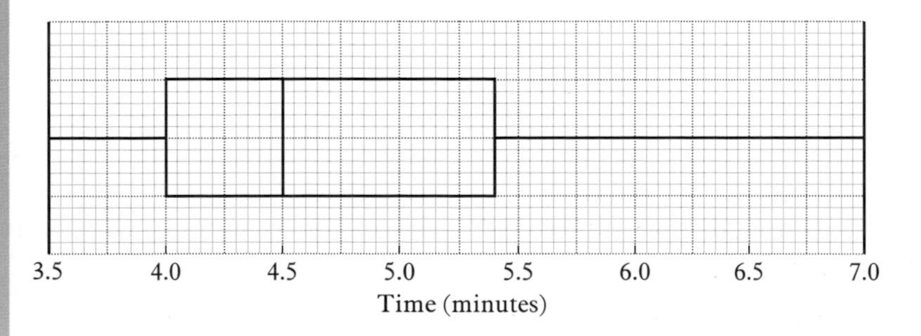

d Use this diagram to make **one** comment about the data for Eastham Runners as compared to that for Weston Harriers. *(1)*

SEG, 1996, Paper 3

4 The histogram represents the weight (in lb) of 100 fish caught by an angler over a one year period.

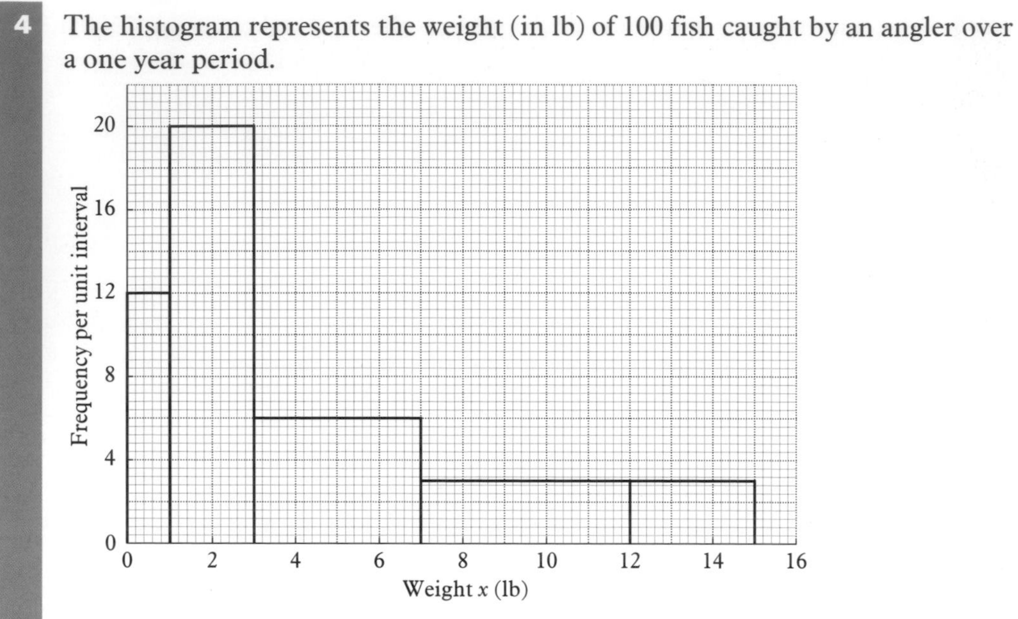

a Use this diagram to copy and complete the following frequency table. *(4)*

Weight x (lb)	Frequency
$0 < x \leqslant 1$	12
$1 < x \leqslant 3$	
$3 < x \leqslant 7$	
$7 < x \leqslant 12$	
$12 < x \leqslant 15$	
	100

Over the same period of time equal numbers of fish were caught and weighed by two other anglers (A and B).
The weights of each set of fish caught were normally distributed with mean and standard deviation recorded as follows:

	Mean weight of fish x (lb)	Standard deviation of weight (lb)
Angler A	12	3
Angler B	14	2

b On graph paper sketch these two distributions labelling each one clearly. *(4)*

To enable comparisons to be made it was agreed to standardise the weights of each distribution.

c **i** What would be the standardised value of a fish caught by A weighing 17 lb? *(2)*

ii A fish caught by B was given a standardised value of $+2.6$.
What was its actual weight? *(3)*

iii Explain whether it is likely that angler A caught (as claimed) a fish weighing 23 lb. Justify your answer. *(2)*

NEAB, 1999

5 The table shows the mean and the standard deviation of the heights of a sample of adult males and a sample of boys aged nine.
The heights of both the adult males and the boys are normally distributed.

	Adult males	Boys aged nine
Mean	180 cm	135 cm
Standard deviation	18 cm	10 cm

David is a boy aged nine whose height is 120 cm.

a How many standard deviations below the mean is David's height? *(1)*

It is believed that the height of a boy aged nine is a good indicator of his adult height.

b Estimate the height that David will be when he is an adult. *(2)*

The diagram shows the distribution of the heights of the boys aged ninc.

c Copy the diagram below and sketch the distribution of the heights of the adult males. *(3)*

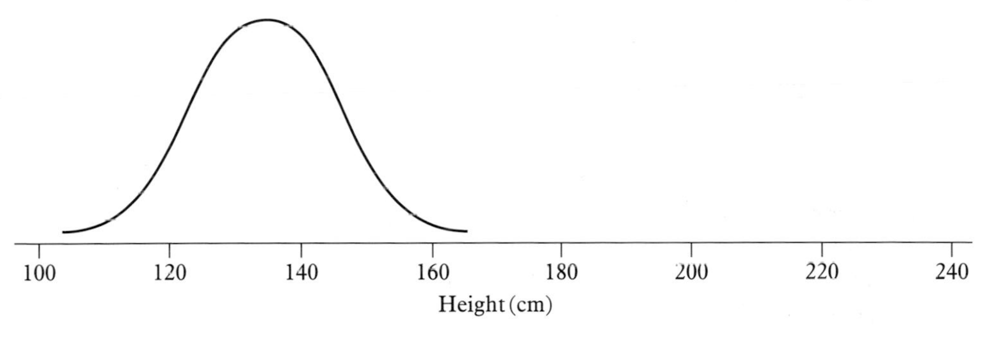

SEG, 1998, Paper 4

6 The number of goals scored by the 11 members of a hockey team in 1993 were as follows:

6 0 8 12 2 1 2 9 1 0 11

a Find the median. *(2)*
b Find the upper and lower quartiles. *(2)*
c Find the interquartile range. *(1)*
d Explain why, for this data, the interquartile range is a more appropriate measure for spread than the range. *(1)*

e The goals scored by the 11 members of the hockey team in 1994 are summarised in the box plot below.

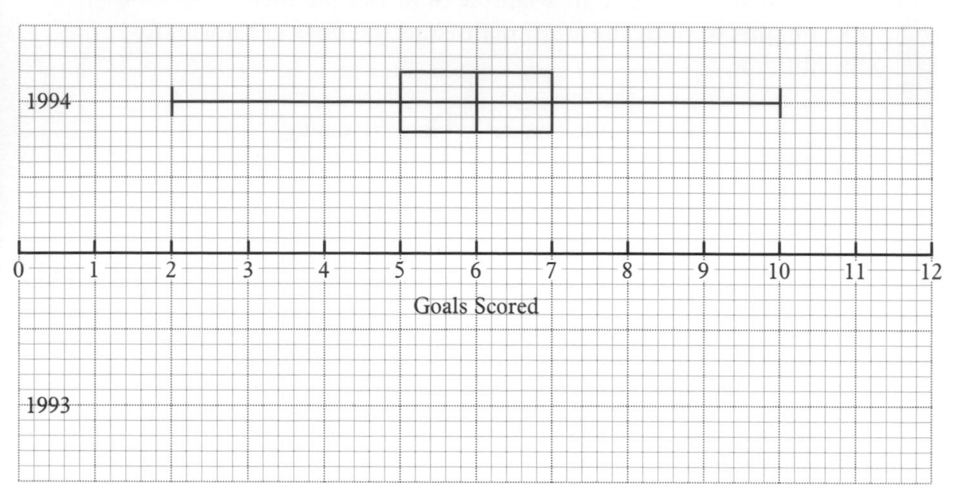

i Copy the diagram above and summarise the results for 1993 in the same way. *(3)*

ii Do you think the team scored more goals in 1994? Explain your reasoning. *(2)*

NEAB, 1996, Paper 2

7 The number of flower buds on a sample of 10 azalea plants were recorded.

25 37 28 16 37 50 48 45 42 49

a Calculate the range of these data. *(1)*

b Calculate the standard deviation of these data.
You may use the statistical functions on your calculator or the following formula. You must show all your working.

Standard deviation $= \sqrt{\dfrac{\Sigma x^2}{n} - \left\{\dfrac{\Sigma x}{n}\right\}^2}$ *(3)*

The number of flower buds on the same 10 plants were recorded two years later. The second set of data had an increased range of 35 but a reduced standard deviation of 7.9.

c Write down two differences which you would expect to see in the values of the second set of data as compared to those of the first set of data. *(2)*

Another type of plant had a positively skewed distribution with a mean of 50 buds per plant.

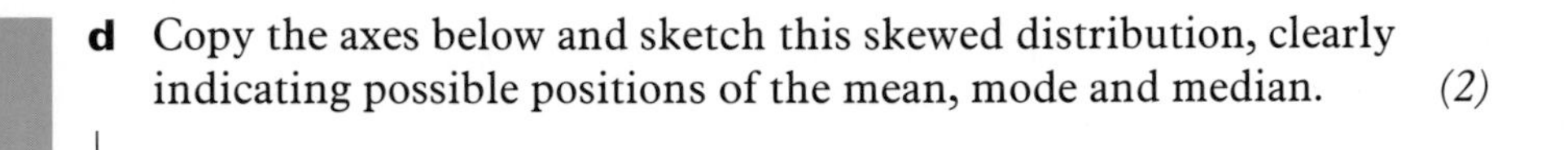

d Copy the axes below and sketch this skewed distribution, clearly indicating possible positions of the mean, mode and median. *(2)*

0 5 10 15 20 25 30 35 40 45 50 55 60 65 70 75 80

SEG, 1998, Paper 1

8 The points awarded (out of a maximum of 6.0) by the 12 judges at a recent ice skating tournament were:

5.7, 5.9, 5.9, 5.7, 3.0, 5.8, 5.7, 5.7, 5.5, 5.8, 5.8, 5.5

a Calculate the mean points scored. *(1)*

b Using the formula $\sqrt{\frac{\Sigma(x-\overline{x})^2}{n}}$ or the statistical functions on your calculator, or otherwise, calculate to two decimal places the standard deviation of the points scored. *(4)*

c The points awarded by the judges are to be standardised by subtracting the mean from each of the values and then dividing the result by the standard deviation.

- **i** What is the standardised value corresponding to a points score of 5.8? *(2)*
- **ii** Which of the original values corresponds to a standardised score of +0.263? *(2)*

d To further the analysis it was decided to exclude any points awarded outside of the range of ±3 on the standardised scale.

- **i** Identify the value to be excluded in this case. *(1)*
- **ii** If the data were analysed without this extreme observation what effect would this have on the original values for the mean and standard deviation calculated in parts **a** and **b**?
  (You should **not** calculate the new values for the mean and standard deviation.) *(2)*

NEAB, 1996, Paper 3

9 The reaction times of a class of 19 girls and 19 boys were measured in hundredths of a second. The summary statistics are shown below.

Reaction times (hundredths of a second)	Median	Lower quartile	Upper quartile	Minimum	Maximum
Girls	14	12	17	10	20
Boys	14	11	18	9	21

a Draw **two** box plots suitable for comparing the reaction times of the boys and girls.

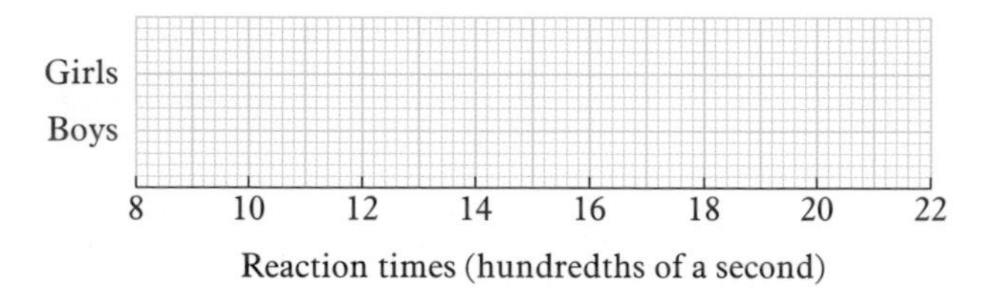

(3)

b Jim was the only boy with a reaction time of 14 hundredths of a second.
How many boys were quicker than Jim? *(1)*

c There were 4 boys with a reaction time greater than Paul's.
What was Paul's reaction time? *(2)*

d There were 14 girls with a reaction time less than Ann's.
What was Ann's reaction time? *(2)*

e A child with a reaction time of more than 18 hundredths of a second is chosesn at random.
Is the child more likely to be a boy or a girl?
Explain your reasoning. *(2)*

Box and whisker diagrams

A **box and whisker diagram** shows the distribution of data. It is sometimes called a box plot. It shows the shape of a distribution. The distribution can be symmetrical, positively skewed or negatively skewed.

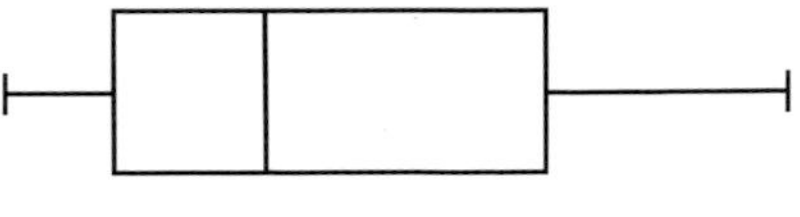

Outliers

Outliers are very small or very large values in a set of data. A value is called an outlier if its distance from the nearest quartile is greater than 1.5 times the interquartile range.

Outliers are often ignored because they can distort the data.

The normal curve

The **normal curve** is a very special and important distribution. It is important as it occurs naturally.
The curve is bell-shaped and symmetrical.
The **mean** and **standard deviation** change the position and shape of the bell.

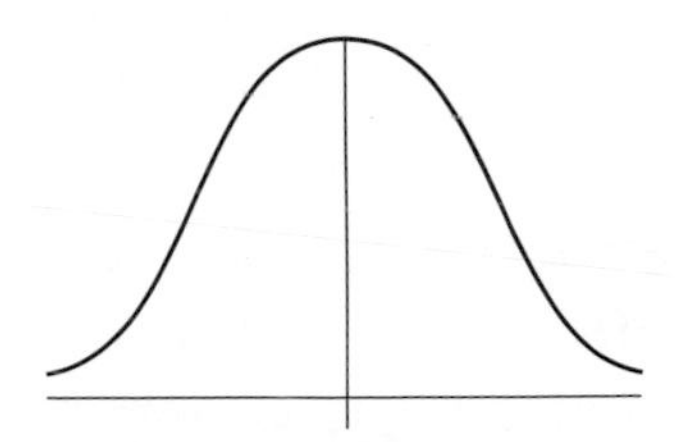

The **standardised score** is the number of standard deviations that a value lies above or below the mean.

$$\text{The standardised score} = \frac{\text{Value} - \text{Mean}}{\text{Standard Deviation}}$$

If the value is to the right of the mean, it has a positive standardised score.
If the value is to the left of the mean, it has a negative standardised score.

For a normal distribution:
68% of all values lie between standardised scores of -1 and $+1$
95% of all values lie between standardised scores of -2 and $+2$
99.7% of all values lie between standardised scores of -3 and $+3$.

Standardised scores are used to compare values from two different normal distributions.
The closer a value is to the mean the more likely it is to happen.

1 Adele works in the local library. As each person returns their books, she asks them how many books they have read in the last month. These are her results.

21	11	2	10	5	4	5	5	6	4	6
12	9	11	3	8	7	4	7	6	4	
14	4	11	5	3	5	5	11	6	10	

a Find the median, the lower quartile and the upper quartile.
b Use these values to draw a box and whisker diagram.
c Calculate m − LQ and UQ − m.
d Describe the distribution.
e Work out the interquartile range.
f Write down any outliers.

2 Rosemary scores 72 in a biology test and 78 in a chemistry test. The marks in biology and chemistry are approximately normally distributed. The mean mark in biology is 76 and the standard deviation is 8. The mean mark in chemistry is 84 and the standard deviation is 6.
a Calculate Rosemary's standardised score in biology.
b Calculate Rosemary's standardised score in chemistry.
c Use the standardised scores to compare her test marks.

3 Max and Chloe are in the school chess team. They have recorded the times that they each took to win their games of chess for the past year. Both sets of times are normally distributed.
Max's times have a mean of 24 minutes and a standard deviation of 12 minutes.
Chloe's times have a mean of 29 minutes and a standard deviation of 10 minutes.
a The time for Max's last win was 26 minutes.
Find the standardised score for this win. Give your answer to 3 s.f.
b The standardised score for Chloe's last win is 0.3
Find the time that she took to win this game.
c One particular game was won in 27 minutes.
Who was most likely to have won this game?
Explain your answer.

9 Errors in statistics

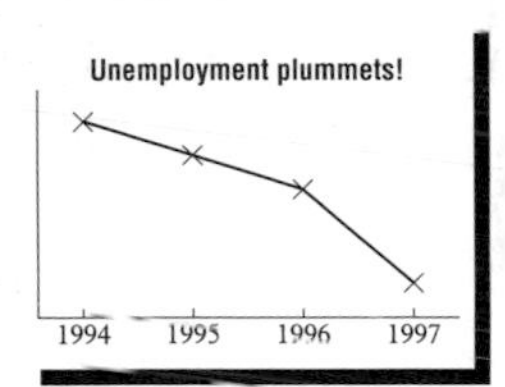

1 Errors in measurement

When you put a plug in a plughole you are happy to assume that it will fit. Plugs have to be made accurately so that you will not be disappointed.
The plug size does not have to be exactly the right size. There is a small error that is allowed. This is called a tolerance.
A plug would probably fit if it is 0.5 mm too small. But it won't fit if it is 2 cm too small.

You cannot measure a length exactly.
The accuracy of a measurement depends on the instrument used for measuring.
Some are more accurate than others.

Parim measures the length of her letter box to the nearest centimetre.
Her length is 29 cm.

Lower bound

28.5 is called the **lower bound**.
It is the smallest number that will round up to 29.

Upper bound

29.5 is called the **upper bound**.
The largest number that will round down to 29 is 29.499 999 99 …
You do not use this value as it's too complicated.
You use 29.5 instead even though 29.5 would round up to 30 to the nearest whole number.
The length of the letter box must lie in the range 29 $\pm$0.5 cm.

Absolute error

The number 0.5 is called the **absolute error**.
It is also sometimes called the maximum absolute error or the maximum possible error.

Example

Write down the lower bound, upper bound and absolute error for each of these:
a 3.6 to 1 d.p. **b** 0.042 to 2 s.f. **c** 700 to the nearest 100.

a lower bound = 3.55
upper bound = 3.65
absolute error = 0.05

b lower bound = 0.0415
upper bound = 0.0425
absolute error = 0.0005

c lower bound = 650
upper bound = 750
absolute error = 50

Exercise 9:1

1 Write down the lower bound, upper bound and absolute error for each of these:

a 7.29 to 2 d.p.
b 86 to 2 s.f.
c 38.1 to 1 d.p.
d 9370 to 3 s.f.
e 400 to the nearest 100
f 6000 to the nearest 1000.

2 **a** Tim has painted a picture. He measures its length as 48 cm to the nearest centimetre. Write down the maximum length that the picture could be.

b Tim has bought a frame. The frame's length is 485 mm to the nearest millimetre.
Will the picture definitely fit into the frame?
Explain your answer.

3 The length of Jane's hall is 14.6 m to 1 d.p.
Jane is going to lay carpet tiles along the length of the hall.
Each tile is 80 cm long. This figure is correct to 2 s.f.

a What is the minimum length of each tile?
b Jane lays 12 tiles in a row.
What is the maximum length that these 12 tiles can be?
c What is the maximum length of the hall?
d What is the maximum number of tiles that Jane will need?

The length, width and height of this box are measured to the nearest centimetre.

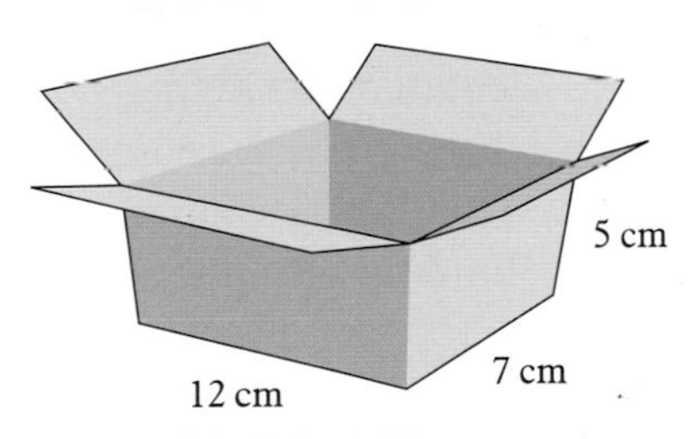

To find the **minimum** volume you use the lower bounds for the length, width and height.
To find the **maximum** volume you use the upper bounds for the length, width and height.

Minimum volume $= 11.5 \times 6.5 \times 4.5 = 336.375$ cm^3
Maximum volume $= 12.5 \times 7.5 \times 5.5 = 515.625$ cm^3

If you calculate the volume using the rounded numbers you get:
$12 \times 7 \times 5 = 420$ cm^3

The absolute error in volume is the difference between this value and either the maximum or minimum volume. Take the larger difference.

The absolute error in the volume is $515.625 - 420 = 95.625$ cm^3.

---

4 Richard has measured the length and height of his garden fence.
The length is 5.8 m. The height is 1.7 m.
Both measurements are given to 1 d.p.

a Write down the maximum length that the fence can be.

b Find the maximum area of one side of the fence.

c Find the absolute error in the area of the fence.

5 Guy has found his birthday present hidden away.
His mother has already wrapped it.
Guy measures the length, width and height to the nearest centimetre.
They are 1.28 m, 75 cm and 34 cm respectively.

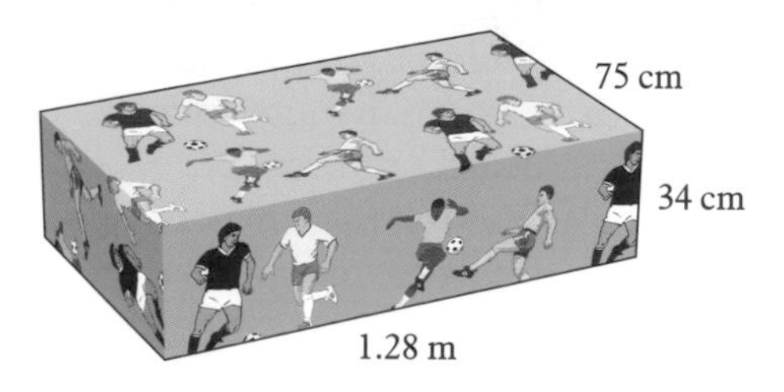

a Write down the maximum possible length of the box.

b Write down the absolute error in the length of the box.

c Find the maximum possible volume of the box.

d Find the absolute error in the volume of the box.

6 Maxine takes 48 s, to the nearest second, to run the length of a field.
The length is 74 m to the nearest metre.

a Write down the maximum possible time for Maxine to run the length of the field.

b Write down the absolute error in the length of the field.

c Find the maximum value of Maxine's speed.

d Find the absolute error in Maxine's speed.

7 David picks grapes at a rate of 7.34 kg every minute.
The weight is accurate to 3 s.f.

a Write down the minimum possible weight that David picks in one minute.

b Find the maximum possible weight that David picks in one hour.

c One day David works for exactly 4 hours and 10 minutes.
Find the minimum possible weight of grapes that he can pick in this time.

Exercise 9:2

1 **a** Liam measured the length of his room. The actual length is 450 cm. Liam's measurement was 452 cm.
(1) Write down the error of Liam's measurement.
(2) Express the error as a percentage of the actual length.

b Liam measured the width of his bedside table. The actual width is 23 cm. Liam's measurement was 25 cm.
(1) Write down the error in Liam's measurement.
(2) Express the error as a percentage of the actual width.

c Look at the answers to **a**(2) and **b**(2).
Explain why the two answers are different even though the error in measurement is the same.

Relative error

The **relative error** gives you a better idea of the size of an error.

$$\text{Relative error} = \frac{\text{error}}{\text{exact value}}$$

Sometimes this is given as a percentage.

Percentage relative error

The **percentage relative error** is the relative error given as a percentage.

$$\text{Percentage relative error} = \frac{\text{error}}{\text{exact value}} \times 100$$

Example A length is measured as 2.06 cm. This is accurate to 3 s.f.
Find: **a** the absolute error **b** the relative error
c the percentage relative error.

a The absolute error is 0.005

b The relative error $= \dfrac{0.005}{2.06} = 0.002\,427\,18\ldots = 0.0024$ to 2 s.f.

c The percentage relative error $= \dfrac{0.005}{2.06} \times 100 = 0.24\%$ to 2 s.f.

---

2 For each of these find: **a** the absolute error
b the relative error
c the percentage relative error.

(1) Measuring the width of a book as 15.3 cm correct to the nearest 0.1 cm.
(2) Measuring a length of time as 48 seconds accurate to 2 s.f.
(3) Measuring a mass as 570 g to 2 s.f.
(4) Measuring a population as 40 million to the nearest million.

3 Mary measures a length of 67 cm to the nearest centimetre.
William measures a length of 460 cm to 2 s.f.
Who has the greater percentage relative error?
Show your working.

Example The lengths of a cuboid are 5, 8 and 10 cm to the nearest centimetre.

a Work out the volume of the cuboid.
b Work out the maximum possible volume.
c Find the percentage relative error for the volume.

a Volume $= 5 \times 8 \times 10 = 400\ \text{cm}^3$

b Maximum possible volume $= 5.5 \times 8.5 \times 10.5 = 490.875\ \text{cm}^3$

c The absolute error in volume $= 490.875 - 400 = 90.875\ \text{cm}^3$

The percentage relative error $= \dfrac{90.875}{400} \times 100 = 22.7187\ldots$

$= 23\%$ to 2 s.f.

---

4 Sheila is going to cover her antique desk with leather.
She has measured the length and width to the nearest centimetre.
These are her measurements: length = 121 cm, width = 76 cm.

a Work out the maximum possible area of the desk.
b Find the relative error of the area.
c Write down the percentage relative error for the area.

5 Peter is going to fill this glass cube of side 85 mm with coloured sand.
The side length is accurate to 2 s.f.

a What is the maximum possible volume of sand that he will need?
b Find the percentage relative error for the volume of sand.

2 Misleading graphs

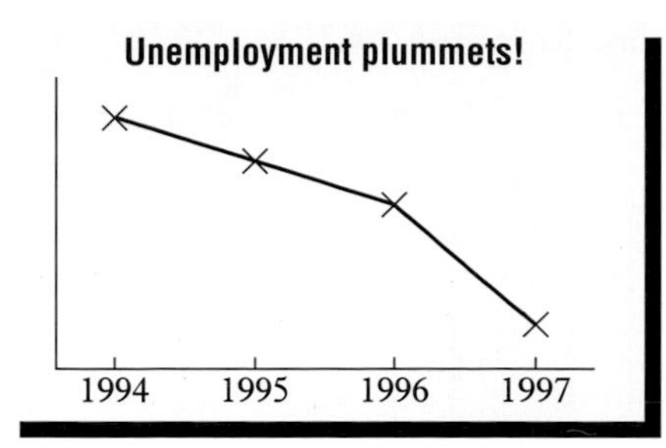

This graph shows that the number of people that are unemployed has fallen very sharply. Or does it?

It has no scale or labels.
Every 1 cm on the graph could stand for 1 person or 10 000 people.

It makes a big difference!

When you are reading a statistical diagram, you should look at it very carefully. You should read the scale and not just go on first impressions.

This section shows how some people use diagrams to give a false impression. Look out for this type of diagram in adverts and on political broadcasts.

Lionel runs a design business.
He wants to borrow some money from the bank to help expand his company.
He has drawn some bar-charts showing his profit for the last 4 years.
He needs to impress the bank manager!

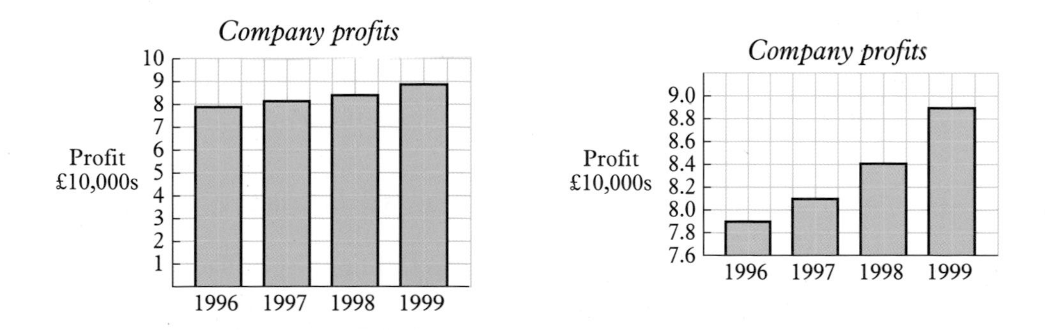

Both bar-charts show the same data but the vertical scales are different.
The chart on the left shows that the profit is hardly changing. The chart on the right makes it look as though the profit is increasing sharply.
Lionel decides to use the graph on the right to show the bank manager!

Exercise 9:3

1 Anthony owns a florist shop.
The bar-chart shows his sales for the last four months.

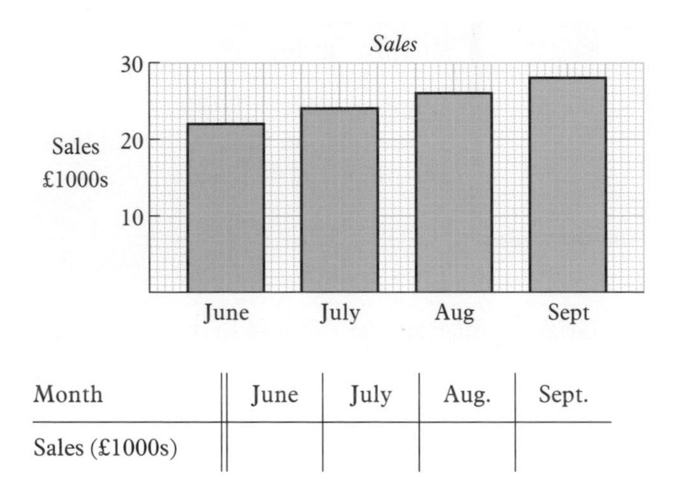

Month	June	July	Aug.	Sept.
Sales (£1000s)				

a Copy the table. Use the bar-chart to fill in the missing numbers.
b Draw another bar-chart to show the data.
Use a vertical scale that starts at £20 000.
c Anthony is selling his shop. Which chart should he use?
Explain your answer.

2 Bianca runs a market stall. She wants to sell the stall and go to college.
These are her sales for the last six months.

Month	Feb.	March	April	May	June	July
Sales in £	4890	4910	4935	4950	5000	5020

Draw a bar-chart for these sales to help Bianca sell her stall.

3 These are the contributions to the European Community budget in 1993 by four countries.

Country	Germany	France	Italy	United Kingdom
Contribution (£ billion)	14.9	9.0	8.0	5.9

a Draw a bar-chart to make the UK's contribution look as large as possible.
b Boris lives in Germany. He wants to show that Germany makes the biggest contribution. Draw a bar-chart for Boris.

This diagram is misleading because of the sloping lines.
They make the company profits look much better than they actually are.

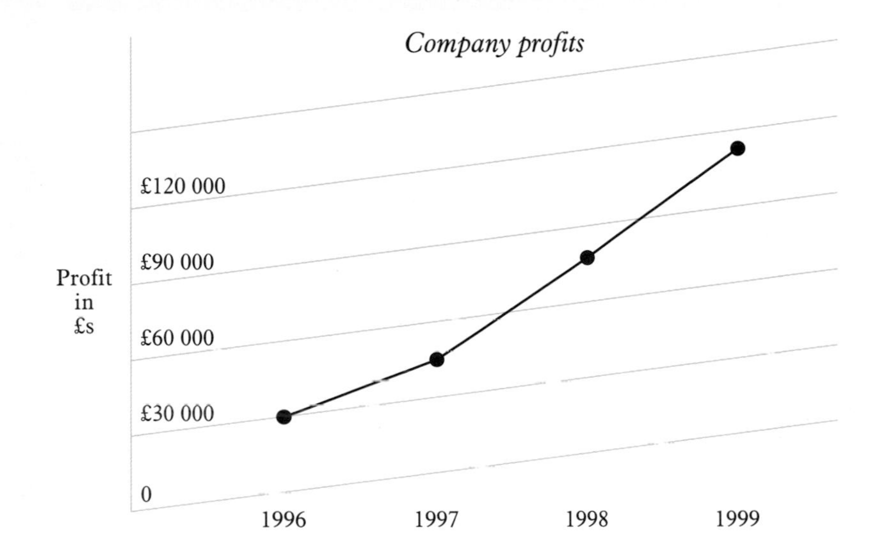

These are some other ways in which diagrams can be misleading:

- Scales do not start at zero.
- The scale is not linear.
  It goes up by different amounts.

1986 1992 1996 1998 1999

- Different sizes of gaps or bars.
- Misleading title.
- Lack of labelling on the axes.
- Lines on the graph drawn too thick.
- Use of shadows.
- Incorrectly plotted points.
- Non-linear time scales.

1996 1997 1998 1999

Look at the diagrams in questions 4–9.
Explain why each one is misleading.

4

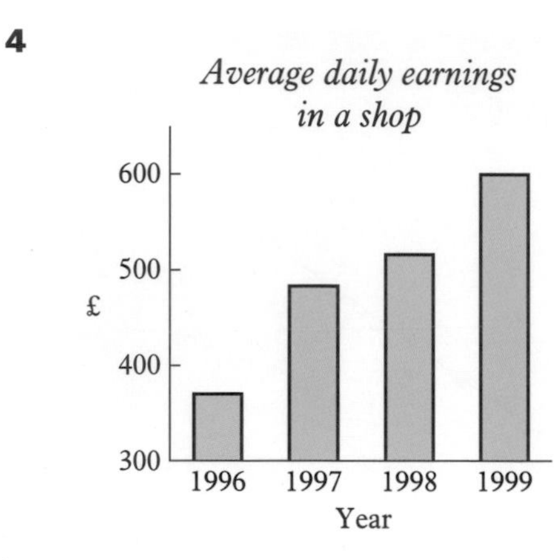

5

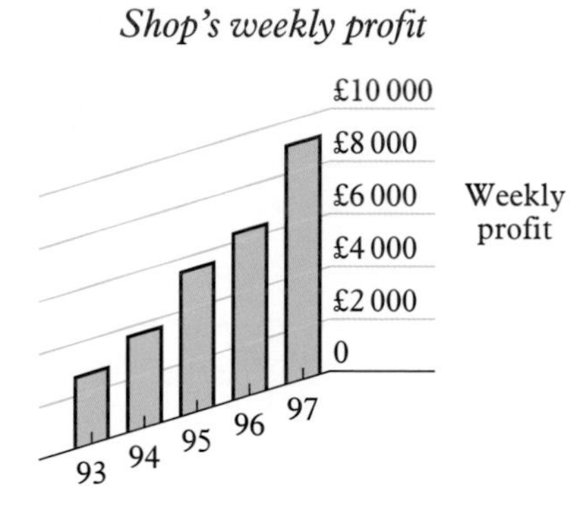

6

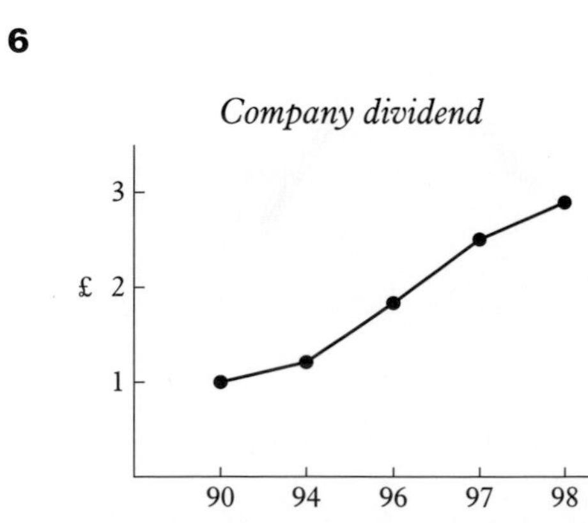

7

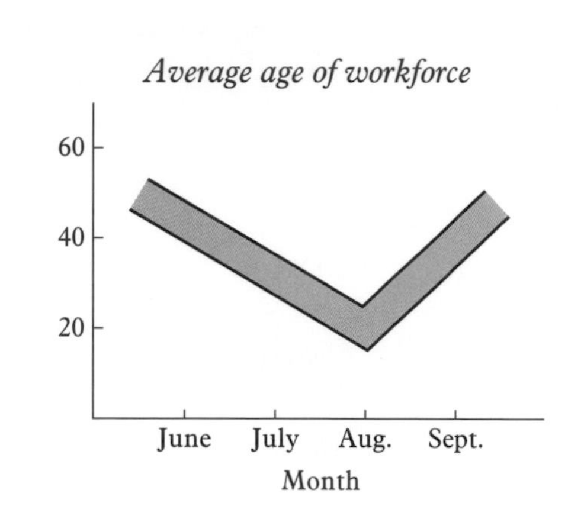

8

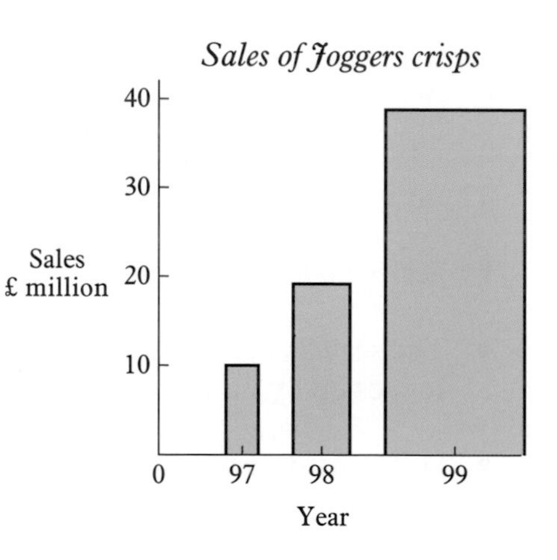

9

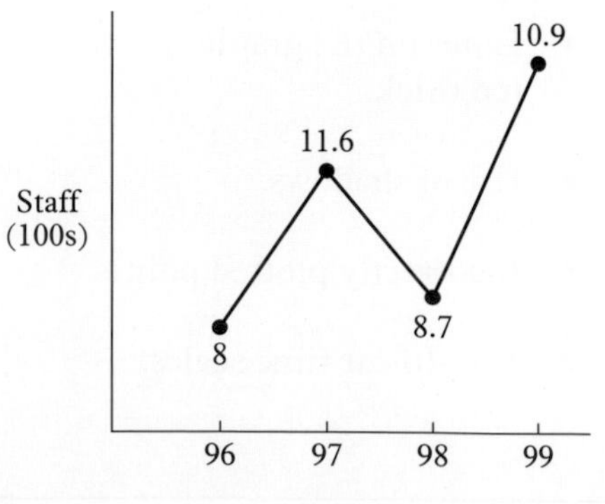

Similar shapes are often used to mislead people.

For any pair of similar shapes A and B:

if the **linear** scale factor for the enlargement from A to B is $\boldsymbol{a}$, then the **area** scale factor for the enlargement from A to B is $\boldsymbol{a} \times \boldsymbol{a} = a^2$.
This means that the **area** of shape B is a^2 times the area of shape A.

For any pair of similar solids C and D:
if the **linear** scale factor for the enlargement from C to D is $\boldsymbol{a}$, then the **volume** scale factor for the enlargement from C to D is $\boldsymbol{a} \times \boldsymbol{a} \times \boldsymbol{a} = a^3$
This means that the **volume** of solid D is a^3 times the volume of solid C.

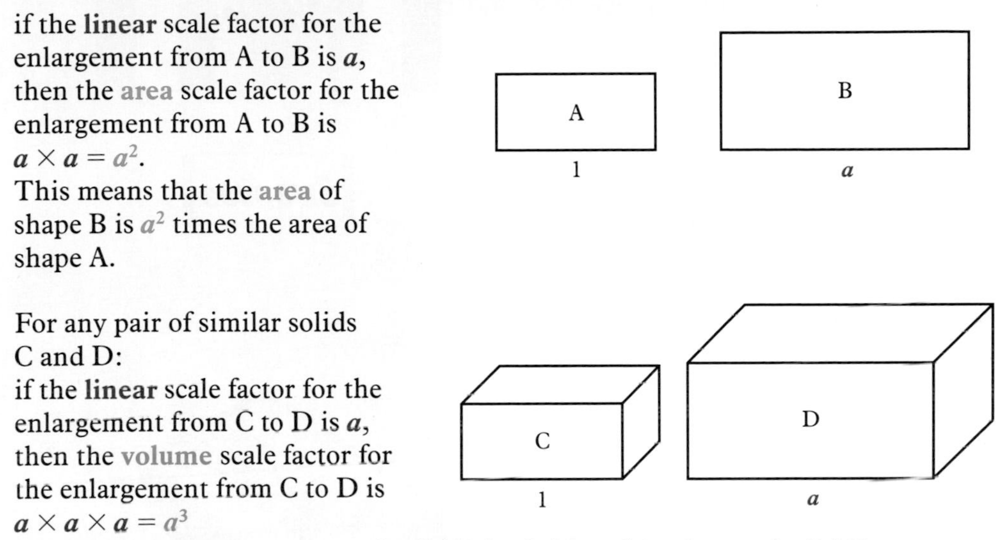

Example Explain why these diagrams are misleading.

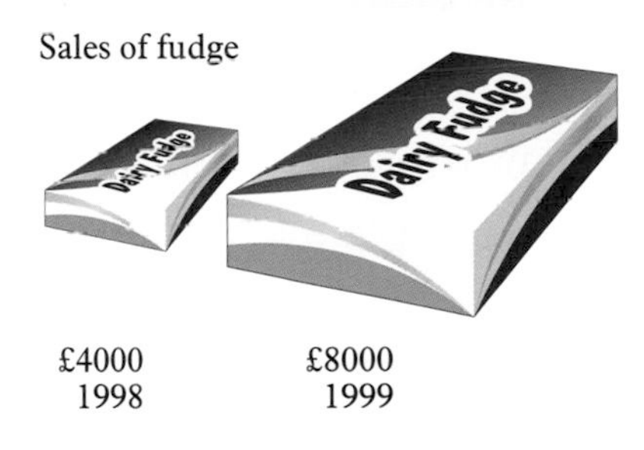

The sales have doubled so the lengths of the first box have been multiplied by 2 to give the second box.

This has the effect of multiplying the volume by $2 \times 2 \times 2 = 8$.

So the volume of the second box is 8 times that of the first box. When you look at the diagrams you can't help thinking that the sales have much more than doubled.

Exercise 9:4

1 A factory supplies scarves to two shops, one in Chester and one in Swansea.

The Chester shop sells twice as many scarves as the Swansea shop.

The diagram shows this data.
Why is it misleading?

2 A supermarket sells two types of special tea.
One is Earl Grey and the other Lapsang Souchong.

It sells three times as much Earl Grey as Lapsang Souchong.

The diagram shows the data.
Why is it misleading?

3 London had twice as many sunny days as Paris in July of 1999.

Pip has drawn this diagram to show this information.

Paris

London

Why is her diagram misleading?

4 A shop sells Welsh butter.
Its sales of this butter have increased by 100% between 1998 and 1999.

They have drawn this diagram to show this information.

Explain why the diagram is misleading.

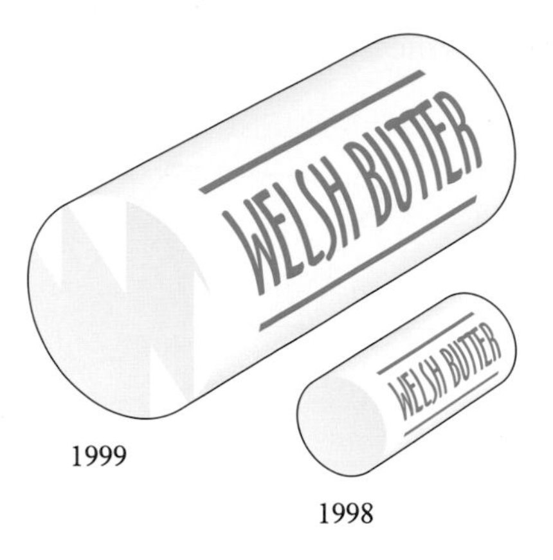

5 A company sells two types of dishwashers, a basic and a premium machine.

It sells three times as many basic machines as premium machines.

a This diagram has been drawn to show this.
Explain why the diagram is misleading.

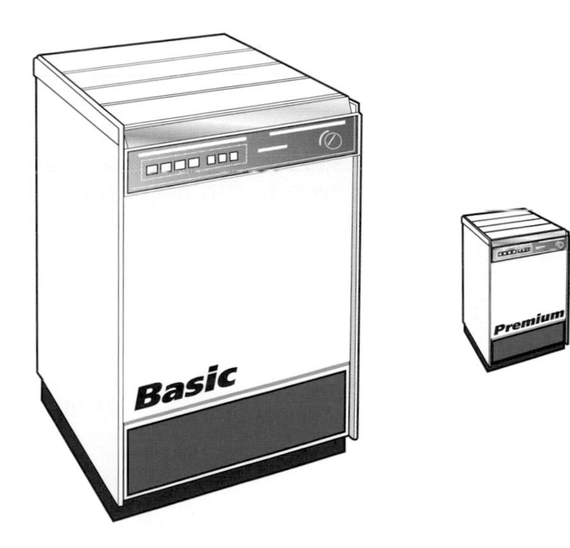

● **b** The sales manager wants cardboard models of the two machines for a sales conference. They must be made in proportion to their sales. The basic model will be three times the volume of the premium one. The height of the premium model must be 50 cm. Calculate the height of the basic model.

1 The measurements on packing cases are given to the nearest centimetre.

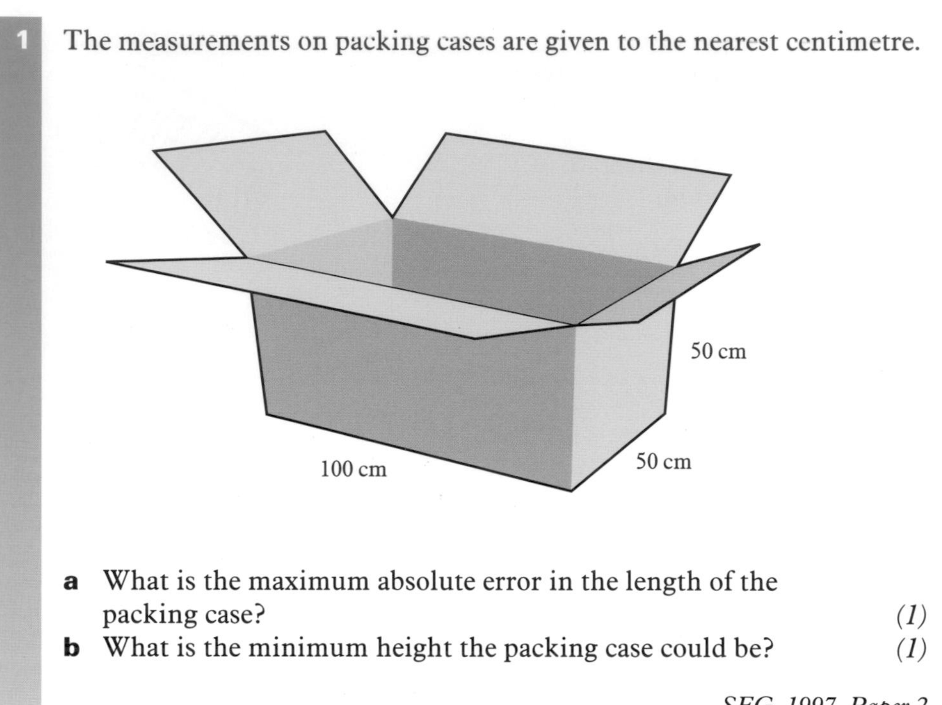

a What is the maximum absolute error in the length of the packing case? *(1)*

b What is the minimum height the packing case could be? *(1)*

SEG, 1997, Paper 2

2 The length of a car is advertised as 444 cm to the nearest centimetre.

a What is the maximum absolute error in this measurement? *(1)*

b What is the shortest length the car could be? *(1)*

A garage stores these cars in rows of ten with exactly 10 cm between each car and exactly 1 m at both ends of each row.

c What is the shortest distance that is needed for each row? *(3)*

SEG, Specimen Papers, Paper 2

3 A pump worked for 1 minute 40 seconds, measured to the nearest 10 seconds.
It pumped 8 litres of water per second, measured to the nearest litre.

a Write down the absolute error in that time. *(1)*

b Calculate the minimum number of litres of water that it pumped. *(3)*

SEG, 1999, Paper 3

4

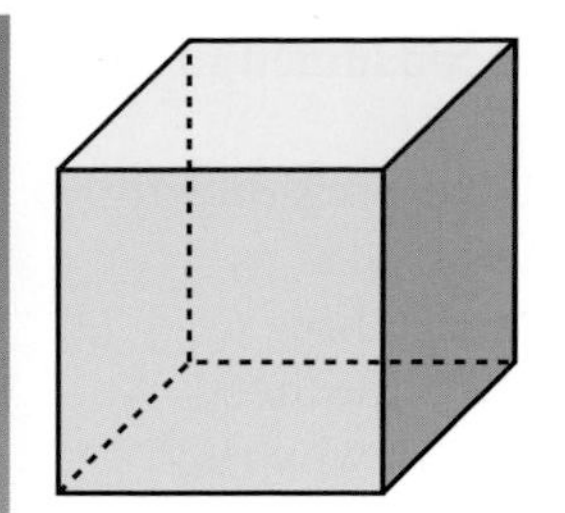

The frame of a cube is made by using thin rods, each of which is 10 cm long.
This measurement is accurate to the nearest centimetre.

a What is the maximum possible length of one of the rods? *(1)*

b What is the maximum possible volume of the cube? *(1)*

c What is the relative error in the volume of the cube? *(3)*

SEG, 1999, Specimen Papers

5 A rocket travels at 61 km per minute for 2.7 minutes. Both measurements are accurate to two significant figures.

a Write down the absolute error in the speed of the rocket. *(1)*

b Find the percentage relative error in the speed of the rocket. *(2)*

c Find the maximum distance that the rocket could have travelled in this time. *(3)*

6 **a** This diagram shows the number of dogs (in millions) owned by people in Britain for the years from 1989 to 1994.

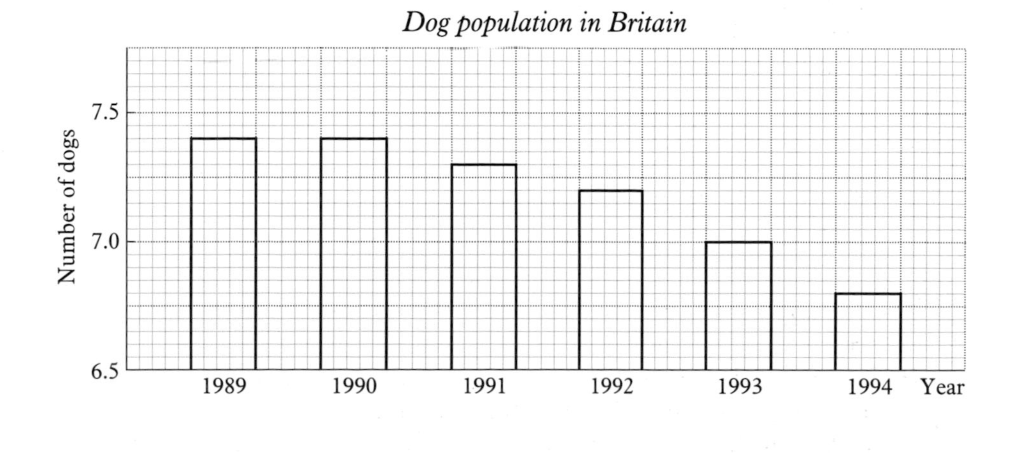

Give **two** reasons why this diagram is misleading. *(2)*

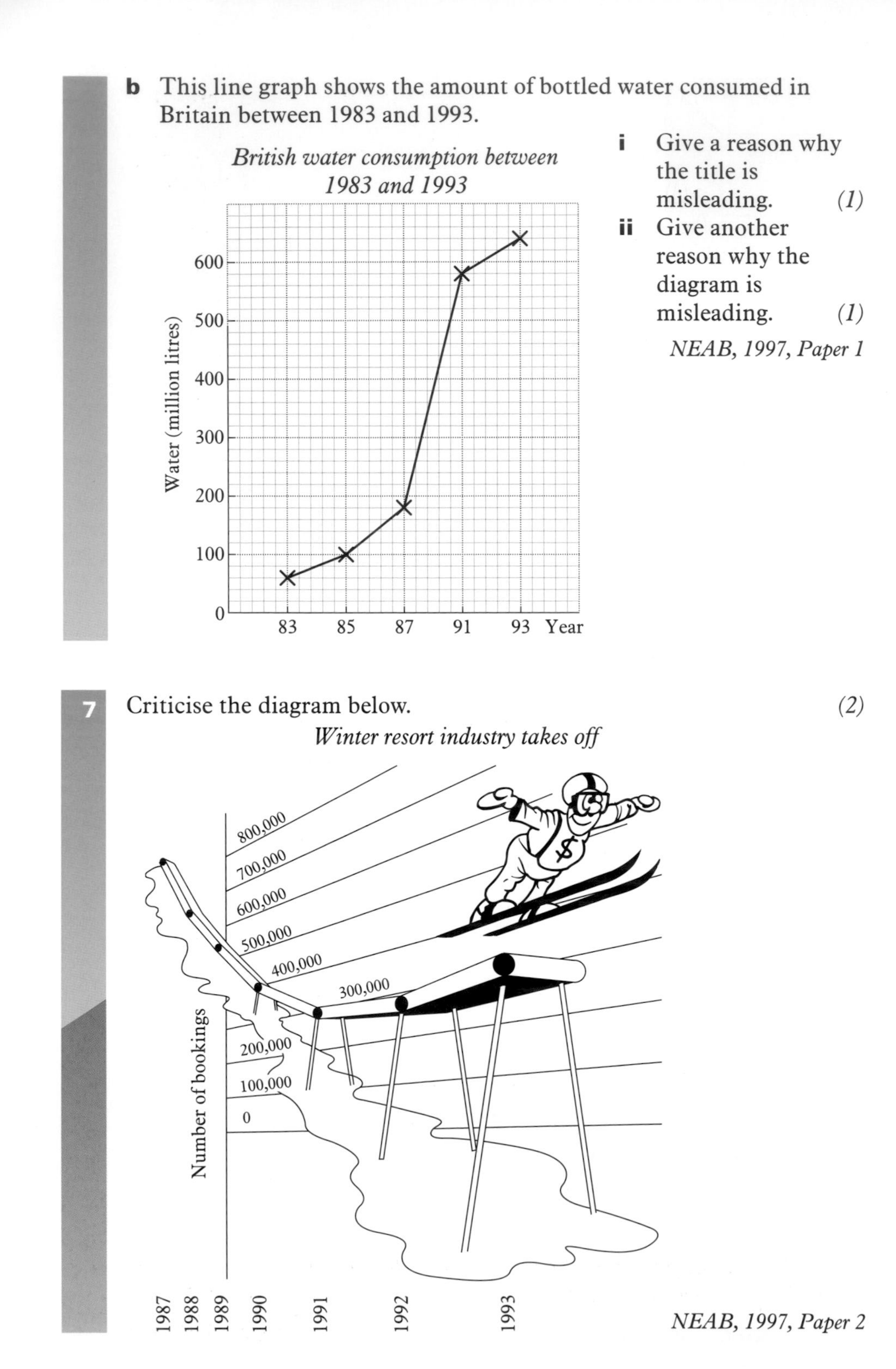

b This line graph shows the amount of bottled water consumed in Britain between 1983 and 1993.

i Give a reason why the title is misleading. *(1)*

ii Give another reason why the diagram is misleading. *(1)*

NEAB, 1997, Paper 1

7 Criticise the diagram below. *(2)*

NEAB, 1997, Paper 2

8 **a** The sets of information **A** and **B**, relate to world population growth.

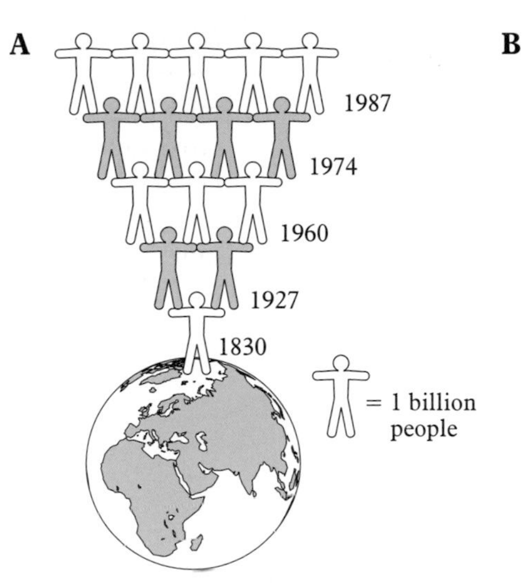

B

Year	Population (billions)
1950	2.55
1960	3.03
1970	3.70
1980	4.45
1990	5.29
1992	5.47
1993	5.56

i Explain why the presentation as shown by diagram **A** could be misleading. *(1)*

ii The population in 1992 is to be represented on diagram **A**. How many symbols should be drawn? *(1)*

b The following diagrams represent different ways in which populations could vary with time.

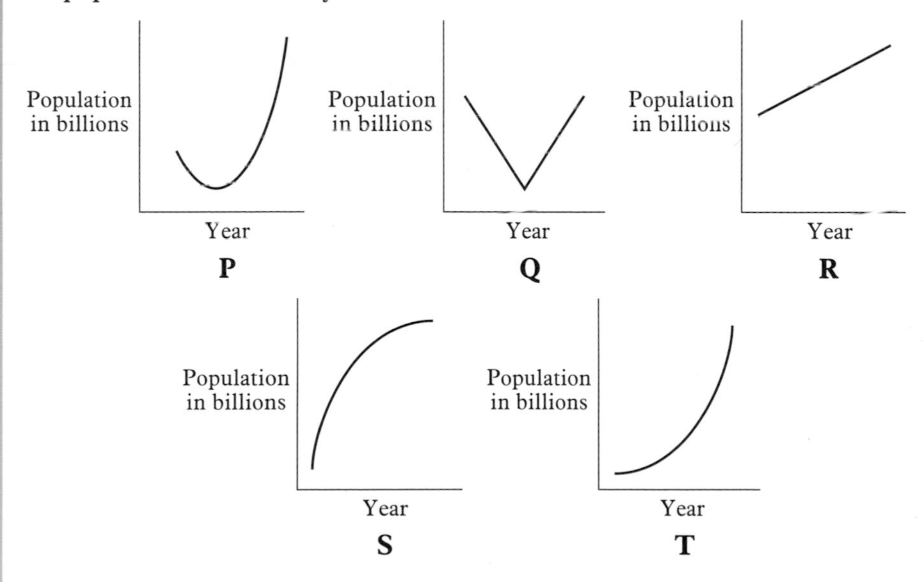

i Which diagram would best represent the years shown in **A**? *(1)*

ii Which diagram would best represent the years shown in **B**? *(1)*

SEG, 1996, Paper 1

9 The diagram below shows the dividend of an American company in dollars.
Give **two** criticisms of this diagram. *(2)*

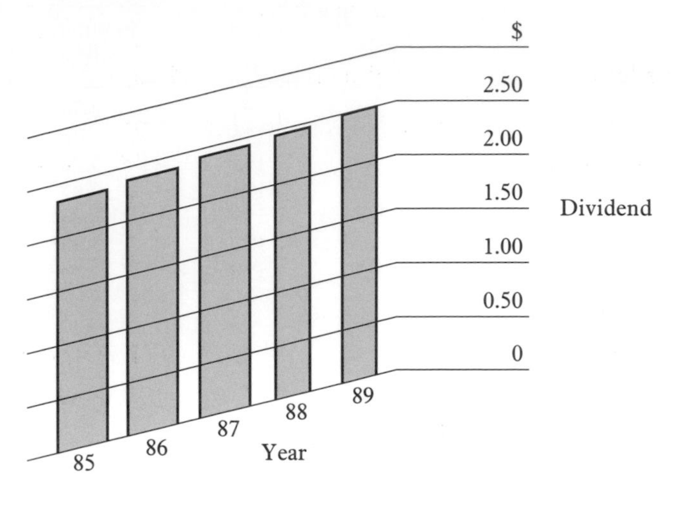

NEAB, 1998, Paper 2

10 A car manufacturer produces two makes of car, Rodeo and Siesta.
They sell twice as many Rodeos as Siestas.

a The diagrams have been drawn to represent this information. Explain why the diagrams are misleading. *(1)*

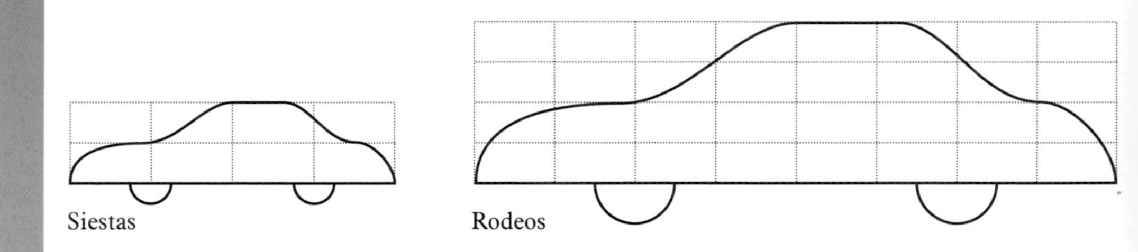

The manufacturer decides to make models of the two cars in proportion to their sales.
The model of the Rodeo will be twice the volume of the model of the Siesta.

The model of the Siesta will be 4 cm long.

b Calculate the length of the model of the Rodeo. *(3)*

SEG, 1998, Paper 4

Parim measures the length of her letter box as 29 cm to the nearest centimetre.

Lower bound 28.5 is called the **lower bound**.

Upper bound 29.5 is called the **upper bound**.

The length of the letter box must lie in the range 29 ± 0.5 cm.

Absolute error The number 0.5 is called the **absolute error**. It is also sometimes called the maximum absolute error or the maximum possible error.

Relative error The **relative error** gives you a better idea of the size of an error.

$$\text{Relative error} = \frac{\text{error}}{\text{exact value}}$$

Percentage relative error The **percentage relative error** is the relative error given as a percentage.

$$\text{Percentage relative error} = \frac{\text{error}}{\text{exact value}} \times 100$$

These are some ways in which diagrams can be misleading:

- Scales do not start at zero
- The scale is not linear
- Lack of labelling on axes
- Use of shadows
- Non-linear time scales
- Different sizes of gaps or bars
- Misleading title
- Lines on the graph drawn too thick
- Incorrectly plotted points
- Use of sloping lines

For any pair of similar shapes A and B:

if the **linear** scale factor for the enlargement from A to B is $\boldsymbol{a}$, then the area scale factor for the enlargement from A to B is $\boldsymbol{a} \times \boldsymbol{a} = a^2$.

This means that the area of shape B is a^2 times the area of shape A.

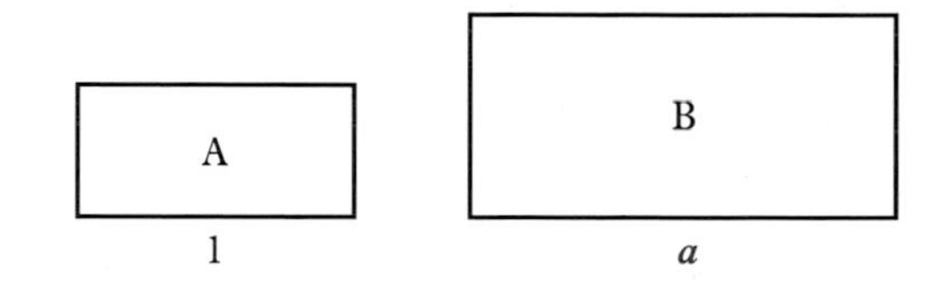

For any pair of similar solids C and D:

if the **linear** scale factor for the enlargement from C to D is $\boldsymbol{a}$, then the volume scale factor for the enlargement from C to D is $\boldsymbol{a} \times \boldsymbol{a} \times \boldsymbol{a} = a^3$

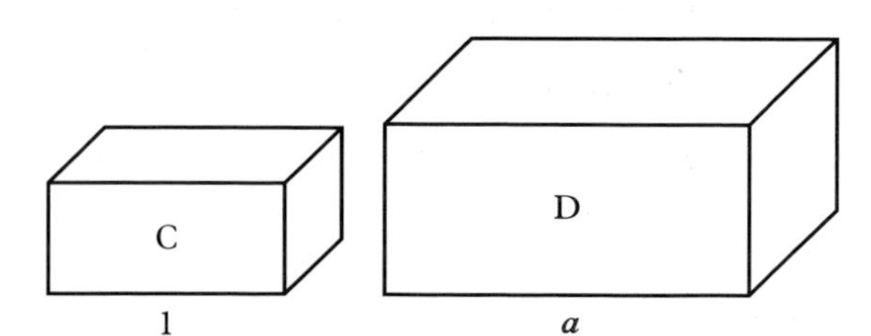

This means that the volume of solid D is a^3 times the volume of solid C.

Working with *Excel*

Misleading diagrams

Bianca runs a market stall. She wants to sell the stall and go to college. These are her sales for the last six months.

Month	February	March	April	May	June	July
Sales in £	4890	4910	4935	4950	5000	5020

Use *Excel* to draw a **bar-chart** to help Bianca sell her stall.

 **Open a new *Excel* document.**

- **Enter the information on Bianca's sales.**
  In Cell A1, type **Month**.
  In Cell A2, type **February**, in Cell A3, type **March** and so on.
  In Cell B1, type **Sales in £**.
  In Cell B2, type **4890**, in Cell B3, type **4910** and so on.

- **Select all the cells you have used.**

- **Use the Chart Wizard button.**

- **Drawing a bar-chart.**
  From **Chart type**, choose **Column**.
  From **Chart sub-type**, choose the top left diagram.

- **Labelling the bar-chart.**
  Click the **Next >** button until you get to the **Chart Options** window.
  In the **Chart title** box, type **Bianca's Sales**.
  In the **Category (X) axis** box, type **Month**.
  In the **Value (Y) axis** box, type **Sales in £**.
  Your screen should look like the one shown. Click **Finish**.
  You can now print this bar-chart.

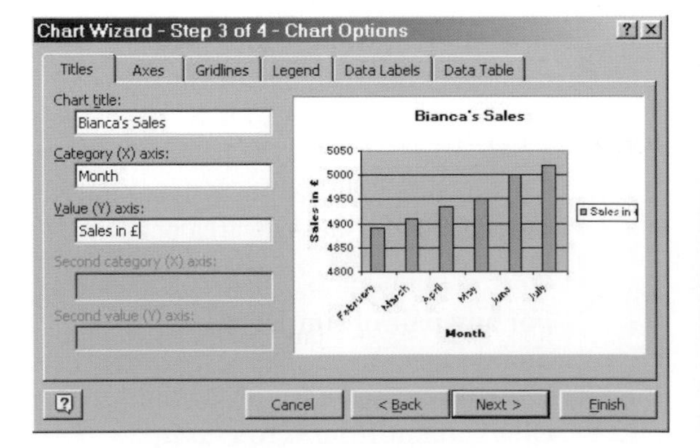

The vertical axis does not start at 0.
Bar-charts compare the heights of the bars.
This is not a representative diagram. The July sales appear to be about twice the March sales.

Changing the vertical axis to start at 0.

Click on a blank area of the bar-chart to select it.
Place the cursor on one of the numbers on the vertical axis.
Double-click.
In the **Format Axis** window, click **Scale**.
In the **minimum** box, enter **0**.
Your screen should look like this.
Click **OK**.
Click on a blank area of the bar-chart to select it.
You can now print it.

Format Axis
Patterns | Scale | Font | Number | Alignment
Value (Y) axis scale
Auto
☐ Minimum: 0
☑ Maximum: 5050
☑ Major unit: 50
☑ Minor unit: 10
☑ Category (X) axis
Crosses at: 4800
☐ Logarithmic scale
☐ Values in reverse order
☐ Category (X) axis crosses at maximum value
OK | Cancel

This bar-chart appears to show a different relationship for Bianca's sales each month. The July sales are very close to the March sales. It is a representative diagram.

Change the vertical axis to start at 4885.

Click on a blank area of the bar-chart to select it before printing it.

This is not a representative diagram.
The July sales appear to be several times the March sales.
This is the bar-chart Bianca would use to sell her stall.

You should always check the values on the axes of a graph.

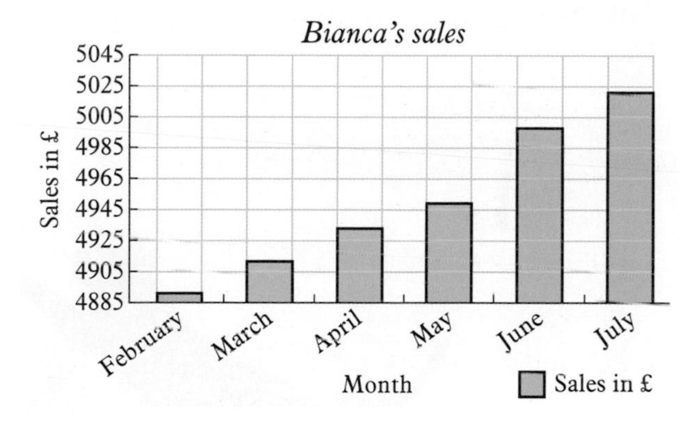

Exercise 9:5

1 These are the sales figures for Weldham Motors for the first six months of the year.

Month	Jan	Feb	Mar	Apr	May	Jun
Cars sold	72	73	75	78	80	85

a The owner wants to sell the company.
Use *Excel* to draw a bar-chart to help the owner sell Weldham Motors.

b The buyer is trying to get the price for Weldham Motors reduced.
Use *Excel* to draw a bar-chart to help the buyer.

1 The sides of this box have been measured to the nearest centimetre.

a What is the maximum possible error in each measurement?

b Write down the minimum height that the box could be.

c Find the maximum area that the front face of the box could be.

d Find the absolute error in the area of the front face.

e Find the maximum possible volume of the box.

f Find the relative error for the volume of the box.

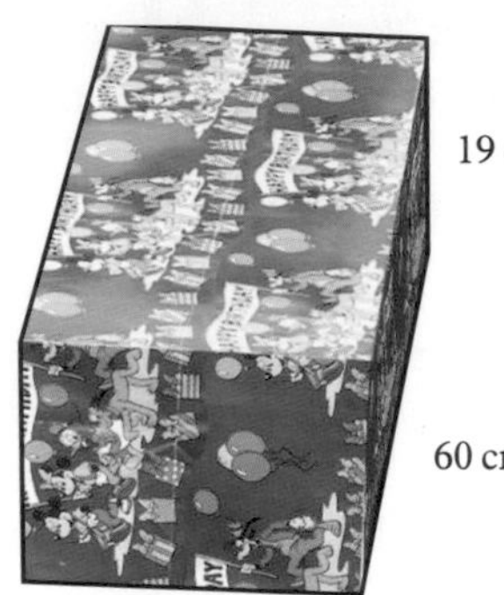

2 Explain how this graph is misleading.

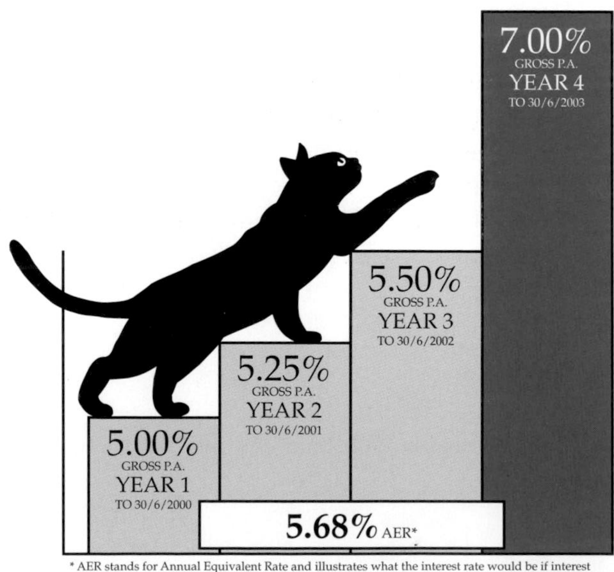

10 Time series

1 Looking for trends
Time series
Trend lines

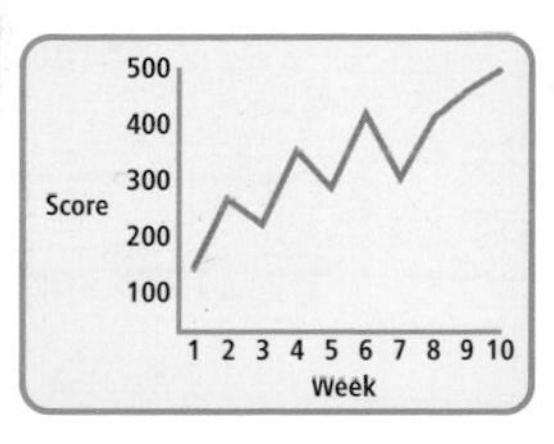

2 Moving averages
Working out moving averages
Seasonal variation
Using moving averages to draw trend lines
Using the trend line to estimate future values
Mean seasonal effect

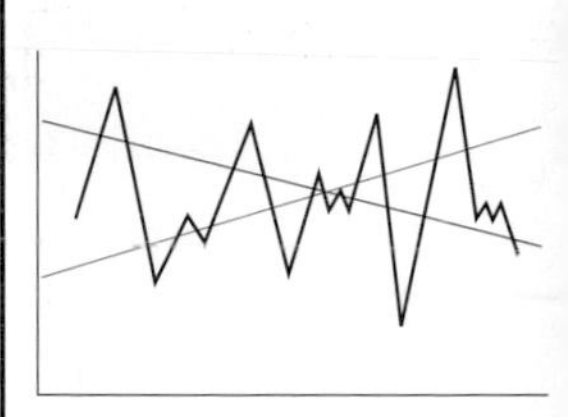

EXAM QUESTIONS

SUMMARY

ICT IN STATISTICS

TEST YOURSELF

1 Looking for trends

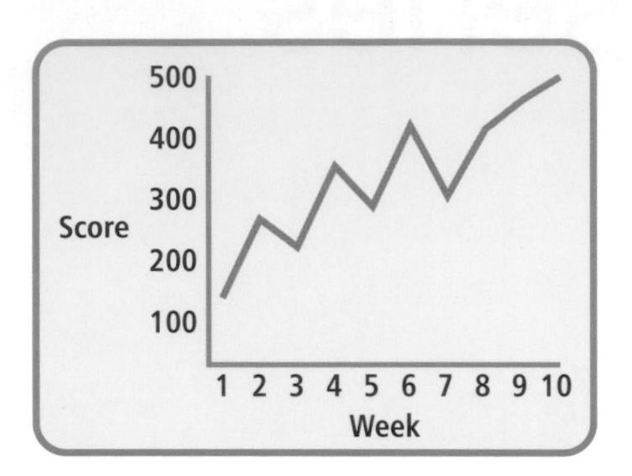

Mike was given a new electronic game 10 weeks ago. Mike recorded his best score for each week and plotted the points in the graph. He uses the graph to predict what score he should get in 5 weeks time. He has decided to try and beat that score.

Statisticians make predictions based on what has happened in the past.

Time series

A **time series** is made up of numerical data recorded at intervals of time. This data is presented in the form of a graph.
The time intervals are plotted on the horizontal axis. These time intervals are the same length. They could, for example, be hours, weeks, months, seasons or years.

These are examples of time series graphs.

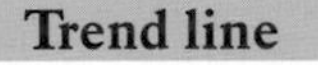

Trend line

In each graph the red line shows the trend. It is called the **trend line** and is drawn by eye.

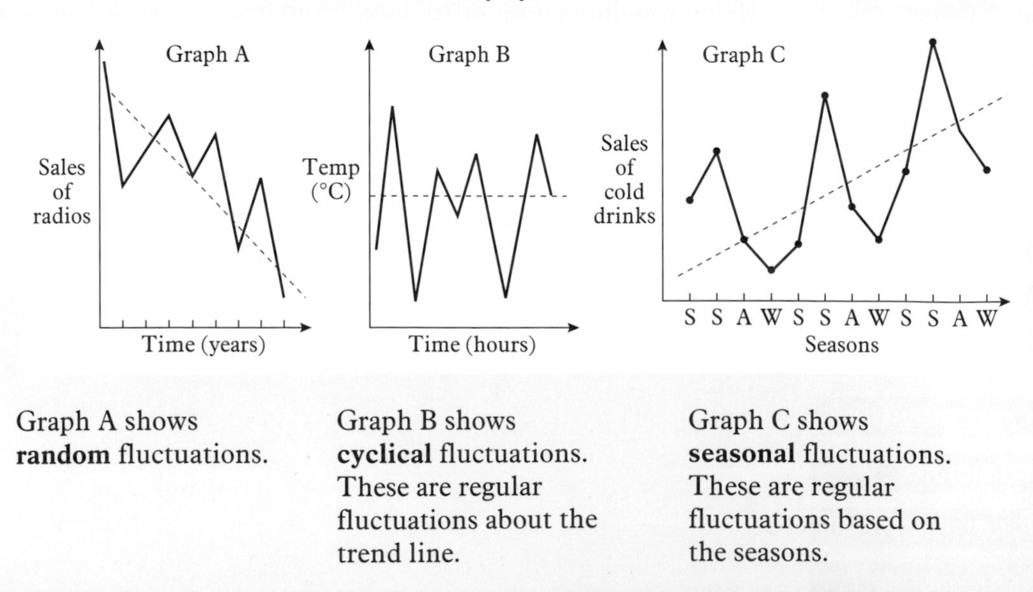

Graph A shows **random** fluctuations.

Graph B shows **cyclical** fluctuations. These are regular fluctuations about the trend line.

Graph C shows **seasonal** fluctuations. These are regular fluctuations based on the seasons.

Exercise 10:1

1 Write down the type of fluctuation you would expect for each of these. Choose from random, cyclical or seasonal.

a Sales of solid fuel.

b Sales of umbrellas.

c Takings in a restaurant.

2 The diagram shows the number of people using Bert's Travel Company in three-monthly periods from 1995–1999.

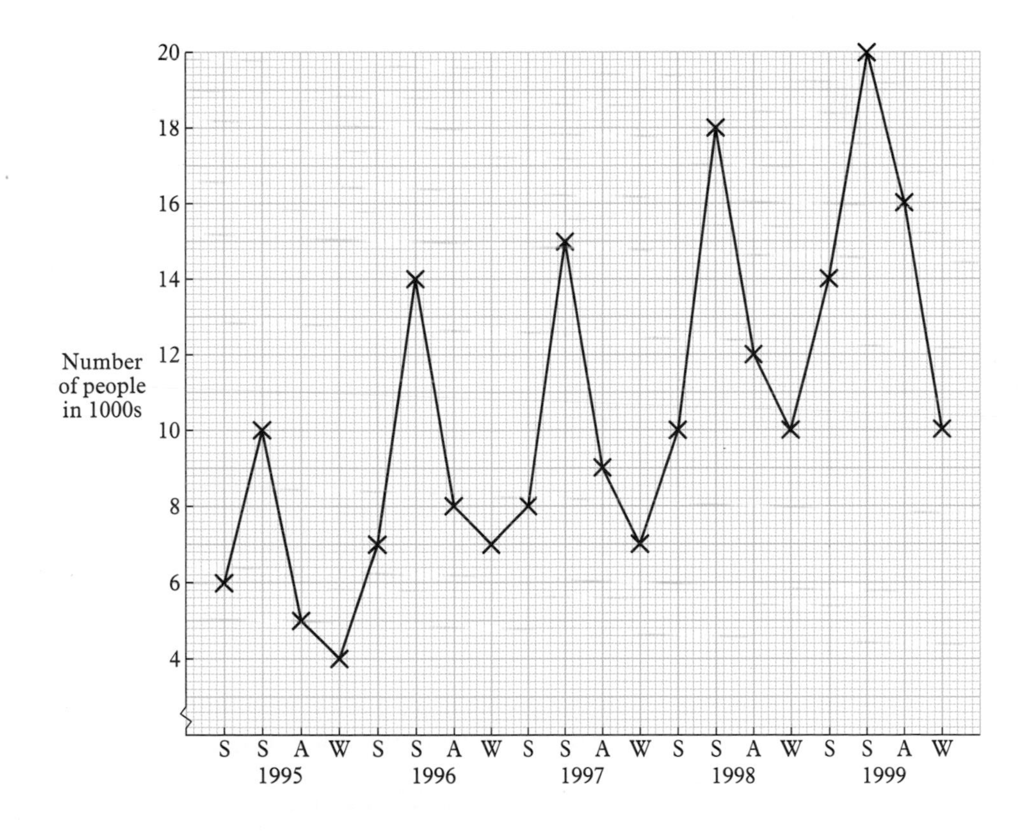

a Describe two features of the diagram.

b In summer 1998, eighteen thousand people used the travel company. How many people used the travel company in Summer 1997?

c How many people used the company in Spring 1996?

d How many people altogether used the company in both Autumn and Winter of 1998?

e Find the percentage increase in the number of people from Summer 1995 to Summer 1998.

3 Serena's company makes tennis racquets.
The table shows the number of racquets her company produced from 1986 to 2000.

Year	1986	1988	1990	1992	1994	1996	1998	2000
Number (100s)	51	80	103	145	180	185	240	270

a Copy these axes onto graph paper.
b Plot the data given in the table.
Join the points with straight lines.
c Draw a trend line on your graph.
d Use your graph to estimate:
(1) the number of racquets made in 1997
(2) the number of racquets that will be made in 2002.
e Which of the two estimates in part **d** will be the least reliable?
Explain your answer.

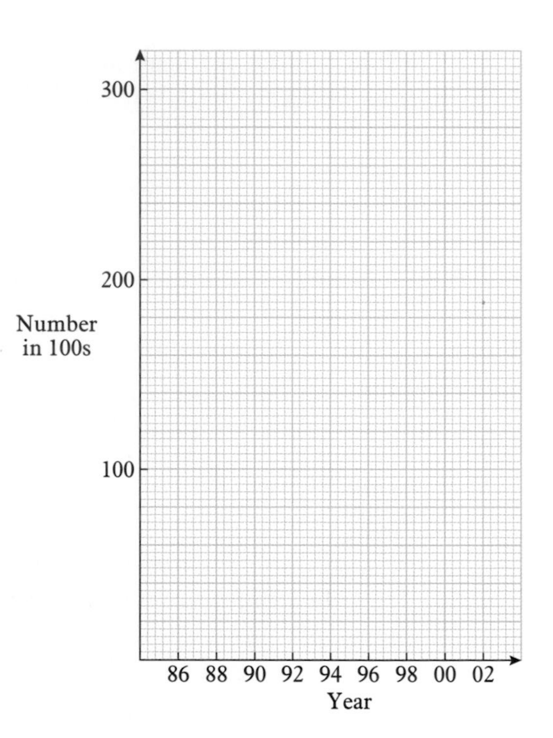

4 The table gives the number of people using James' Wedding Car Service for a five-year period. The data is given in three-monthly periods.

Year	Quarter	Number	Year	Quarter	Number
1996	1st	5	1998	1st	10
	2nd	11		2nd	16
	3rd	19		3rd	29
	4th	12		4th	12
1997	1st	8	1999	1st	9
	2nd	14		2nd	18
	3rd	23		3rd	34
	4th	13		4th	16

a Copy these axes onto graph paper.

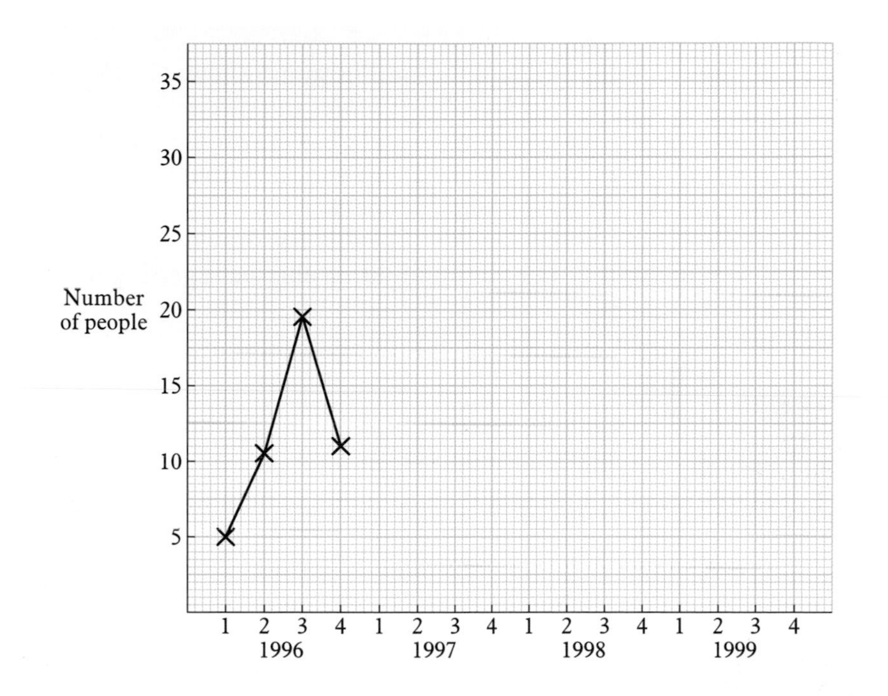

b The points for 1996 have already been plotted. Complete the graph.
c Describe two features of your graph.
d Draw a trend line on your graph.
e Describe the seasonal variation. Explain the cause of the variation.

5 The table gives the sales, in millions of pounds, for a new company in its first eight years of trading.

Year	1	2	3	4	5	6	7	8
Sales (£m)	3.4	7.1	7.8	11.3	14.9	19.2	23.4	29

a Draw a graph to show this data.
b Draw a trend line on your graph.
c Use your trend line to predict the sales in the 10th year.
d How reliable is your answer to part **c**?

2 Moving averages

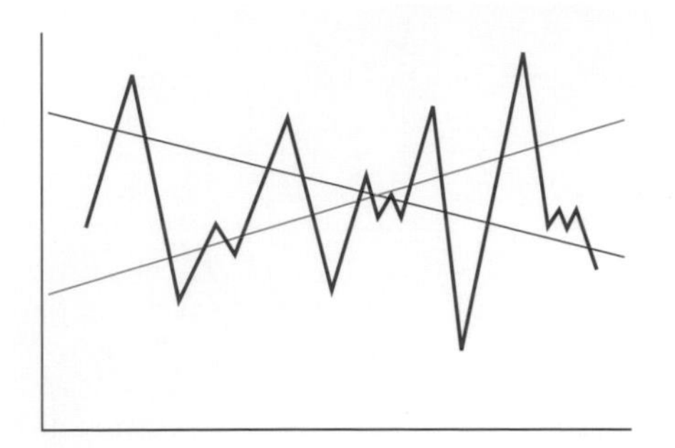

The graph shows Gary's weekly scores in cricket.
He has drawn the blue trend line to show that he is getting better.
Paula has drawn the red line to show that he is getting worse!

Trend lines can sometimes be very difficult to draw.
You can use a series of different average values to help you draw a trend line.

Moving average

Moving averages are averages worked out for a given number of items of data as you work through the data.

A three-point moving average uses 3 items of data at a time.
A four-point moving average uses 4 items of data and so on.

Example Work out the three-point moving average for this data.

3, 5, 7, 15, 23, 34

Take the first three values. 3, 5, 7
Work out the average. $(3 + 5 + 7) \div 3 = 5$

Now miss out the 1st value and use the 15. 5, 7, 15
Work out the average. $(5 + 7 + 15) \div 3 = 9$

Now miss out the 2nd value and use the next value. 7, 15, 23
Work out the average. $(7 + 15 + 23) \div 3 = 15$

Miss out the 3rd value and include the last value. 15, 23, 34
Work out the average. $(15 + 23 + 34) \div 3 = 24$

The three-point moving average is the series of values: 5, 9, 15, 24.

Exercise 10:2

1 Work out the three-point moving average for this data.

8, 9, 4, 15, 14, 7, 12

2 Work out the four-point moving average for this data.

5, 10, 11, 14, 13, 22, 27, 26

3 Work out the five-point moving average for this data.

2, 5, 6, 3, 9, 7, 15, 21, 8

4 Work out the six-point moving average for this data.

5, 4, 8, 7, 9, 9, 5, 10, 14, 13, 15, 3, 11, 4

5 John has recorded the numbers of people using the Barton train each day for the last 15 days. This is his data.

45, 31, 50, 36, 48, 38, 46, 39, 49, 56, 51, 44, 39, 56, 53

Work out the five-point moving average for John's data.

6 Penny has written down the number of books borrowed from the school library on 20 consecutive school days. This is her data.

12, 27, 18, 15, 20, 33, 26, 21, 17, 13, 20, 22, 30, 25, 15, 17, 9, 19, 28, 36

Giving your answer correct to 1 d.p., find the moving average:

a of order 4

b of order 5.

7 Liam has written down the money he earned each month for last year. Here are his amounts in pounds (£):

134, 165, 148, 204, 256, 172, 155, 142, 199, 178, 120, 133

Find a four-point moving average for Liam's data. Give your answer to the nearest penny.

Moving averages show the general trend. They eliminate seasonal variation. You can use a table to help work out moving averages.

Example The table shows the profits, in millions of pounds, of a company over three years.

	1997	1998	1999
1st quarter	138	126	130
2nd quarter	151	167	155
3rd quarter	228	240	248
4th quarter	87	103	97

Since there are four quarters it is appropriate to use a four-point moving average. You set up a new table.

Year	Quarter	Profit	Four-point moving average
1997	1	138	
	2	151	
			151
	3	228	
			148
	4	87	
			152
1998	1	126	
			155
	2	167	
			159
	3	240	
			160
	4	103	
			157
1999	1	130	
			159
	2	155	
			$157\frac{1}{2}$
	3	248	
	4	97	

Moving averages can be used to help you draw a trend line.

You plot each moving average on your graph.

The first value covers the first four quarters. You plot **151** in the middle of these quarters. It is plotted half-way between the 2nd and 3rd quarters.

The second value covers quarters 2, 3, 4, 1 so you plot **148** between the 3rd and 4th quarters.

The third value covers quarters 3, 4, 1, 2 so you plot **152** between the 4th and 1st quarters.

Continue in this way until you have plotted all values of the moving average.

Draw a line of best fit through these points.

This is the trend line for the data.

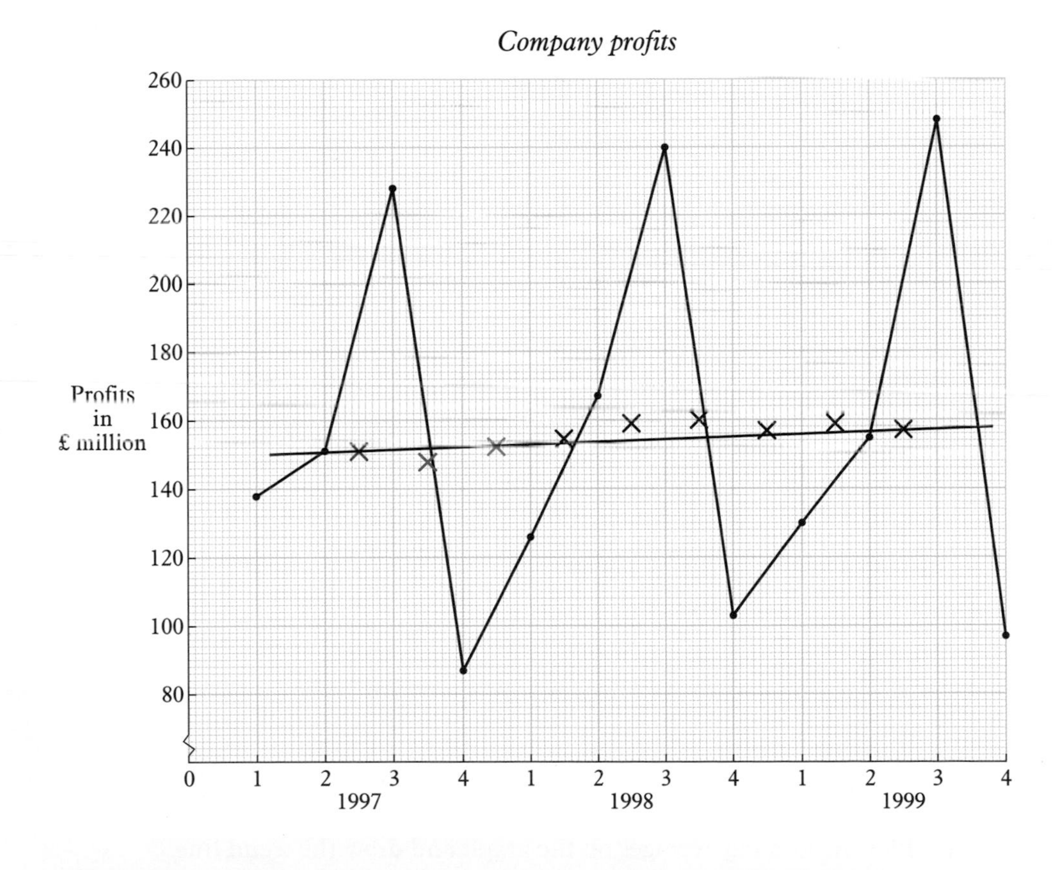

Exercise 10:3

1 The table shows the number of computers a shop sold each quarter over the last three years.

	1998	1999	2000
1st quarter	315	340	351
2nd quarter	571	590	592
3rd quarter	446	470	491
4th quarter	963	989	

a Draw a graph to show this data.

b Use a table to calculate the four-point moving average for the data.

c Plot the moving averages on the graph and draw the trend line.

d Why is it difficult to use your trend line to predict the number of computers sold in the 4th quarter of year 2000?

2 The table gives the annual profit, in thousands of pounds, of Sam's business for the last 15 years.

Year	Profit
1985	18
1986	24
1987	25
1988	21
1989	17

Year	Profit
1990	25
1991	23
1992	22
1993	28
1994	27

Year	Profit
1995	29
1996	24
1997	27
1998	30
1999	29

a Draw a graph to show this data.

b Use a table to calculate the five-point moving average for the data.
You plot the first five-point moving average at 1987, the second at 1988 and so on.

c Plot the moving averages on the graph and draw the trend line.

d Use your trend line to predict the profit in the year 2000.

Example This graph shows a seasonal fluctuation.

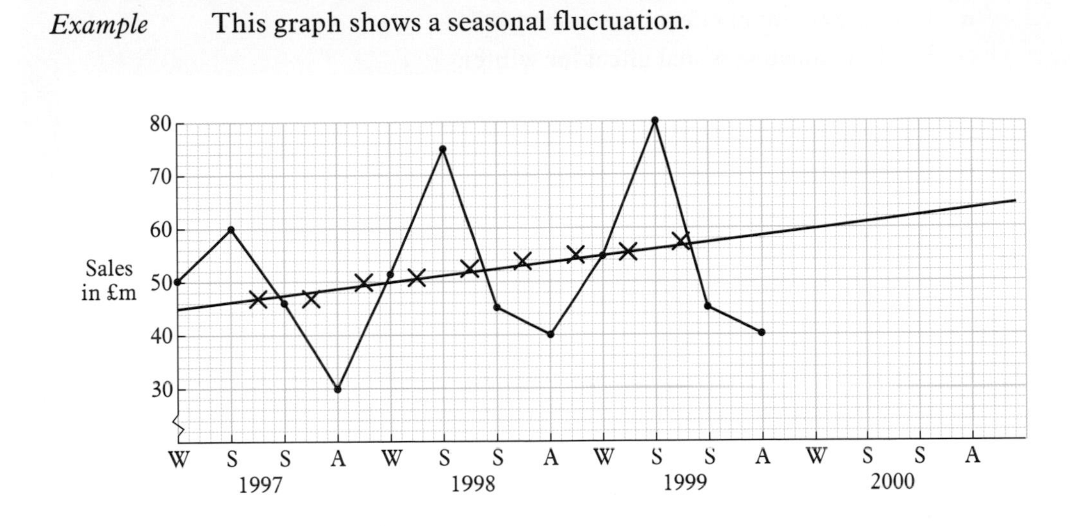

The effect of the season can be estimated at any point on the graph.
Look at the point for autumn of 1997.
The observed value of the graph is £30 million.
The value from the trend line is £48.5 million.
The estimated seasonal effect is **observed value − trend value** = £30 − £48.5 million
= −£18.5 million

The estimated seasonal effect on sales in autumn 1997 is a decrease of £18.5 million from the general trend.

You can also estimate the mean seasonal effect for the autumn quarter.
Work out the seasonal effect for the autumn of each year and then find the mean.

Autumn, Year	Seasonal effect
1997	30 − 48.5 = −18.5
1998	40 − 54 = −14
1999	40 − 59 = −19

The mean of the autumn seasonal effect is −(18.5 + 14 + 19) ÷ 3 = −£17 (2s.f.)
This is a decrease of £17 million from the general trend.

You can use this value to predict the sales for autumn 2000.
The trend line gives a value of £64 million.
You now include the mean seasonal effect.
The predicted sales total is £64 − £17 = £47 million.

3 Use the graph in the example to answer this question.

a Find the seasonal effect for winter 1997.

b Find the mean seasonal effect for winter.

c Predict the sales for the winter of 2000.

d Find the seasonal effect for spring 1998.

e Find the mean seasonal effect for spring.

f Predict the sales for the spring of 2000.

4 Use the graph that you drew in question **1** to answer this question.

a Find the seasonal effect for the 4th quarter in 1998.

b Find the mean seasonal effect for the 4th quarter.

c Predict the sales in the fourth quarter of 2000.

d Find the seasonal effect for the 1st quarter in 1999.

e Find the mean seasonal effect for the 1st quarter.

f Predict the sales in the first quarter of 2001.

5 George wants to predict his gas bills for next year.
He pays four times a year.
The table shows his last 12 quarterly bills in pounds (£).

	Winter	Spring	Summer	Autumn
1998	62	46	35	50
1999	73	51	36	56
2000	78	59	40	57

a Draw a graph to show this data.

b Calculate an appropriate moving average for this data.

c Plot the moving averages on the graph and draw the trend line.

d Find the average seasonal effect for:
(1) winter (2) spring (3) summer (4) autumn.

e Predict George's four bills for 2001.
Show your working.

1 The diagram shows the number of passengers travelling on all UK airline services in three-month periods from 1971–1979.

UK airline passengers

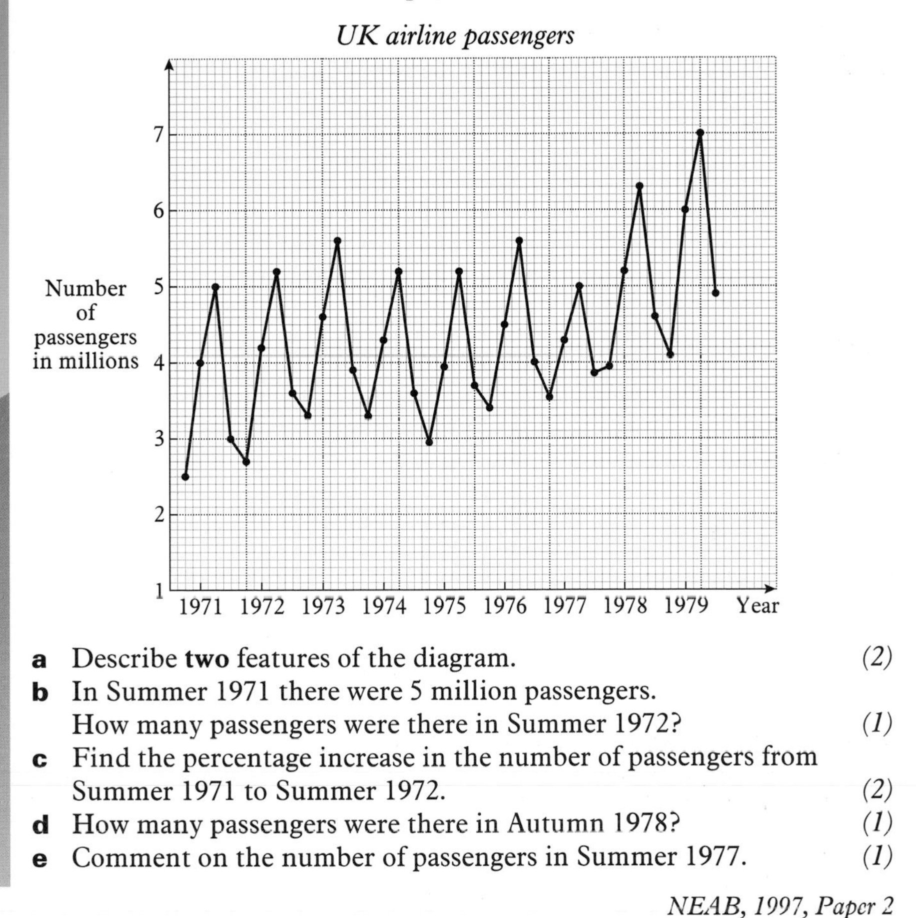

a Describe **two** features of the diagram. *(2)*

b In Summer 1971 there were 5 million passengers.
How many passengers were there in Summer 1972? *(1)*

c Find the percentage increase in the number of passengers from Summer 1971 to Summer 1972. *(2)*

d How many passengers were there in Autumn 1978? *(1)*

e Comment on the number of passengers in Summer 1977. *(1)*

NEAB, 1997, Paper 2

2 The table shows the number of cars produced in Britain, in the years shown, from 1984 to 1996.

Year	1984	1986	1988	1990	1992	1994	1996
Production of cars (thousands)	900	1000	1250	1290	1290	1500	1690

a Plot this information on graph paper. *(3)*

b Draw a trend line on your graph. *(1)*

c Use your graph to estimate:

i the number of cars produced in 1985 *(1)*

ii the number of cars that will be produced in 1998. *(1)*

NEAB, 1998, Paper 1

3 Leena wanted to estimate how much her next electricity bill might be.
She found her last 11 bills.
The table shows her last 11 quarterly bills in pounds (£).

	Spring	Summer	Autumn	Winter
1992	95	60	110	155
1993	108	66	120	170
1994	110	80	135	

These data have already been plotted on the graph.

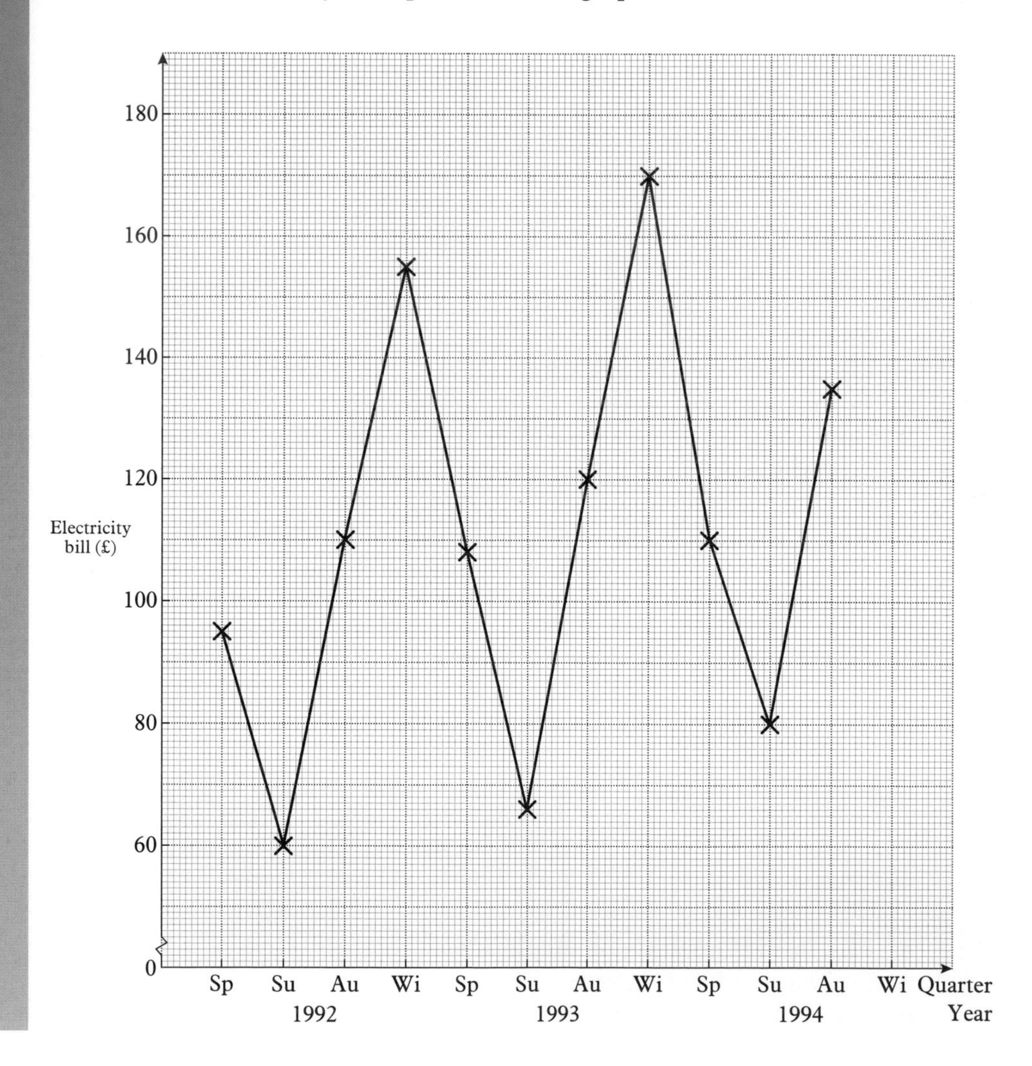

a Copy the table. Calculate the four-point moving averages for these data and enter them in the table. *(3)*

	Quarterly Bill (£)	Four-point moving average
Spring 92	95	
Summer 92	60	
Autumn 92	110	
Winter 92	155	
Spring 93	108	
Summer 93	66	
Autumn 93	120	
Winter 93	170	
Spring 94	110	
Summer 94	80	
Autumn 94	135	

b Copy the graph. Plot the moving averages on the graph and draw the trend line. *(3)*

c Use your trend line to predict the bill for Winter 1994. *(3)*

SEG, 1998, Paper 4

4 Below is a table showing the quarterly electricity bills paid for a school during a period of three years. The Headmistress, in a drive to save money, wishes to display these data in order to show the rising cost over the years.

		YEAR		
		1987	1988	1989
QUARTER	1st	£9000	£9550	£9990
	2nd	£5000	£5450	£6100
	3rd	£4000	£4850	£5350
	4th	£8900	£9850	£10400

a On graph paper, plot the above figures joining the points with straight lines. *(4)*

b Suggest a reason for the seasonal variation shown by your graph. *(1)*

c i Copy the table. Calculate suitable moving averages for these data and use them to complete the table. Plot these values on your graph. *(4)*

Year	Quarter	Bill paid £	
1987	1	9000	
	2	5000	
	3	4000	
	4	8900	
1988	1	9550	
	2	5450	
	3	4850	
	4	9850	
1989	1	9900	
	2	6100	
	3	5350	
	4	10400	

ii What is the purpose in plotting moving averages? *(1)*

d i On your graph, draw a trend line by eye. *(1)*

ii Use your graph to estimate the electricity bill for the school during the first quarter of 1990. *(2)*

SEG, 1999, Specimen Papers

5 The table below shows the number of units of electricity used by a householder during eight successive quarters in 1995 and 1996.

Year	Electricity consumption (units) Quarter 1	Quarter 2	Quarter 3	Quarter 4
1995	1450	1080	730	1280
1996	1630	1220	930	1460

a Draw, on graph paper, a time-series graph to represent the amounts of electricity used over the eight quarters. *(2)*

b **i** What are the main seasonal trends shown by your graph? *(1)*
ii What is the most likely explanation for them? *(1)*

c Copy and complete the following table. *(3)*

Year	Quarter	Consumption	Four-point moving average
	1	1450	
	2	1080	
1995			1135
	3	730	
	4	1280	
	1	1630	
	2	1220	
1996	3	930	
	4	1460	

d Plot the moving averages on your graph. *(2)*

e Draw by eye the trend line on the graph. *(1)*

f The seasonal effects for the quarters 1 and 2 are as follows:

Seasonal effect	
Quarter 1	Quarter 2
+390	−77.5

Use this information and the trend line to provide seasonally adjusted forecasts for Quarter 1 and Quarter 2 of 1997. *(4)*

NEAB, 1997, Paper 3

6 The table shows the amounts, in £1000s, deposited in a bank on each of 12 weekdays before Christmas.

	Mon	Tues	Wed	Thurs	Fri
Week beginning 6/12	250	190	200	215	230
Week beginning 13/12	280	215	235	240	255
Week beginning 20/12	305	245			

a On graph paper, draw a graph and plot the daily amount deposited. *(3)*

b Copy this table. Calculate the five-point moving averages for the data and complete the table. *(4)*

Weekday	Deposits	Moving totals	Five-point moving average
Mon	250		
Tue	190		
Wed	200		
Thur	215		
Fri	230		
Mon	280		
Tue	215		
Wed	235		
Thur	240		
Fri	255		
Mon	305		
Tue	245		

c Explain why a five-point moving average is appropriate. *(1)*

d Plot the moving averages on the graph and draw the trend line. *(2)*

e Calculate the average daily variation for Wednesday. *(2)*

f Use the daily variation in order to estimate the sum deposited on the next Wednesday. *(2)*

SEG, 1997, Paper 4

Time series A **time series** is made up of numerical data recorded at intervals of time. This data is presented in the form of a graph.

Trend line In each graph below, the red line shows the trend. It is called the **trend line** and is drawn by eye.

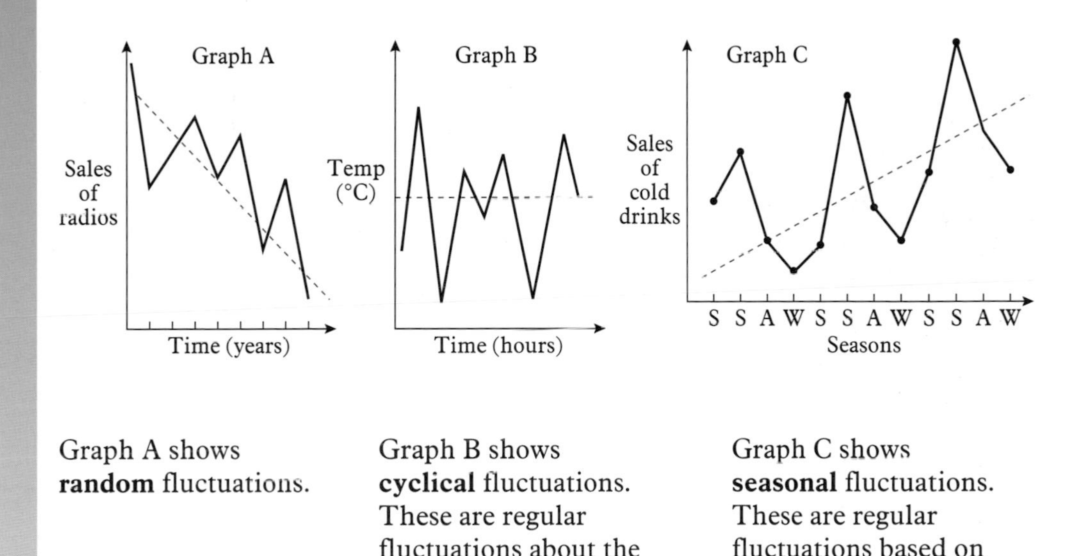

Graph A shows **random** fluctuations.

Graph B shows **cyclical** fluctuations. These are regular fluctuations about the trend line.

Graph C shows **seasonal** fluctuations. These are regular fluctuations based on the seasons.

Moving average **Moving averages** are averages worked out for a given number of items of data as you work through the data.

A three-point moving average uses three items of data each time.
A four-point moving average uses four items of data and so on.

Moving averages show the general trend. They eliminate seasonal variation.

The effect of the season can be estimated at any point on the graph.
The estimated seasonal effect is **observed value − trend value**.

You can also estimate the mean seasonal effect for a given season.
Work out the seasonal effect for each year in the data and then find the mean of these seasonal effects.
You can use seasonal variations to help you predict values.
Read the predicted value from the trend line and then include the average seasonal effect.

Working with *Excel*

Drawing moving averages and trend lines

The table gives the annual profit, in thousands of pounds, of Sam's business for the last 15 years.

Year	Profit	Year	Profit	Year	Profit
1985	18	1990	25	1995	29
1986	24	1991	23	1996	24
1987	25	1992	22	1997	27
1988	21	1993	28	1998	30
1989	17	1994	27	1999	29

Use *Excel* to: **a** calculate the **five-point moving average** for the data
b draw a graph showing the **trend line**.

- **Enter the information on yearly profit into a new *Excel* document.**
  In Cell A1, type Year, in Cell B1, type Profit and in Cell C1, type 5 point Moving Average.
  In Cell A2, type 1985, in Cell A3, type 1986 and in Cell A4, type 1987.
  Select Cells A2, A3 and A4.
  Place the cursor on the bottom right-hand corner of Cell A4.
  Hold down the left mouse button and drag down the column to enter the year up to 1999.

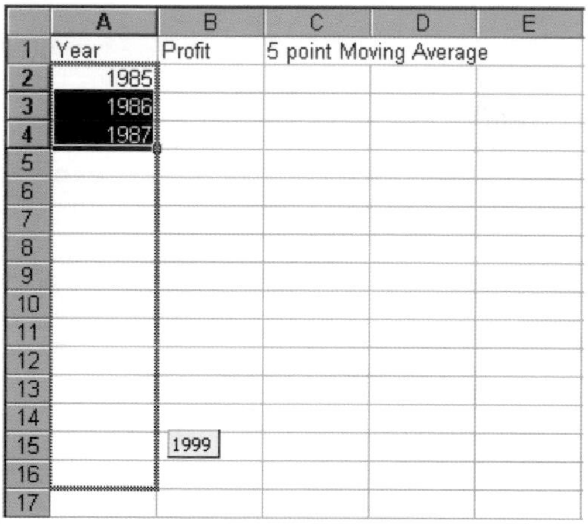

	A	B	C	D	E
1	Year	Profit	5 point Moving Average		
2	1985				
3	1986				
4	1987				
5					
6					
7					
8					
9					
10					
11					
12					
13					
14					
15		1999			
16					
17					

  The yellow box shows when you reach 1999. This copies the pattern into each cell.

  In Cell B2, type 18, in Cell B3, type 24 and so on.

- **Calculate the five-point moving average.**
  In Cell C4, insert the formula: =(**B2**+**B3**+**B4**+**B5**+**B6**)/5 and press the Enter key.
  This calculates the average of the five cells.

  Select Cell C4.
  Place the cursor over the bottom right-hand corner of Cell C4 and drag down to Cell C14.
  This copies the formula for the five-point moving average.

- **Select all the cells you have used.**

- **Use the Chart Wizard button.**

Draw a graph showing a trend line.
From Chart type, choose XY (Scatter).
From Chart sub-type, choose the bottom left diagram.

Label the graph.
Click the Next> button until you get to the Chart Options window.
In the Chart title box, type Yearly Profit.
In the Value (X) Axis box, type Year.
In the Value (Y) axis box, type Profit in £1,000s.
Your screen should look like the one shown. Click Finish.

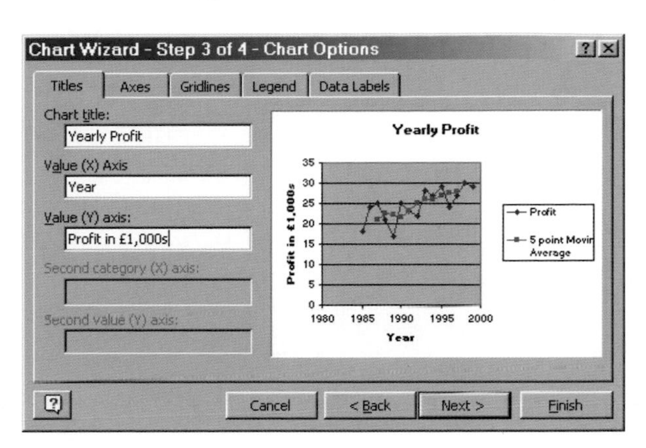

Change the vertical axis to start at 15.

Change the horizontal axis to start at 1985.
Click on a blank area of the graph to select it before printing it.

Exercise 10:4

1 Akuna has kept a record of the money she spends on food each month for the last 12 months. The table shows her data. The amounts are in pounds (£).

Month	Jan	Feb	Mar	Apr	May	Jun	Jul	Aug	Sept	Oct	Nov	Dec
Amount	83	91	88	90	87	93	91	95	93	94	96	105

Use *Excel* to: **a** calculate a three-point moving average for the data
b draw a graph showing the trend line.

2 Larry is a travelling salesman. He keeps a record of the mileage that he does each working day. This is his data.

Day	Mon	Tues	Wed	Thurs	Fri
Mileage	196	250	158	247	191

Day	Mon	Tues	Wed	Thurs	Fri
Mileage	127	156	130	144	129

Use *Excel* to: **a** calculate a three-point moving average for the data
b draw a graph showing the trend line.

1 The graph shows the quarterly sales, in £10 000s, of garden equipment at the Green Fingers Garden Centre over a period of four years.

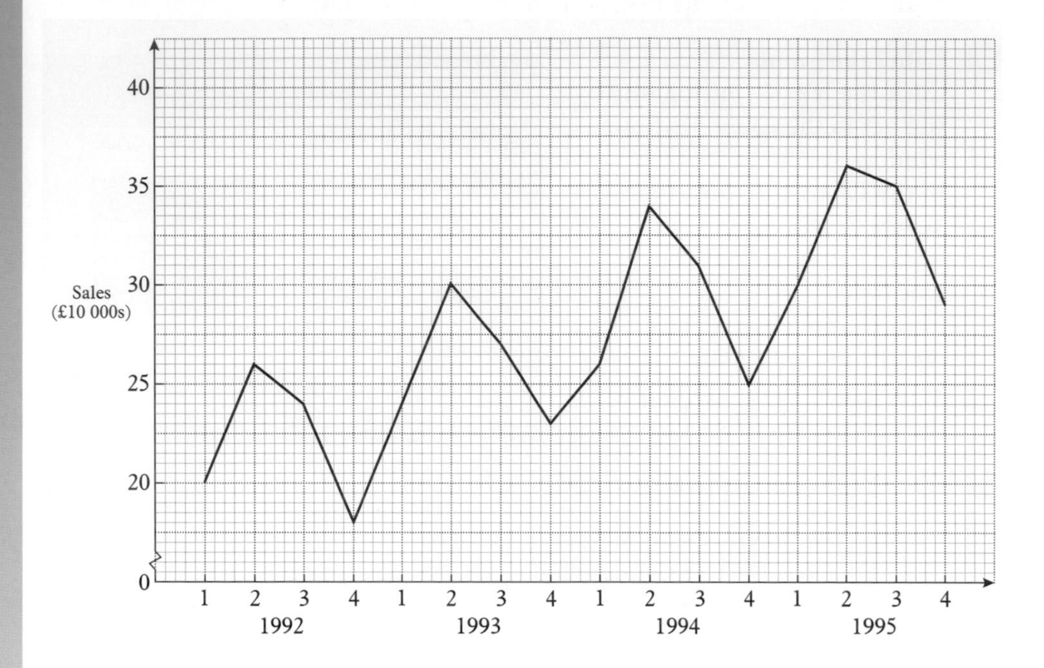

a Suggest a reason for the seasonal variation shown by the graph. *(1)*

b Copy the table. Calculate the four-point moving averages for this data and enter these values. *(3)*

c Copy the graph. Plot these moving averages on the graph. *(2)*

d On your graph, draw a trend line by eye. *(1)*

Year	Quarter	Sales £10 000s	Four-point moving average
1992	1	20	
	2	26	
	3	24	
	4	18	
1993	1	24	
	2	30	
	3	27	
	4	23	
1994	1	26	
	2	34	
	3	31	
	4	25	
1995	1	30	
	2	36	
	3	35	
	4	29	

SEG, 1996, Paper 2

11 Weighting

1 Index numbers
Calculating index numbers
Price index and percentage relative
Using indices to find values
Weighted index numbers
Retail Price Index (RPI)

2 Birth and death rates
Crude death rates
Crude birth rates
Standard population
Standardised death rate

EXAM QUESTIONS

SUMMARY

TEST YOURSELF

1 Index numbers

Max is an estate agent. He is writing a report on the cost of housing in different areas and how the prices have changed over the last three years. Max uses index numbers to show these differences.

Index number An **index number** shows how a quantity changes over a period of time.

Price index The **price index** shows how the price of something changes over a period of time.

The table shows how the price of one house has changed over the years 1997 to 1999.

Year	1997	1998	1999
Price (£)	62 000	66 960	73 780

Max needs to choose a price to compare all the other prices with.
This price is called the **base value**.
He chooses the price in 1997 as the base value.
The year 1997 is called the **base period**.
The base value price **£62 000** is given an index number of **100**.

You use the percentage increase in price to find the index numbers for the other two years.

For 1998 the index number $= \frac{66\,960}{\mathbf{62\,000}} \times 100 = \mathbf{108}$.

The index number 108 shows that the price has increased by 8%.

For 1999 the index number $= \frac{73\,780}{\mathbf{62\,000}} \times 100 = \mathbf{119}$.

The index number 119 shows that the price has increased by 19% since the base period.

The plural of index is indices. The table shows these indices.

Year	1997	1998	1999
Index	100	108	119

Exercise 11:1

1 The table shows the prices for a model of new car for the four years 1996 to 1999.

Year	1996	1997	1998	1999
Price (£)	12 300	12 792	13 407	13 776

a Taking 1996 as the base period, work out the index numbers for all four years. Put your answers in a table.
b Look at the prices in 1997 and 1998 only.
Taking 1997 as the base period, work out an index number for 1998.

2 Rashid bought a computer in 1997 for £1200.
The values of the computer in 1998 and 1999 are £840 and £420.
Taking 1997 as the base period, work out the index numbers for 1998 and 1999. Put your answers in a table.

An index number is sometimes called the **percentage relative**.

Example These are the prices of a house in 1990 and 1998.

Year	1990	1998
Price (£)	54 700	58 529

a Calculate the percentage relative for the house price in 1998.
Use 1990 as the base year with index 100.
b Write down the percentage increase in price.

a The percentage relative for 1998 $= \frac{58\,529}{54\,700} \times 100 = 107$.

b The index 107 shows that the price has increased by 7%.

3 The costs of a wedding in 1994 and 1998 are given in the table.

Year	1994	1998
Cost (£)	6700	7906

a Calculate the percentage relative for the cost of a wedding in 1998. Take 1994 as the base year with index 100.
b Write down the percentage increase in cost.

4 Morgan's garage sells second-hand cars. The average price of cars sold for each of the years 1997 to 2000 is given in the table.

Year	1997	1998	1999	2000
Price (£)	3600	3996	4284	4428

Take 1997 as the base year with an index of 100.
a Express the price in 1998 as an index number.
b Do the same for each of the years 1999 and 2000.

5 The table shows how the workforce of a car factory has changed in the years 1998 to 2000.

Year	1998	1999	2000
Workforce	600	657	570

a Take 1998 as the base year. Express the workforce in each of 1999 and 2000 as an index number.
b Write down the percentage increase in the workforce between 1998 and 1999.
c Find the decimal number that is missing in this sentence:
The workforce increased by a factor of ... between 1998 and 1999.
d Find the percentage decrease in the workforce between 1999 and 2000.

6 The table shows the increase in population of Markton.

Year	1998	1999
Population	5100	5508

a Find the percentage relative for the population in 1999.
Take 1998 as the base year.
b The population increased by 5% between 1999 and 2000.
Find the population in 2000.
c Write down the percentage relative for the population in 2000 taking 1999 as the base year.

The table shows the indices for the cost of Sam's bus ticket for the years 1999 and 2000.

Year	1999	2000
Index	100	120

Sam's ticket cost £1.30 in 1999 so **£1.30** is the base value.
You can use the indices to find the cost of the ticket in 2000.
Let C be the cost of the ticket in 2000. Then

$$\text{Cost of the ticket in 2000} = \mathbf{1.30} \times \frac{120}{100} = £1.56$$

Exercise 11:2

1 The table shows the index for the cost of a tin of biscuits.

Year	1999	2000
Index	100	105

a The tin of biscuits cost £8.80 in 1999.
Find the cost in 2000.

b What does the index number of 105 for 2000 tell you about the cost of a tin of biscuits?

2 The table shows the indices for Pam's house insurance costs.

Year	1997	1998	1999
Index	100	110	145

a The cost of insurance in 1997 was £120.
Find the cost of Pam's insurance in: (1) 1998
(2) 1999.

b Write down the percentage increase in costs between 1997 and 1998.

c Calculate the percentage increase in costs between 1998 and 1999.

d Compare the increase in the costs between 1997 to 1998 and 1998 to 1999.

3 These are the indices for the cost of raw materials and wages in a nursing home.

Year	1990	1995	2000
Raw materials	100	100	125
Wages	100	118	152

a Raw materials cost £78 000 in 1990.
Find the cost of raw materials in 2000.

b The wages bill was £280 000 in 1990.
Find the wages bill for 1995.

c Compare the cost of raw materials in 1990 and 1995.

d Compare the change in the costs of raw materials and wages over the 10-year period.

The table shows the indices for the cost of servicing a boiler. The cost of servicing is £45 in the year 2000. You can use the indices to find the cost in 1995.

Year	1995	2000
Index	100	125

Let the cost in 1995 be C.

Then $C \times \frac{125}{100} = £45$ so $C = £45 \times \frac{100}{125} = £36$

The cost in 1995 was £36

4 The table gives the indices for the cost of renting a student flat.

Year	1990	1994	1998
Index	100	104	117

a Between which two years was the rise bigger?
b The rent was £37.96 per week in 1994. Find the rent per week in 1990.
c Calculate the rent per week in 1998.
d Write down the percentage increase in rent over the eight-year period.

5 The table gives the indices for Jane's salary and living costs over 10 years.

Year	1990	1995	2000
Salary	100	106	109
Living costs	100	107	115

a Jane's salary was £152.64 per month in 1995. Find Jane's salary in 1990.
b Find Jane's salary in 2000.
c Jane's living costs were £136.96 in 1995. What were her living costs in 2000?
d Write down the percentage increase in Jane's living costs between 1990 and 1995.
e Compare the changes in Jane's salary and living costs over the 10 years.
f Find the percentage increase in Jane's salary between 1995 and 2000.

6 The table shows the indices for Ted's investments at the end of each year.

Year	1995	1996	1997	1998	1999
Index	100	112	96	103	110

a When did Ted's investments fall in value?

b His investments were worth £2304 at the end of 1997.
How much were they worth at the end of 1999?

c When were Ted's investments at their highest value?

● **d** What was the highest value of Ted's investments?

e Write down the percentage increase in the investments between 1995 and 1999.

Index numbers give the percentage change compared to a single base year.
To look at differences from year to year you use **chain base numbers.**
The changes in values are always worked out using the previous year as the base.

Example The table shows the values of an investment account at the end of the year for the last 4 years.

Year	1998	1999	2000	2001
Value (£)	5000	6280	6500	5710

Find the chain base numbers for each year.
Give each number correct to 1 decimal place.

Index for 1999 $= \frac{6280}{5000} \times 100 = 125.6$ **The base year is 1998**

Index for 2000 $= \frac{6500}{6280} \times 100 = 103.5$ **The base year is 1999**

Index for 2001 $= \frac{5710}{6500} \times 100 = 87.8$ **The base year is 2000**

The chain base numbers are:

Year	1998	1999	2000	2001
Chain	100	125.6	103.5	87.8

This shows that the value increased by 25.6% in the first year, then it increased by 3.5% in the second year. It then decreased by 12.2% in the third year.

7 The table shows the value of an investment portfolio at the end of the year for the last five years.

Year	1997	1998	1999	2000	2001
Value (£)	4000	4621	4831	4406	3221

a Find the chain base numbers for the four years. Give the numbers to 1 decimal place.
b Describe what the chain base numbers show.

8 Tom owns three garages. The chain base numbers are shown for the profits made by each garage over the last 4 years.

Year	1999	2000	2001	2002
Hadleys	100	108	110	107
Marchmont	100	156	109	118
Carmend	100	102	104	103

Describe the change in profits of each garage over these 4 years.

9 Emma owns two shops. A newsagents and a corner shop.

a The value of the newsagents shop at the end of each of the last 5 years is shown in the table.

Year	1998	1999	2000	2001	2002
Value (£)	56 000	62 000	69 000	73 000	87 000

Use the chain base method to calculate index numbers for these 5 years.

b The value of the corner shop was £72 000 at the end of 1998. The standard index numbers for the value of this shop are shown in the table.

Year	1998	1999	2000	2001	2002
Index number	100	131	128	125	126

Find the value of the corner shop at the end of the remaining 4 years.

c Use the chain base method to calculate index numbers for the corner shop.

d Use both sets of chain base numbers to compare the values of the two shops over the 5 year period.

You have already met weighted means in Chapter 3.

The weighted mean $= \dfrac{\Sigma wx}{\Sigma w}$ where w is the weighting given to each value of x.

Example Danny sits three Maths papers. The papers have different weightings.

Paper	Weighting, w	Danny's mark, x
1	60	65
2	30	74
3	10	90

Danny's weighted mean mark $= \dfrac{\Sigma wx}{\Sigma w}$

$$= \frac{60 \times 65 + 30 \times 74 + 10 \times 90}{60 + 30 + 10}$$

$$= 70.2$$

Exercise 11:3

1 Rupert and Mary also sat the three Maths papers. These are their scores.

Paper	Weighting	Rupert	Mary
1	60	55	68
2	30	71	70
3	10	98	59

a Find the weighted mean mark for: (1) Rupert (2) Mary.
b Who performed the best out of Danny, Rupert and Mary?
• **c** Cheb's mean mark is 58.6 He scored 54 in Paper 1 and 60 in Paper 2. Find Cheb's score for Paper 3.

2 Emma and Sarah have entered a sports competition. They are awarded marks for four different sports. The sports have different weightings. These are the results.

Sport	Emma	Sara	Weighting
Running	9	8	45
Squash	5	7	20
Swimming	4	5	70
Long jump	6	3	10

Work out the mean score for: **a** Emma **b** Sara.

3 The table shows information collected about the employees of a factory for the years 1998 and 1999.

Type of work	1998 Number of employees	1998 Weekly wage (£)	1999 Number of employees	1999 Weekly wage (£)
Production	520	710	487	734
Cleaning	49	328	53	338
Catering	11	514	9	560

a Find the total wage bill for (1) 1998 (2) 1999

b Use your answers to part **a** to calculate an index for the total wage bill in 1999. Take 1998 as the base year.

c The catering staff had a rise of nearly 9% between 1998 and 1999. Explain why the index shows a decrease in the total wage bill.

Weighted index number

An index number can include a number of different items. The index number must take into account the proportions of the different items. These are called weightings. The final index number is called a **weighted index number**.

The table shows the indices for the basic costs of a factory for two years.

Year	1998 (base year)	1999	Weighting
Production	100	105	210
Services	100	103	40
Rent	100	109	12

The weightings reflect the proportion of money spent on each item.
The increase of 5% on production costs in 1999 would have a greater effect on the total costs than the 9% increase on rent.
This is because more money is spent on production than on rent.

You use the weightings to find the weighted index for costs.

$$\text{Weighted index for costs} = \frac{\Sigma w \times \textbf{index}}{\Sigma w} = \frac{210 \times 105 + 40 \times 103 + 12 \times 109}{210 + 40 + 12}$$

$$= 104.877\,786\,2\ldots$$

$$= 105 \text{ to the nearest whole number}$$

This **weighted index** is actually the mean of the three indices 105, 103 and 109 with the weightings taken into account. It is sometimes called the **weighted mean index**, or the **weighted average**.

4 A firm uses three raw materials to make its products.
The table shows the indices for the costs of each raw material in 1995 and 1997.

Raw material	1995	1997	Weighting
Wood	100	112	250
Plastic tubes	100	103	120
Resin	100	105	30

Calculate a weighted index number for the total cost of materials in 1997.

5 The table shows the running costs of a school over two years.

	1997 index	1998 index	Weighting
Salaries	100	104	76
Repairs	100	172	3
Books	100	165	5

a Give a reason why the weighting of salaries is so high.
b Calculate the weighted mean index.
c Explain why your answer is closer to the salaries index of 104 than to the other two indices.

6 The table shows the change in price over two years of five food items.

Item	1997	1999
Bread	62p	71p
Steak	£4.25	£4.95
Baked beans	23p	9p
Eggs	54p	57p
Apples	35p	39p

a Calculate the price index for each of the items. Take 1997 as the base year.
b These are the weightings of the five items for two families.

Item	Weightings – Jones	Weightings – Williams
Bread	35	41
Steak	26	3
Baked beans	8	16
Eggs	32	0
Apples	15	5

c Work out the mean price index for:
(1) the Jones family
(2) the Williams family.
d Explain why your two answers to part **c** are different.

The Retail Price Index, RPI, is the best known index number. It is known as the 'cost of living' index. It looks at how people spend their income. The RPI is based on the results of a survey on expenditure of a number of households.

The RPI measures the change in the cost of a representative basket of goods and services. It was started in 1987 and this acts as the base year.

The weighting of each item comes from the survey and the total of the weightings is 1000. The indices are worked out for each year. These were the indices in September 1993.

	Group	Weighting, w	Index
1	Food	154	111.3
2	Catering	49	118.0
3	Alcoholic drink	83	114.7
4	Tobacco	36	106.4
5	Housing	175	138.0
6	Fuel and light	54	109.0
7	Household goods	71	110.0
8	Household services	41	113.0
9	Clothing and footwear	73	111.0
10	Personal goods and services	37	115.0
11	Motoring expenditure	128	120.0
12	Fares and other travel costs	23	126.0
13	Leisure goods	47	107.0
14	Leisure services	29	117.0
	Total weighting =	1000	

The weighted index can be worked out using the formula $= \dfrac{\Sigma w \times \text{index}}{\Sigma w}$

The weighted index for the 14 groups is called the 'All groups' index.

Exercise 11:4

Use the data in the table above for this exercise.

1 Use the formula given above to calculate the 'All groups' index for September 1993.

2 These are the values for the groups Food and Motoring expenditure.

Group	Weighting	Index
Food	154	111.3
Motoring expenditure	128	120.0

a Calculate a weighted index number for these two groups only.
b Why do you think you get a different value from the 'All groups' index?

3 These are the values for the groups Alcoholic drink and Tobacco.

Group	Weighting	Index
Alcoholic drink	83	114.7
Tobacco	36	106.4

Calculate a weighted index for these two items only.

4 Calculate a weighted index number for each of these sets of groups:
a Household goods and Household services
b Leisure goods and Leisure services
c Food, Catering, Fuel and light
d the first six groups.

5 The Parkinson family do not have a car.
a Work out a new index excluding group 11.
b What has happened to the index by removing group 11? Explain why this happened.

6 The Guha family always use their car for travel. They do not use other forms of transport.
a Work out a new index excluding group 12.
b What has happened to the index by removing group 12? Explain why your answer is different from question **5** part **b**.

Sometimes it is necessary to leave out just part of a group instead of the whole group.

Group 12 covers spending on 'Fares and other travel costs'.
Suppose that spending on fares has a weighting of 18 and an index number of 158.
You can omit spending on fares but leave in the other travel costs.

The 'All groups' index $= \dfrac{\Sigma w \times \text{index}}{\Sigma w} = \dfrac{117\,790}{1000}$

You now subtract the value of $\Sigma w \times$ index for the fares. This is 18×158.
The total weighting has also changed from 1000 to $1000 - 18$.
So the new index without fares is

$$\frac{117\,790 - 18 \times 158}{1000 - 18} = 117.052\,953\ldots$$

$$= 117.1 \text{ to 1 d.p.}$$

This is still $\dfrac{\Sigma w \times \text{index}}{\Sigma w}$ for the indices that are left.

7 Group 9 covers spending on 'Clothing and footwear'.
Clothing has a weighting of 61 and an index number of 109.7
Work out the index number for all groups except the part of group 9 for clothing.

8 Group 10 covers spending on 'Personal goods and services'.
Personal goods has a weighting of 19 and an index number of 111.5
Work out the index number for all groups except the part of group 10 for personal goods.

• **9** Use subtraction to find the index for each of these:

a all groups except group 5
b all groups except group 3
c all groups except groups 13 and 14
d all groups except groups 3 and 4.

2 Birth and death rates

Sometimes there's a very good reason for a high death rate …

Crude death rate

The **crude death rate** is normally given as the number of deaths per thousand of population.

$$\text{Crude death rate} = \frac{\text{number of deaths}}{\text{total population}} \times 1000$$

The formula can be adapted for births as well as deaths.

$$\text{Crude birth rate} = \frac{\text{number of births}}{\text{total population}} \times 1000$$

If the population is small, the rates can be given per 100 of population or another smaller number.

The table shows the births and deaths in 1997 for the village of Lukehampton:

Population	Births	Deaths
7692	37	45

$$\text{Crude birth rate} = \frac{37}{\mathbf{7692}} \times 1000 = 4.8 \text{ (2 s.f.) per 1000 of population.}$$

$$\text{Crude death rate} = \frac{45}{\mathbf{7692}} \times 1000 = 5.9 \text{ (2 s.f.) per 1000 of population.}$$

Exercise 11:5

1 In 1998 the population of Dartville was 11 473. There were 51 deaths. Calculate the crude death rate for 1998.

2 **a** The table shows the births and deaths in 1995 for the city where Ken lives.

Population	Births	Deaths
65 821	467	960

Work out: (1) the crude birth rate (2) the crude death rate.

b The table shows the births and deaths in 1995 for the town where Sally lives.

Population	Births	Deaths
28 106	324	297

Work out: (1) the crude birth rate (2) the crude death rate.

c Use your answers to parts **a** and **b** to compare the two populations.

3 **a** The squirrel population of a forest was 729 last year.
There were 63 deaths.
Find the crude death rate giving your answer per 100 of population.

b Another forest had a population of 15 029 squirrels. Assuming the crude death rate was the same for both forests, estimate the number of deaths in the second forest.

4 The town of Severly had a crude death rate of 7.8 deaths per 1000 population in 1995.

a The population was 50 482 in 1995. How many deaths were there?

b In 1998 the population was 67 204. How many deaths would you expect if the death rate was the same for both years?

c Give a reason why the death rate might not be the same for both years.

5 The crude death rate in 1999 for a population of 593 bears was 23.7 per 100 of population.
The crude birth rate in 1999 was 22.9 per 100 of population.

a In 1999, how many bears: (1) were born (2) died?

b If this pattern continues over the next few years what will happen to the population of bears?

To compare the death rates of two towns you need to look at the proportion of people in the different age groups.

Otto has collected information on his town from the library.
The table shows the data he found on ages.

Age	Number
0–19	1056
20–39	6294
40–59	8458
60 and over	4606

Otto wants to get a sample of 1000 people that represents this population.
He needs a stratified sample so that the size of each group is represented.
The sample of 1000 people selected in this way is called a standard population for Otto's town.

Standard population

The **standard population** consists of 1000 people and is a representative sample of the whole population.

Total population $= 1056 + 6294 + 8458 + 4606 = 20\,414$

Age group	Standard population for Otto's town
0–	$\frac{1056}{20\,414} \times 1000 = 52$
20–	$\frac{6294}{20\,414} \times 1000 = 308$
40–	$\frac{8458}{20\,414} \times 1000 = 414$
60–	$\frac{4606}{20\,414} \times 1000 = 226$

Make sure the numbers add up to 1000: $52 + 308 + 414 + 226 = 1000$ ✓

Exercise 11:6

1 The population of Great Larmly is made up as follows:

Age group	0–25	26–50	51–75	76–100
Number	6391	13 407	8467	2810

Find the standard population for Great Larmly.
Make sure that your numbers add up to 1000.

2 The table gives the population, in millions, for 1991, of different age groups and the predicted population for the year 2051.

Year	Under 16	16–39	40–64	65–79	80 and over
1991	11.7	20.4	16.5	6.9	2.1
2051	10.5	16.7	18.1	8.9	5.5

a Find the standard population for each year.
b Compare the two standard populations.

To decide whether the death rate of a town is high you need to take into account the proportions of the population in different age groups. You would expect a higher death rate as the proportion of elderly people increased.
You need to use the standard population.

The table gives information on the population of Darncup town.
The standard population of the whole country is given.

Age (years)	Population	Deaths	Standard population
0–19	2000	14	320
20–39	3500	28	260
40–59	6000	18	240
60 and over	2800	252	180

Standardised death rate

The **standardised death rate** is calculated using the standard population of the whole country.

To work out the standardised death rate you first find the death rate for each age range.

For 0–19: Death rate per 1000 $= \frac{14}{2000} \times 1000 = 7$ and so on.

You then use these values and the standard population to find the number of expected deaths in the standard population.

Age	Death rate per 1000	Standard population	Deaths per 1000 of standard population
0–	$\frac{14}{2000} \times 1000 = 7$	320	$\frac{7}{1000} \times 320 = 2.24$
20–	$\frac{28}{3500} \times 1000 = 8$	260	$\frac{8}{1000} \times 260 = 2.08$
40–	$\frac{18}{6000} \times 1000 = 3$	240	$\frac{3}{1000} \times 240 = 0.72$
60 and over	$\frac{252}{2800} \times 1000 = 90$	180	$\frac{90}{1000} \times 180 = 16.2$
			Total 21.24

The standardised death rate is the total of the expected deaths in the standard population.

The standardised death rate for Darncup is **21.24** deaths per 1000 people.

The crude death rate would have been $\frac{312}{14\,300} \times 1000 = 21.82$ (2 d.p.)

The standardised death rate is a better measure because it takes into account the spread of ages and compares it with the standard population.

---

3 Chelsi has collected this data on her town.

Age	Population	Deaths	Standard population
0–15	4 700	29	240
16–30	6 100	85	210
31–45	10 300	107	190
46–60	9 400	131	200
60+	7 000	170	160

Work out:

a the crude death rate

b the standardised death rate.

4 Gordon has collected this data for the towns Harpdon and Walston.

Harpdon

Age	Population	Deaths	Standard population
0–15	1300	17	120
16–40	6000	18	600
41+	3800	45	280

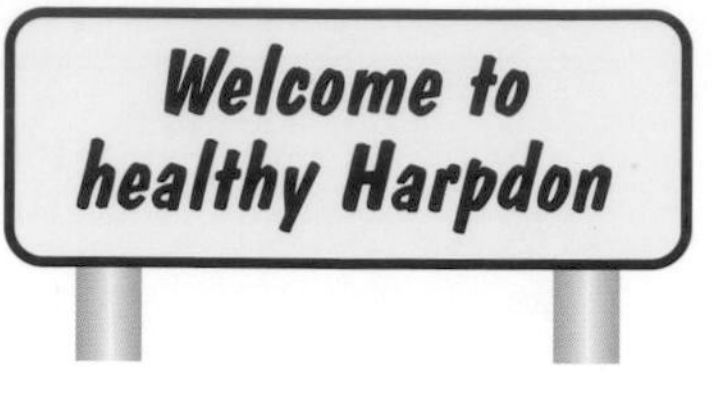

Walston

Age	Population	Deaths	Standard population
0–15	900	14	120
16–40	5600	29	600
41+	2100	21	280

a Work out the standardised death rate for Harpdon.
b Work out the standardised death rate for Walston.
c Which town is the healthier place to live?
Explain your answer.

5 These were the death rates for three towns in 1998.

Town	Angville	Barton	Grantmore
Death rate per 1000	19.3	27.1	20.6

One of the towns has a large number of retired people. Which town is it?
Explain your answer.

Standard populations are sometimes given as percentages.
You change the percentages to numbers of people out of 1000.

Age	Standard population
0–	18%
20–	30%
40–	36%
60 and over	16%

becomes

Age	Standard population
0–	180
20–	300
40–	360
60 and over	160

18% of 1000 = 180
30% of 1000 = 300
and so on

Check that the numbers in the standard population add up to 1000:
180 + 300 + 360 + 160 = 1000 ✓

---

6 Katy is researching the health of two different areas.
Here is some of her data.

Age group	Area A		Area B		Standard population
	Population	Deaths	Population	Deaths	
0–	2000	4	1100	7	25%
15–	4200	8	2400	13	35%
30–	7000	10	1700	5	20%
45–	6000	5	2000	4	10%
60–	3900	52	700	17	6%
75–	2500	27	350	11	4%

a Calculate the standardised death rate for each area.
b In which area would you prefer to live?
Give a reason for your answer.
c What is the advantage of Katy using the standardised death rate to compare areas rather than the crude death rate?

Standardised rates are not only used for births and deaths. They can also be used for any situation where the population is divided into age groups e.g. unemployment rates, accident rates.

---

7 The table shows the rates of unemployment in the standard population of the whole country.

Age group (years)	under 30	30–	45–	60 and over
Standard population	250	350	300	100
Percentage unemployed	6	7	11	15

a Show that the standardised rate of unemployment is 87.5.

These are the unemployment figures for the village of Barnworth.

Age group (years)	under 30	30–	45–	60 and over
Population	300	350	300	50
Percentage unemployed	5	8	12	120

b Calculate the crude unemployment rate for Barnworth.
c Show that the standardised unemployment rate is 96.5.

1 The table shows the indices for bicycle insurance costs.
The base year for these indices is 1994.

Year	1994	1995	1996
Index	100	105	108

a What does the index number of 105 for 1995 tell you about insurance costs? *(1)*

The cost of insuring a particular type of bicycle in 1995 was £31.50

b **i** How much was the insurance in 1996? *(3)*

ii What was the actual increase in cost for this insurance from 1994 to 1996? *(1)*

In 1997 the insurance cost for this bicycle was £34.80

c Calculate the index number for 1997. *(2)*

SEG, 1998, Paper 2

2 The table shows the percentage relatives and weight of certain commodities in 1995, taking 1993 as the base year.

	Percentage relatives		Weight
	1993	1995	
Mortgage	100	110	0.4
Heat and lighting	100	130	0.2
Clothing	100	125	0.1
Food	100	115	0.25
Other items	100	120	0.05

a Give **one** reason why a weighted average is sometimes more appropriate than the ordinary arithmetic average. *(1)*

b Given that the clothing bill in 1993 was £400, how much would it have been in 1995? *(2)*

c Use the information in the table above to calculate a retail price index for 1995. *(3)*

SEG, 1996, Paper 4

3 The price of a holiday in 1995 and 1996 was

Year	1995	1996
Price	£310	£380

a Calculate the percentage relative for the price of the holiday in 1996, taking 1995 as the base year with index number 100. *(2)*

b Taking 1995 as the base year with an index of 100, the cost of a different holiday is divided into accommodation and travelling as shown in the table below:

Percentage relative	1995	1996	Weight
Accommodation	100	130	70%
Travel	100	120	30%

Calculate the combined percentage relative for the holiday in 1996. *(2)*

SEG, 1999, Specimen Papers

4 Fabro plc uses four raw materials, A, B, C, and D, in the manufacture of a product. The ratio, by weight, of the four materials, A, B, C and D, needed to produce each item is 2:4:12:1, respectively. One kilogram of material D is used in the manufacture of one item. The costs of these raw materials in the years 1995–97 were as follows.

Cost per kilo (£)

Raw material	1995	1996	1997
A	2.50	2.50	3.00
B	1.00	1.20	1.50
C	4.00	4.50	4.50
D	5.00	5.00	6.00

a Show that the total raw material cost for **one** item in 1995 was £62.00 *(2)*

b Calculate the raw material cost of producing **one** equivalent item in 1997. *(2)*

c Use the results obtained in parts **a** and **b** to calculate, to one decimal place, a raw material cost index for 1997 using 1995 as base. *(2)*

d An index of Fabro's labour costs is as follows.

Year	1995	1996	1997
Labour cost index	120	130	160

Calculate the percentage increase in these costs from 1995 to 1997. *(2)*

e Compare the change in labour costs between 1995 and 1997 with the change in raw material costs. *(2)*

NEAB, 1998, Paper 3

5 The following data were collected from the wages section of a large manufacturing company.

Employee status	1990 No. of employees	1990 Weekly wage (£)	1995 No. of employees	1995 Weekly wage (£)
Non-manual – men	15	190	40	294
Manual – men	490	172	120	275
Non-manual – women	20	123	100	202
Manual – women	90	100	420	156

a Calculate an index based on the total wage bill for 1995 compared to that of 1990. *(4)*

b Calculate, using the 1990 numbers of employees a weighted index to measure the increase in wages paid by the company between 1990 and 1995. *(4)*

c An equivalent index measuring the increase in wages over the two periods, weighted by the 1995 numbers of employees, gave a value of 158.1
Suggest a reason for the difference in this index and the one calculated in part **b**. *(2)*

NEAB, 1997, Paper 3

6 The following data have been taken from the General Index of Retail Prices and relate to September 1993.

	Group	Weight	Index
1	Food	154	111.3
2	Catering	49	118.0
3	Alcoholic drink	83	114.7
4	Tobacco	36	106.4
5	Housing	175	138.0
6	Fuel and light	54	109.0
7	Household goods	71	110.0
8	Household services	41	113.0
9	Clothing and footwear	73	111.0
10	Personal goods and services	37	115.0
11	Motoring expenditure	128	120.0
12	Fares and other travel costs	23	126.0
13	Leisure goods	47	107.0
14	Leisure services	29	117.0

(Source: *Monthly Digest of Statistics*)

The 'All groups' index number for September 1993 is 117.79

a Calculate, to one decimal place, a weighted index number to represent expenditure on motoring and travel (i.e. Groups 11 and 12). *(3)*

b Comment on the difference between the 'All groups' index and your answer in **a**. *(1)*

c Included in Group 5 (Housing) is expenditure on mortgage interest payments which has a weighting of 60 and an index number of 168.2

i Calculate, to two decimal places, an index number representing 'All groups' excluding expenditure on mortgage interest. *(4)*

ii Explain briefly the effect this has had on the 'All groups' index. *(1)*

NEAB, 1996, Paper 3

7 The table shows the age distribution of a town X, the number of deaths for each age group and the percentage of each age group in the whole country.

Age group	Population in 1000s	Number of deaths	Standard population
Under 10	20	324	15%
10–24	32	125	25%
25–44	40	268	25%
45–64	23	426	20%
65 and over	12	560	15%

a Calculate the crude death rate for town X. *(3)*

The crude death rate for a town Y is 19.2 per thousand.

b How would you expect the age distribution of town Y to differ from town X? *(1)*

c Calculate the standardised death rate for town X. *(4)*

d Explain why the standardised death rate is a better representation than the crude death rate. *(1)*

SEG, 1998, Paper 3

8 Kate, in her Humanities project, is comparing the two towns of Dormingly and Garthside. At the moment she is looking into health and has collected these data.

	Dormingly		Garthside		
Age group	Population	No. of deaths	Population	No. of deaths	Standard population
0–	3000	6	1500	10	25%
15–	4000	8	2500	12	35%
30–	8000	12	2000	5	20%
45–	6000	6	1000	2	10%
60–	4000	56	500	12	6%
75–	4000	50	500	12	4%

a Calculate the standardised death rate for Dormingly. *(4)*

b What is the advantage of Kate quoting the standardised death rate for a town rather than the crude death rate? *(1)*

c Garthside has a standardised death rate of 6.45 per 1000 people. In which of these towns would you prefer to live? Give a reason for your answer. *(1)*

SEG, 1999, Specimen Papers

9 A local newspaper suggested that Westhope is a healthier place to live than Martrent. Data for the two towns is given below.

	Westhope		Martrent		Standard population
Age group	No. of deaths	Population	Death rate per 1000	Percentage of population	Percentage of population
0–	6	1500	8	40%	25%
20–	4	2000	2	25%	30%
40–	24	1600	10	30%	30%
60–	100	2000	90	5%	15%

a Find the crude death rates per thousand for each age group in Westhope. *(4)*

b Calculate the standardised death rates for the two towns. *(4)*

c With reference to your answer in **b** is the newspaper correct in its suggestion that Westhope is a healthier place to live than Martrent? Give a reason for your answer. *(2)*

NEAB, 1999

10 In December 1996 the cost of petrol was 62.5 pence per litre.
In December 1999 the cost was 78.7 pence per litre.

a Using 1996 as the base year, express the cost of petrol in 1999 as an index number, to one decimal place. *(2)*

With 1999 as the base year, the index number for December 2001 was 91.2.

b Find the cost of petrol per litre in December 2001. *(2)*

NEAB, 1999

11 The table below shows the prices of petrol and bread in 1989 and 1999.

	Price (pence)		Price index (1999 relative to 1989)
Article	1989	1999	
Gallon of petrol	192	320	X
Loaf of bread	64	80	Y

a i Calculate the value of X, the price index of a gallon of petrol. *(1)*

ii Calculate the value of Y, the price index of a loaf of bread. *(2)*

b For each year write down the total cost of a gallon of petrol and 2 loaves of bread. *(2)*

c Calculate the price index for the total cost of a gallon of petrol and 2 loaves of bread. *(1)*

NEAB, 2000

12 In a village a record was kept of the ages of those people who died in 1992.
The data are shown on the stem and leaf diagram.

Age at death (years)

Stem	Leaf
0	6
1	1
2	3 5
3	0 0 1
4	5 5 6 9 9
5	3 7 8 9
6	7 8 9

Key: 2|3 denotes 23 years

a What is the range of their ages? *(2)*

b How many people died in this village in 1992? *(1)*

At the start of the year there were 750 people in the village.

c Calculate the death rate for this village. *(2)*

The same year seven people died in another village.
The death rate for this village was 22.

d Calculate the number of people who lived in this village at the start of the year. *(2)*

SEG, 1999, Paper 3

Weighted index number An index number can include a number of different items. The index number must take into account the proportions of the different items. These are called weightings. The final index number is called a **weighted index number**.

The table shows the indices for the basic costs of a factory for two years.

Year	1998 (base year)	1999	Weighting
Production	100	105	210
Services	100	103	40
Rent	100	109	12

The weightings reflect the proportion of money spent on each item.
The increase of 5% on production costs in 1999 would have a greater effect on the total costs than the 9% increase on rent.
This is because more money is spent on production than on rent.

You use the weightings to find the weighted index for costs.

$$\text{Weighted index for costs} = \frac{\Sigma w \times \textbf{index}}{\Sigma w} = \frac{210 \times 105 + 40 \times 103 + 12 \times 109}{210 + 40 + 12}$$

$$= 104.877\,786\,2\ldots$$

$$= 105 \text{ to the nearest whole number}$$

This **weighted index** is actually the mean of the three indices 105, 103 and 109 with the weightings taken into account. It is sometimes called the **weighted mean index**, or the **weighted average**.

Crude death rate The **crude death rate** is normally given as the number of deaths per thousand of population.

$$\text{Crude death rate} = \frac{\text{number of deaths}}{\text{total population}} \times 1000$$

To decide whether the death rate of a town is high you need to take into account the proportions of the population in different age groups. You would expect a higher death rate as the proportion of elderly people increased.
You need a standard to compare the population with to see if there is a higher proportion of elderly people.

Standard population The **standard population** consists of 1000 people and is a representative sample of the whole population.

Standardised rates are not only used for births and deaths. They can also be used for any situation where the population is divided into age groups e.g. unemployment rates, accident rates.

1 The table gives the cost of 1 litre of petrol in 1989 and 1999.

Year	Cost of 1 litre of petrol
1989	41.2 p
1999	72.9 p

a With 1989 as the base year, express the cost of 1 litre of petrol in 1999 as an index number. Give your answer to 3 significant figures.

b With 1989 as the base year, the index number for the cost of 1 litre of petrol in 1995 is 128.
Calculate the cost of 1 litre of petrol in 1995. Give your answer to the nearest tenth of a penny.

c What does the index number 128 tell you?

2 Pupils have to sit three Maths papers in a contest.
The weighting of the papers is 2:3:5.
These are the marks for Simon and Charles.

	Weighting	Simon	Charles
Paper 1	2	70	40
Paper 2	3	45	50
Paper 3	5	40	

a Calculate the weighted mean mark for Simon.

b Charles' weighted mean mark was 40.5
What mark did Charles score on Paper 3?

3 Tom has collected this data for the town of Calvinham.

Age	Population	Deaths	Standard population
0–15	2800	19	240
16–30	8300	104	210
31–45	8900	77	190
46–60	7400	121	200
60+	4900	192	160

Work out:

a the crude death rate for Calvinham

b the standardised death rate for Calvinham.

12 Finding probabilities

1 Rules for finding probabilities

Independent events
Probability of independent events
Probability of an event happening more than once
Mutually exclusive events
Exhaustive events

2 Probability trees

Tree diagrams
Tree diagrams for dependent events

EXAM QUESTIONS

SUMMARY

TEST YOURSELF

1 Rules for finding probabilities

All the balls in the machine have an equal chance of being the first ball out.
What has happened in previous draws has no effect on the next draw.

Independent events

Two events are **independent** if the outcome of one has no effect on the outcome of the other.

If you roll a fair dice and toss a coin you can get 1, 2, 3, 4, 5 or 6 on the dice and either a head or a tail with the coin.

Whatever you get on the dice has no effect on what you get with the coin. The two events are independent.

Exercise 12:1

Look at the events in questions **1**–**5**.
Write down whether each pair of events is independent.

1 Nita throws a dice and scores a 5.
Vicram throws a dice and scores a 2.

2 A bag contains balls of different colours.
Diane picks out a red ball and keeps it.
Ashley then picks out a blue ball at random.

3 Penny picks a black card at random from a pack of cards and then replaces it. Marika then picks out a red card at random from the pack.

4 Andy drops a drawing pin and it lands point down.
He picks up the pin and drops it again and it lands point up.

5 David's car breaks down on the way to college.
David is late for his first lecture on the same day.

Rudi uses this spinner and dice.
The sample space diagram shows all the possible outcomes.

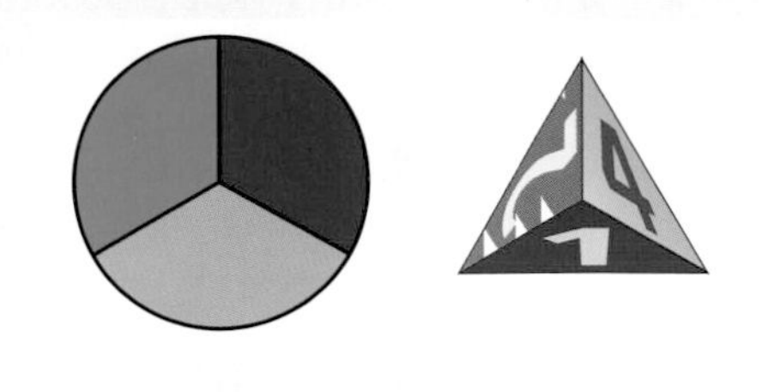

		Dice			
		1	2	3	4
Spinner	R	R,1	R,2	R,3	R,4
	B	B,1	B,2	B,3	B,4
	G	G,1	G,2	G,3	G,4

Using the sample space, you can work out
$P(\text{a red and a number less than 3}) = \frac{2}{12} = \frac{1}{6}$
You can also work out this probability using independent events.

Probability of independent events

If two **events** A and B are **independent** then the **probability** of them both happening is called P(A and B)

$P(\text{A } \mathbf{and} \text{ B}) = P(\text{A}) \times P(\text{B})$

So for Rudi:
$P(\text{a red } \mathbf{and} \text{ a number less than 3}) = P(\text{red}) \times P(\text{a number less than 3})$
$= \frac{1}{3} \times \frac{1}{2} = \frac{1}{6}$

This is the same answer as you get using the sample space diagram but this method is much quicker to use.

Example Donna throws an ordinary dice and uses this spinner. Find the probability that she gets a green and a 6.

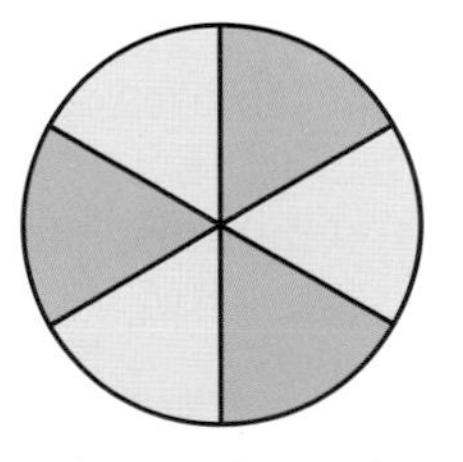

$P(\text{green}) = \frac{1}{2}$ and $P(6) = \frac{1}{6}$
The two events are independent so:
$P(\text{green and } 6) = \frac{1}{2} \times \frac{1}{6}$
$= \frac{1}{12}$

Exercise 12:2

1 The probability that Ken will win his darts match against Barry is $\frac{2}{5}$.
The probability that Sally will win her darts match against Jude is $\frac{1}{4}$.
Find the probability that both Ken and Sally win their matches.

2 The probability that Jane forgets her bus fare on any day is $\frac{1}{3}$.

a Find the probability that she forgets her bus fare today and tomorrow.

b Find the probability that she forgets her bus fare today but does not forget it tomorrow.

3 A new torch is switched on and left on continuously.
The probability of the battery lasting 5 hours is 0.3
The probability of the bulb lasting 5 hours is 0.9

a Find the probability that both the battery and the bulb will last 5 hours.

b Find the probability that both will fail in the first 5 hours of use.

4 Graham has five male friends and three female friends.

a He needs some help with his homework so he rings one of his friends at random. Find the probability that this friend is female.

b He needs help with his homework the next night.
Graham again rings a friend at random.
Find the probability that he rings a female friend on both nights.

5 Robert is a car salesperson. He has a 30% chance of selling a car in any one hour period at work.

a Write this probability as a decimal.

b Write down the probability that he will not sell a car in any one hour.

c Find the probability that he will sell a car in the first hour of work but not in the second hour.

You can use independent events to find the probability of an event happening more than once.
You multiply the probability of it happening once by itself.

Example This spinner is used three times.
Find the probability of getting three reds.

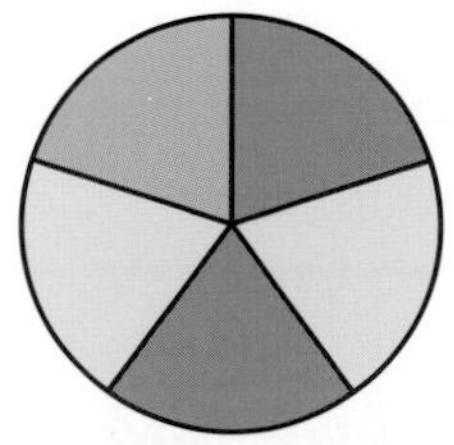

The probability of a red on one spin is $\frac{2}{5}$.
Each spin is independent, so multiply the probabilities.

$P(\text{three reds}) = P(\text{red on 1st spin}) \times P(\text{red on 2nd spin}) \times P(\text{red on 3rd spin})$

$= \frac{2}{5} \times \frac{2}{5} \times \frac{2}{5} = \frac{8}{125}$

---

6 A coin is thrown three times.
Find the probability of getting three heads.

7 The spinner in the example on page 332 is used four times.
Find the probability of getting four blues.

8 The probability of a client being late for an appointment is 0.15
a Find the probability that the next three clients will all be late
b Find the probability that none of the next three clients will be late.

9 James throws this tetrahedral dice five times.
The faces are numbered 1, 2, 3, and 4.
Find the probability that James gets:
a a four on the first throw
b all fours on the first three throws
c all fours on all five throws.

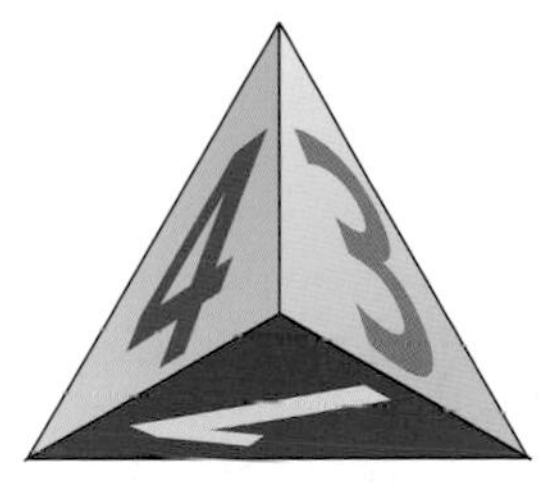

10 Walker's garage is offering a free sports car to any customer who can get 7 sixes on 7 rolls of a dice.
Find the probability of a customer winning the car.
Give your answer as a decimal.

Mutually exclusive

Events are **mutually exclusive** if they cannot happen at the same time.

When a dice is thrown the events 'getting an even number' and 'getting a 3' cannot happen at the same time.
This is because 3 is not an even number. The events 'getting an even number' and 'getting a 3' are mutually exclusive.

Probability of mutually exclusive events

For two **mutually exclusive events** A and B, the **probability** that either event A **or** event B will occur can be found by **adding** their probabilities together.

$$P(\text{A or B}) = P(\text{A}) + P(\text{B})$$

$$\begin{aligned} P(\text{getting an even number or a 3}) &= P(\text{getting an even number}) + P(\text{getting a 3}) \\ &= \tfrac{1}{2} + \tfrac{1}{6} \\ &= \tfrac{2}{3} \end{aligned}$$

Exercise 12:3

1 State whether each pair of events is mutually exclusive.

a When tossing a coin:
'getting a head' and 'getting a tail'.

b When picking a card at random from a pack:
'getting a heart' and 'getting a red card'.

c When picking a ball at random from a bag of coloured balls:
'getting a white ball' and 'getting a black ball'.

d When picking a person at random from a crowd:
'picking a boy' and 'picking a person with black hair'.

2 A tin of sweets contains 5 toffees, 6 fruit chews and 9 chocolates.
Carol chooses a sweet at random from the tin.
Find the probability that Carol chooses:

a a toffee
b a chocolate
c a toffee or a fruit chew
d a fruit chew or a chocolate.

3 Jane rolls a 12-sided dice. The sides are numbered 1 to 12.
Write down the probability that Jane gets:

a 8
b an odd number
c 3 or 4
d a multiple of 3
e a number more than 5
f a number less than 6.

Exhaustive events

Events are **exhaustive** if all possible outcomes are included.

When throwing a dice:
The events 'getting an odd number' and 'getting an even number' are exhaustive.

The events 'getting an odd number' and 'getting a 4' are not exhaustive because 2 and 6 are not included.

---

4 State whether each pair of events is exhaustive.

a When throwing a coin:
'getting a head' and 'getting a tail'.

b When choosing a card at random from a pack of 52 playing cards:
'getting a red card' and 'getting a spade'.

c When picking a ball from a bag containing 3 red, 2 black and 5 pink balls:
'picking a red ball' and 'not picking a red ball'.

Exercise 12:4

In this exercise you need to decide whether the events are independent or mutually exclusive before you work out the probability.

1 The probabilities that Ken and Pam will go to Spain for their holiday are 0.7 and 0.4 respectively.
Find the probability that:
a both will go to Spain
b only Pam will go to Spain
c only one will go to Spain
d neither will go to Spain.

2 Guy picks a cube at random from a box of coloured cubes.
The probability that he picks a red cube is 0.24
The probability that he picks a purple cube is 0.59
Find the probability that Guy picks:
a a red or a purple cube
b a cube that is neither red nor purple.

3 Corinne prints sports emblems on to T-shirts. She has four designs and five colours.
The four designs are: golf, tennis, football, snooker.
The five colours are: red, blue, green, black, orange.
She chooses the design and colour at random.
Find the probability that she chooses:
a a tennis design
b an orange design
c a football design in red
d a golf design in green or black.

4 The probabilities that James, Katy and Matthew will pass their French oral are 0.5, 0.75 and 0.9 respectively.
Find the probability that:
a both James and Katy pass
b James fails but Katy and Matthew pass
c all three pass
d all three fail.

5 **a** The events A and B are independent.
$P(\text{A}) = 0.25$
$P(\text{A and B}) = 0.075$
Find $P(\text{B})$.
b The events R and Q are mutually exclusive
$P(\text{R or Q}) = 0.77$, $P(\text{R}) = 0.24$. Find $P(\text{Q})$.

2 Probability trees

Squirrels always want to eat the bird food. There are many paths that this squirrel can take across the branches of the tree. Only one of these paths will lead to the bird food.

Tree diagrams

You can use **tree diagrams** to show the outcomes of two or more events.
Each branch represents a possible outcome of one event.
The probability of each outcome is written on the branch.
The final result depends on the path taken through the tree.

Example The probability that a new car will develop a fault in the first year is 0.7
Two new cars are chosen at random.

a Show all the possible outcomes on a tree diagram.
b Use the diagram to find the probability that both will develop a fault.
c Find the probability that only one will develop a fault.

a The first set of branches of the tree shows what can happen to the first car. The second set of branches shows what can happen to the second car.

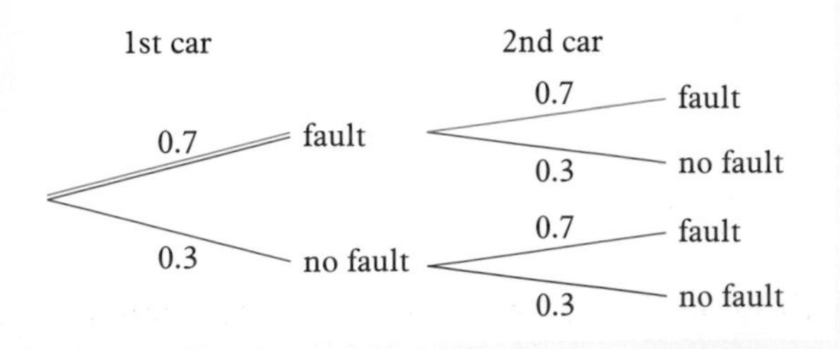

Each path through the tree gives a different outcome.
The tree has four paths so there are four outcomes.

b The blue path gives the outcome 'both cars will develop a fault'.
You **multiply** the probabilities on the branches of this path since both the first car **and** the second car have faults.
This gives you the probability of this final outcome.
The probability that both cars will develop a fault $= 0.7 \times 0.7 = 0.49$

c Each red path gives an outcome of one of the two cars developing a fault.

The top red path gives the probability of the first car developing a fault $= 0.7 \times 0.3 = 0.21$

The lower red path gives the probability of the second car developing a fault $= 0.3 \times 0.7 = 0.21$

Now you **add** the probabilities of each path to find the probability that only one car develops a fault, i.e. the first car **or** the second car $= 0.21 + 0.21 = 0.42$

You **multiply** the probabilities along the branches.
You **add** the probabilities when more than one path is used.

Exercise 12:5

1 The probability that a garage serves more than 100 customers during one day is 0.6

a Copy this tree diagram showing all the possible outcomes for two days. Fill it in.

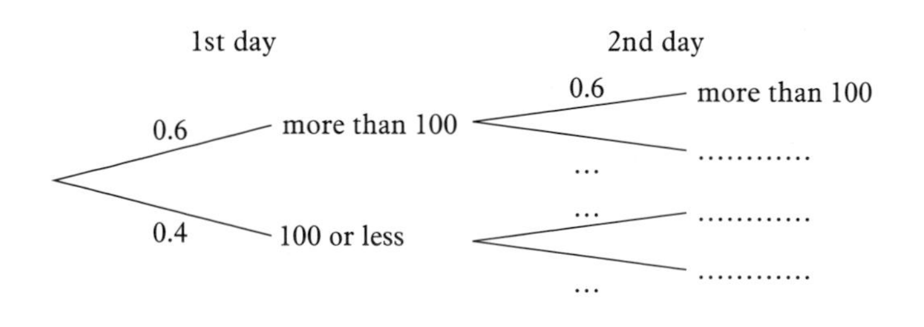

Find the probability that there were more than 100 customers on:
b both days **c** the first day only **d** only one day.

2 Vicky rolls a coloured dice twice. Two faces are red and four are blue.

a Copy this tree diagram showing all the possible outcomes. Fill it in.

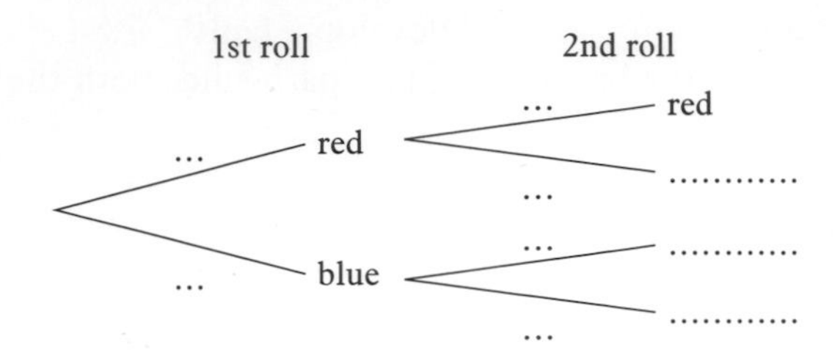

Find the probability that Vicky gets:

b two reds

c two blues

d only one red

e red on the 1st roll and blue on the 2nd roll.

3 A company making lamps uses switches from two companies, Ant and Bakers in the ratio 2:3.
4% of the switches from Ant are faulty and 5% of the switches from Bakers are faulty.
A lamp is chosen at random.

a Copy this tree diagram showing all the possibilities for the switch on the lamp. Fill it in.

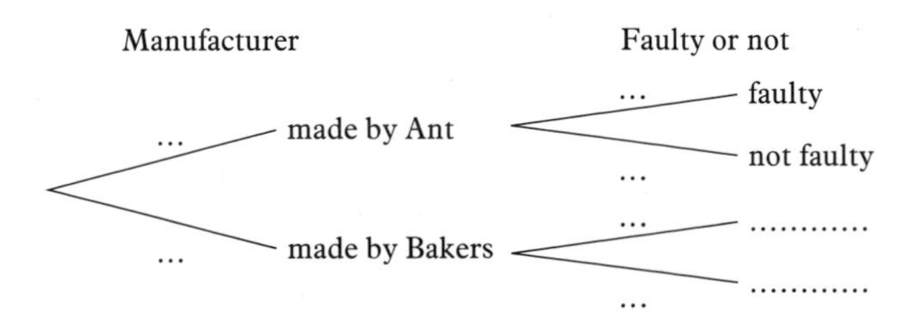

Find the probability that the switch:

b is made by Ant and is faulty

c is not faulty.

4 Simon uses this spinner three times.

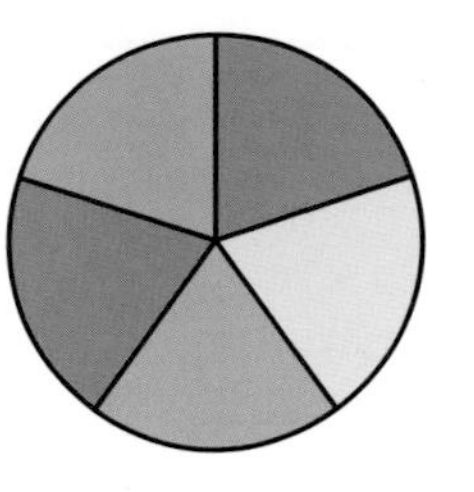

a Copy the probability tree.
Fill it in.

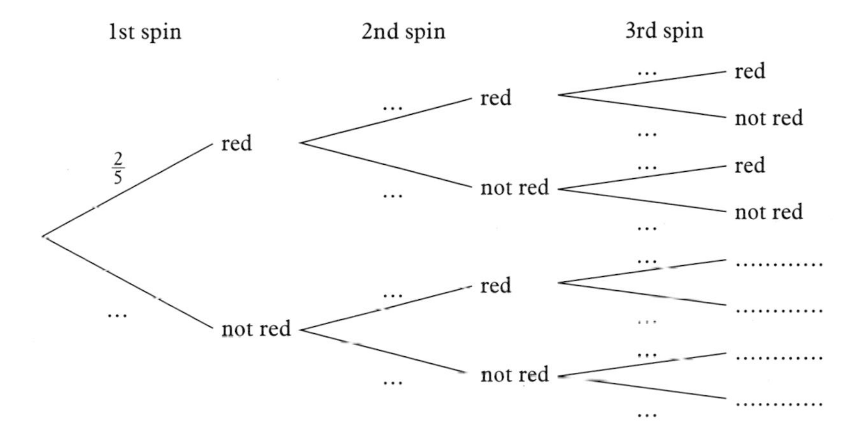

Find the probability that Simon gets:

b three reds

c only one red

d exactly two reds

e a red on the second spin only.

5 Belinda rolls a dice twice.

a Copy this tree diagram. Fill it in.

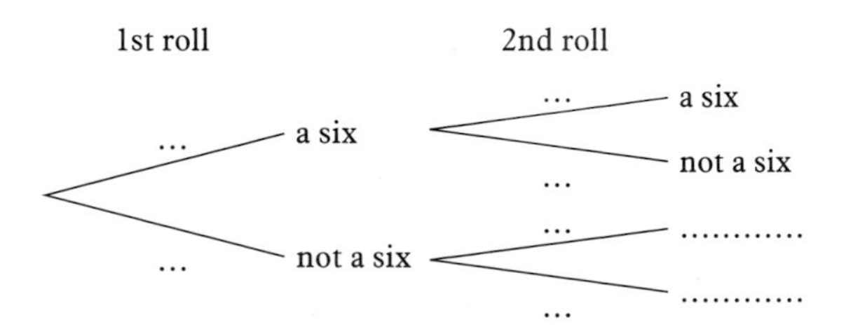

Find the probability that Belinda gets:

b just one six

c two sixes

d a six on the second roll only.

You have been drawing tree diagrams for independent events.
Sometimes the events are dependent. You can still use a tree diagram to work out probabilities.

Example Ria drives through two sets of traffic lights on her way to work.
The probability that the first set of lights will be red is 0.3
If the first set of lights is red, the probability that the second set will be red is 0.8
If the first set of lights is not red then the probability that the second set will be red is 0.4
Use a tree diagram to find the probability that:

a both sets are red **b** only one set is red.

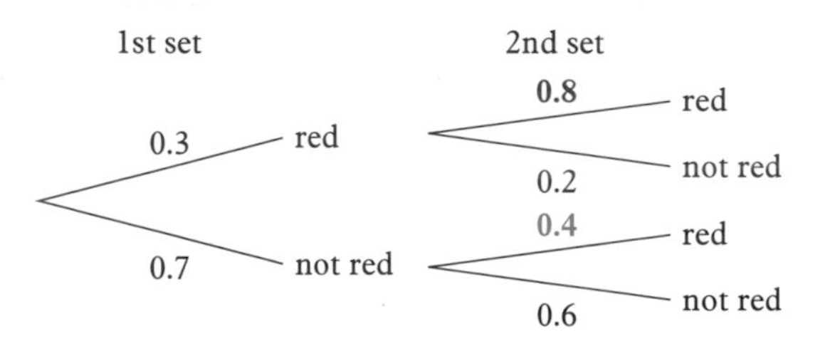

In the top two paths the 1st set was red.
This means that the probability that the 2nd set is red is **0.8**
In the bottom two paths the 1st set was not red.
This means that the probability that the 2nd set is red is **0.4**

a $P(\text{both sets are red}) = 0.3 \times 0.8 = 0.24$
b $P(\text{only one set is red}) = (0.3 \times 0.2) + (0.7 \times 0.4) = 0.06 + 0.28 = 0.34$

Exercise 12:6

1 The school team has two tennis matches to play.
The probability that they win the first match is $\frac{2}{5}$.
If they win the first match the probability that they win the second is $\frac{3}{4}$.
If they lose the first match the probability that they win the second is $\frac{1}{3}$.

a Copy this tree diagram. Fill it in.

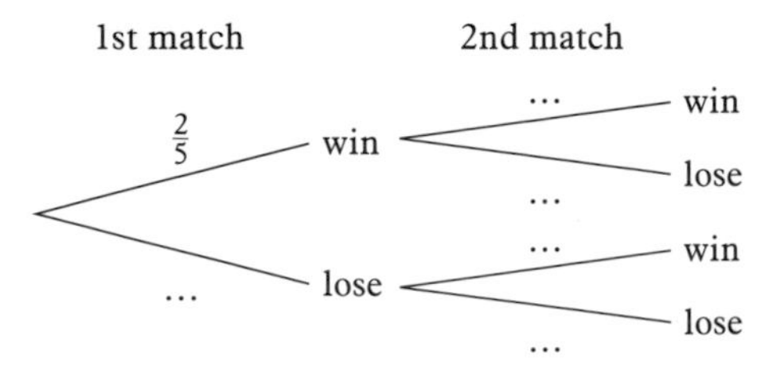

Find the probability that the team:

b loses both matches **c** wins only one match.

2 A chess team consists of seven girls and five boys.
Two of them are chosen at random to represent the team.

a Copy the tree diagram.
Fill it in.

1st person chosen 2nd person chosen

... girl ... girl
... boy
... boy ... .
... .

Find the probability that:

b both are girls

c both are the same sex

d only one boy is chosen

e the 1st is a boy and the 2nd a girl.

3 Tim goes through 3 sets of traffic lights.
All 3 sets of traffic lights have the same probabilities for red, amber and green.

a The diagram shows part of the probability tree.
Copy the diagram and complete it.

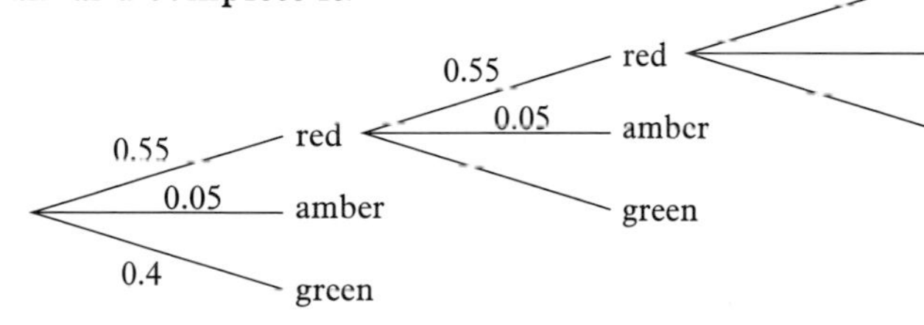

Find the probability that:

b all three sets show red

c only the first set is red.

Sometimes one of the paths of a tree diagram stops before the others.

Example Stan is taking his French oral test. The probability of him passing is 0.8
If he fails the test he has to resit. He can only have two tries.

a Draw a tree diagram to show the possible outcomes.

b Find the probability that he fails both tests.

a

1st test 2nd test

0.8 pass

0.2 fail 0.8 pass
0.2 fail

The top path stops because Stan has passed his test.
There is no need for him to resit.

b $P(\text{Stan fails both tests}) = 0.2 \times 0.2 = 0.04$

---

4 Martin has to pass a test before he can join the school choir.
The probability that he passes is $\frac{5}{8}$. He is allowed two tries.

a Copy the tree diagram. Fill it in.

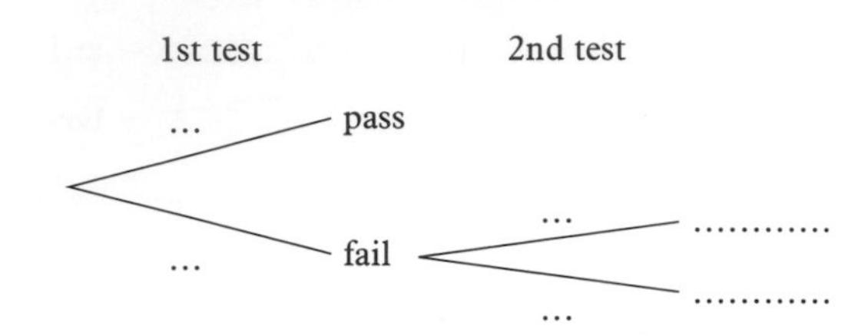

Find the probability that Martin:

b passes on the 2nd try

c fails both tests.

5 Jenny has a 45% chance of getting a place on a gym course.
If she does not get a place on the first course, then she can try for a second course.
Use a tree diagram to find the probability that Jenny:

a gets a place on the 1st course

b does not get a place on either course.

6 Keith has to throw a six with a dice to win a game.
He is allowed to throw the dice three times.
Use a tree diagram to find the probability that Keith:

a wins on the 2nd try

b wins on the 3rd try

c does not win

● **d** wins.

Questions will sometimes use the words "with replacement" or "without replacement"

"with replacement" means the events are independent
"without replacement" means the events are dependent.

Example A box contains 3 red and 5 blue cubes:

a Paul takes out a cube at random, replaces it, then takes out a second cube.
Find the probability that Paul picks out 2 red cubes.

b Sue uses the same box of counters.
She takes out 1 cube at random but does not replace it.
She then picks out a second cube.
Find the probability that Sue picks out 2 red cubes.

a

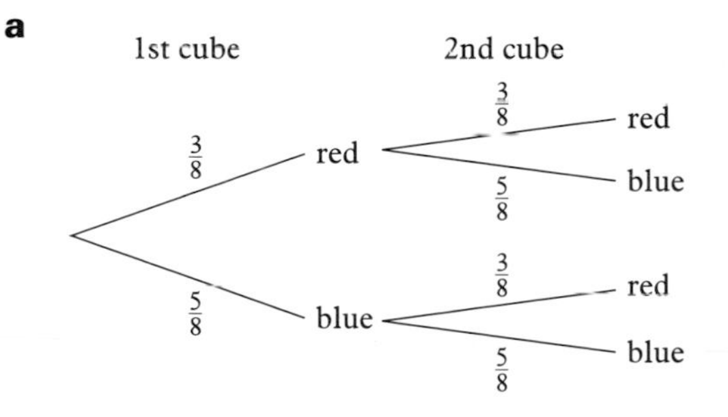

$P(\text{Paul picks out 2 red cubes}) = \frac{3}{8} \times \frac{3}{8} = \frac{9}{64}.$

b

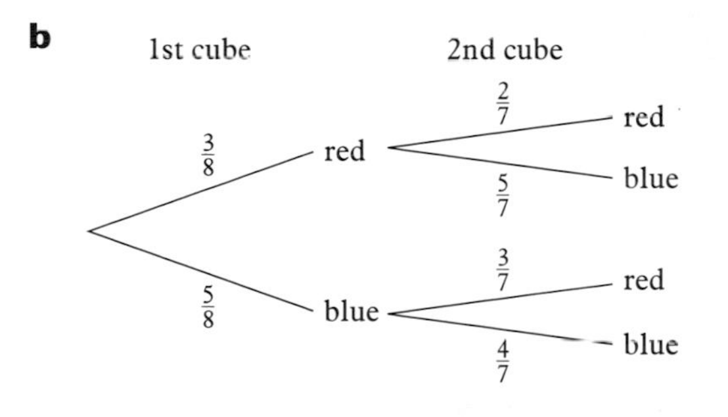

$P(\text{Sue picks out 2 red cubes}) = \frac{3}{8} \times \frac{2}{7} = \frac{6}{56} = \frac{3}{28}.$

Sometimes without replacement has to be assumed.

e.g. when a chocolate chosen at random is eaten
when a battery chosen at random is tested to destruction
when two things are chosen at the same time.

7 A bag contains 6 green and 4 purple beads.

a Jo takes out a bead at random, replaces it, then takes out a second bead. Find the probability that Jo picks out 2 purple beads.

b Anoop uses the same bag of beads.
He takes out 1 bead at random but does not replace it.
He then picks out a second bead.
Find the probability that Anoop picks out 2 green beads.

8 Sam has a drawer containing 3 pairs of socks. One pair is black, one pair is white and the other pair is red. Same takes out two socks at random. Find the probability that:

a both socks are red **b** both socks are the same colour.

9 A fruit bowl contains 4 apples, 2 oranges and 3 bananas.
Colin eats one piece of fruit at random from the bowl.
His sister Cara then eats a piece of fruit at random.
Find the probability that:

a Cara eats an apple **b** both eat apples **c** neither eat apples.

Example A bag contains 4 yellow and 6 red balls.
Two balls are picked out at random.
Find the probability that *at least one* ball is red.

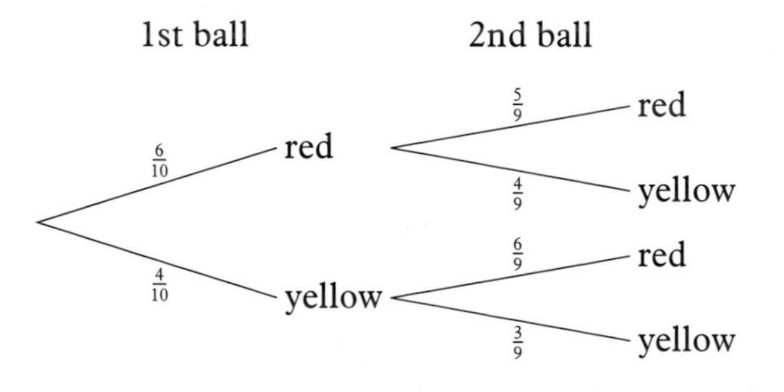

The events P '*at least one* ball is red' and P' '*no* balls are red' are mutually exclusive and nothing else is possible.

Either *no* balls are red or *at least one* of them is red.

$$P(\text{at least one ball is red}) = 1 - P(\text{no balls are red})$$
$$= 1 - \left(\tfrac{4}{10} \times \tfrac{3}{9}\right)$$
$$= 1 - \tfrac{12}{90}$$
$$= \tfrac{13}{15}$$

Exercise 12:7

1 James spins two coins.
Find the probability that James gets at least one head.

2 Amy throws a biased dice three times.
For each throw the probability of Amy not scoring a 6 is $\frac{2}{3}$.
Find the probability that she gets at least one 6 in three throws.

3 Graham is making tiaras with semi-precious stones.
The probability of a stone having a flaw is 0.2
He uses 3 stones to make a tiara.
Find the probability that, of the stones that he uses,
a none is flawed
b only one is flawed
c two are flawed
d at least one is flawed.

4 Tom plants 3 sunflower seeds.
The probability of a seed growing is 0.9
Find the probability that:
a no seeds grow
b at least one seed grows.

5 Julia plays three matches in a tournament.
She can win, lose or draw each match.
The probability that she wins is 0.6, the probability that she loses is 0.3
Find the probability that:
a she draws a match
b she wins all three matches
c she does not win all three matches.

6 A box contains 3 red, 2 blue and 5 green counters.
Sara picks out a counter at random, notes the colour and replaces it.
She does this three times.
a Find the probability that all three counters are the same colour.
b Use your answer to part **a** to find the probability that all three counters are not the same colour.

1 The sweets in a jar are red, yellow or orange.
The probability that a sweet, chosen at random, will be red is $\frac{1}{4}$ and the probability that it will be yellow is $\frac{2}{5}$.
If I choose one sweet at random what is the probability

a that it will be red or yellow, *(2)*
b that it will be orange, *(2)*
c that it will be white? *(1)*

There are 60 sweets in the jar.

d Calculate the number of red sweets. *(2)*
e $\frac{1}{3}$ of the red sweets have soft centres and the rest have hard centres.
How many red sweets have hard centres? *(2)*

NEAB, 1997, Paper 1

2 A fair eight-sided dice had six red faces and two white faces.

a The dice is thrown once.
What is the probability of obtaining
i a red face, *(1)*
ii a white face? *(1)*

b The dice is thrown twice.
What is the probability that two red faces are obtained? *(2)*

c The dice is thrown four times.
i What is the probability that red is obtained every time? *(2)*
ii Is it likely or unlikely that red is obtained every time? *(1)*

NEAB, 1998, Paper 2

3 Last year 1800 cars were repaired in a local garage. The repair record for each car indicates whether or not it runs on leaded or unleaded fuel and whether it has a large engine size (1600 cc or above) or small (below 1600 cc). 64% of all the cars ran on leaded fuel. Of the 340 cars of large engine size, 80% ran on leaded fuel.

a Copy and complete the table opposite, entering the number of vehicles in each category. *(4)*

Fuel type \ Engine size	Large	Small	Totals
Leaded			
Unleaded			
Totals			1800

b Using the table, calculate the probability that a repair record chosen at random is for a car that:
i runs on leaded fuel and has a small engine size, *(1)*
ii runs on leaded fuel or has a small engine size, *(2)*
iii runs on unleaded fuel, given that it has a small engine size. *(2)*

NEAB, 1998, Paper 3

4 In an examination there were two questions each of which required the candidates to select the one correct answer from a choice of five answers.
A candidate decides to guess the answers to these questions.

a What is the probability of guessing the **first one** correctly? *(1)*

b What is the probability of getting the first one wrong? *(1)*

c Copy and complete the tree diagram to show the outcome of guessing the answers to two questions. *(3)*

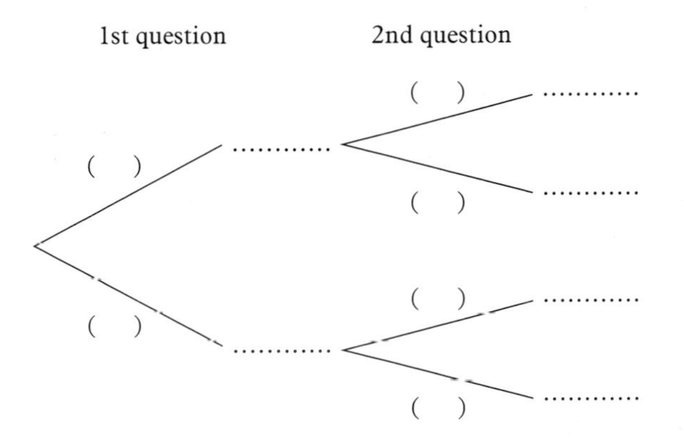

d Calculate the probability that **both** questions are guessed correctly. *(2)*

e Calculate the probability that **only one** of the two questions is guessed correctly. *(2)*

In another test, **thirty** similarly structured questions were all guessed by the candidate.

f How many would you expect him to get right? *(2)*

SEG, 1996, Paper 2

5 A company produces chocolate.
60% of its production is milk chocolate and the rest is plain chocolate.
Of the milk chocolate bars 75% are large and the others are small.
Of the plain chocolate bars 70% are large and the others are small.

a Draw a clearly labelled tree diagram to represent these probabilities. *(3)*

b A bar of chocolate is selected at random.

i Calculate the probability that it a small bar of milk chocolate. *(2)*

ii Calculate the probability that it is a small bar of chocolate. *(3)*

c If 270 small bars of chocolate were selected at random, how many of them would you expect to be milk chocolate? *(3)*

SEG, 1998, Paper 4

6 Peter is playing football and tennis.
He has one match in each sport to play.
The probability that he will win the football match is 0.3; the probability that he will draw is 0.5
He has a 60% chance of winning the tennis match, otherwise he will lose.

a Copy and complete the tree diagram. *(3)*

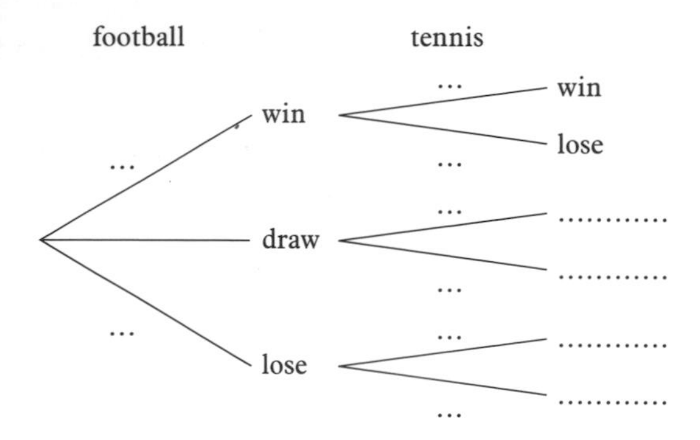

Find the probability that Peter:

b wins both matches *(1)*

c loses both matches *(1)*

d wins only one match *(2)*

e wins the tennis match only *(2)*

7 Jim has to go through one set of traffic lights and one level crossing on his way to work. The probability that he will have to stop at the lights is $\frac{5}{12}$.
He has a 25% chance of having to stop at the level crossing.

a Draw a tree diagram showing all the possible outcomes. *(3)*

Find the probability that Jim:

b stops at both *(1)*

c does not have to stop at either *(1)*

d only has to stop at the crossing *(2)*

e only has to stop at one of them. *(2)*

8 Steven has two bags of cubes. The first bag has red and green cubes in the ratio 3:5. The second bag has green and white cubes in the ratio 2:1.
Steven picks up a cube at random from each bag.

a Draw a tree diagram to show all the possible outcomes. *(3)*

Find the probability that:

b both cubes are green *(1)*

c the 1st is green and the 2nd white *(1)*

d neither cube is green *(2)*

e only one cube is green. *(2)*

9 A city bus company carries out a survey of the commuting habits of city-centre workers. The survey reveals that 40% of commuters travel by bus, 25% by train and the remainder by private vehicle.

Of those who travel by bus, 60% have a journey of less than 5 miles and 30% have a journey of between 5 and 10 miles.

Those travelling by train include 30% who have a journey of between 5 and 10 miles and 60% who travel more than 10 miles.

Of those using private vehicles, 20% travel less than 5 miles with an equivalent percentage travelling more than 10 miles.

a Use the above information to copy and complete the following tree diagram.

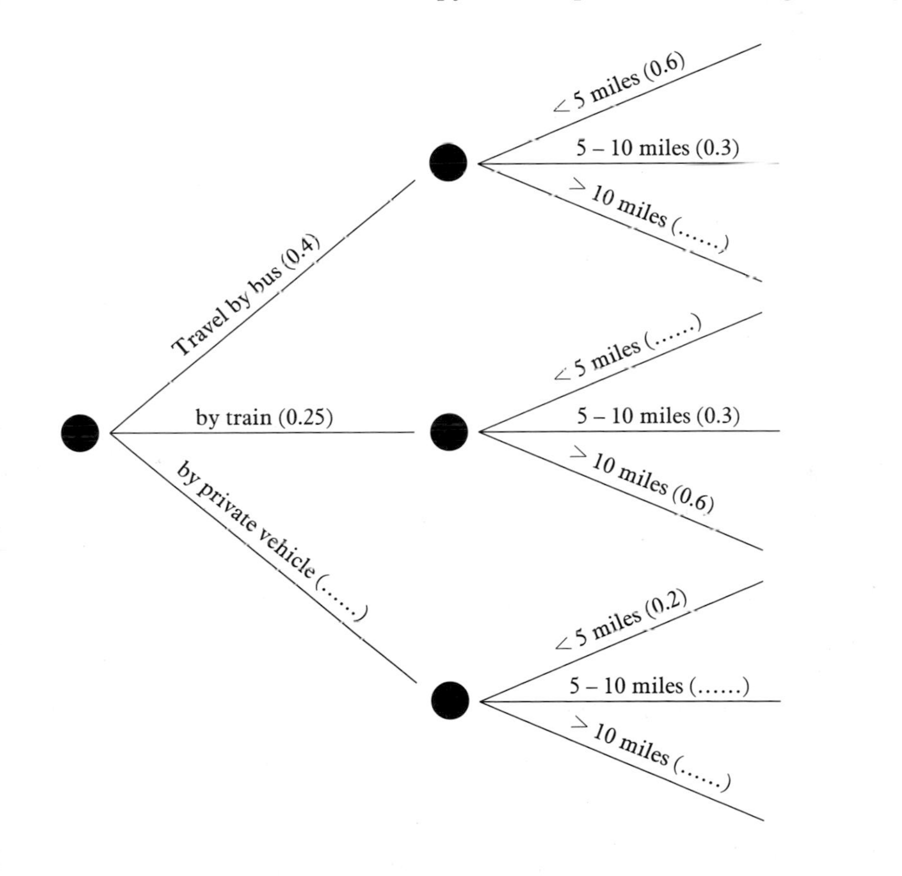

(3)

b Hence determine the probability that a commuter chosen at random

i travels by bus for a journey of less than 5 miles, *(2)*

ii has a journey of more than 10 miles, *(3)*

iii uses a private vehicle, given that the commuter travels between 5 and 10 miles. *(4)*

NEAB, 1997, Paper 3

10 A school debating team has six members, four girls and two boys. Two members are to be chosen at random to lead the debate. The names of the six members are written on pieces of paper and placed in a hat. Two pieces are then chosen at random.

a Complete the tree diagram by filling in the probabilities on the branches.

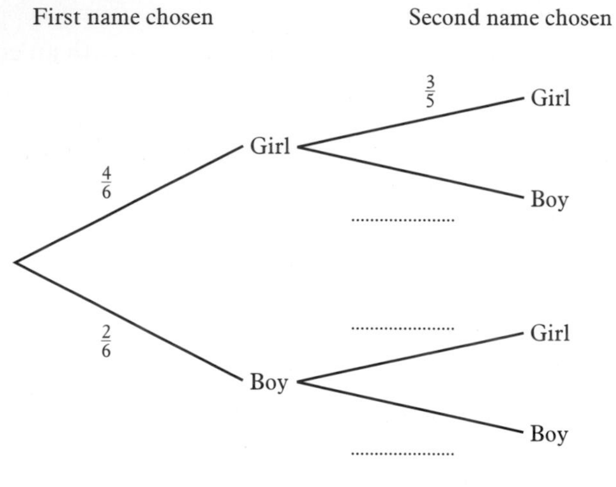

(3)

b Find the probability that both girls are chosen. *(2)*

c Find the probability that at least one boy is chosen. *(2)*

Independent events

Two events are **independent** if the outcome of one has no effect on the outcome of the other.

If you roll a fair dice and toss a coin you can get 1, 2, 3, 4, 5 or 6 on the dice and either a head or a tail with the coin.

Whatever you get on the dice has no effect on what you get with the coin.

Probability of independent events

If two **events** A and B are **independent** then the **probability** of them both happening is called $P(\text{A} \textbf{ and } \text{B})$.

$P(\text{A} \textbf{ and } \text{B}) = P(\text{A}) \times P(\text{B})$

Mutually exclusive

Events are **mutually exclusive** if they cannot happen at the same time.

When a dice is thrown the events 'getting an even number' and 'getting a 3' cannot happen at the same time.
This is because 3 is not an even number.
The events 'getting an even number' and 'getting a 3' are mutually exclusive.

Probability of mutually exclusive events

For two **mutually exclusive events** A and B, the **probability** that *either* event A *or* event B will occur can be found by *adding* their probabilities together.

$P(\text{A} \textbf{ or } \text{B}) = P(\text{A}) + P(\text{B})$

Exhaustive events

Events are **exhaustive** if all possible outcomes are included.

Tree diagrams

You can use tree diagrams to show the outcomes of two or more events.
Each branch represents a possible outcome of one event.
The probability of each outcome is written on the branch.
The final result depends on the path taken through the tree.
You **multiply** the probabilities along the branches.
You **add** the probabilities when more than one path is used.

1 The probability that Jack will go to college is $\frac{3}{4}$.
The probability that Andrew will go to college is $\frac{3}{5}$.
Find the probability that they will both go to college.

2 Joy is responsible for checking the final product on an assembly line making toy cars.
A car has a 2% chance of being faulty.
Two cars are chosen at random.
Find the probability that:

a both cars are faulty
b only one car is faulty
c neither car is faulty
d only the first car is faulty.

3 A bag contains four red, three blue and eight white balls.
Two balls are picked at random.

a Copy and complete the tree diagram.

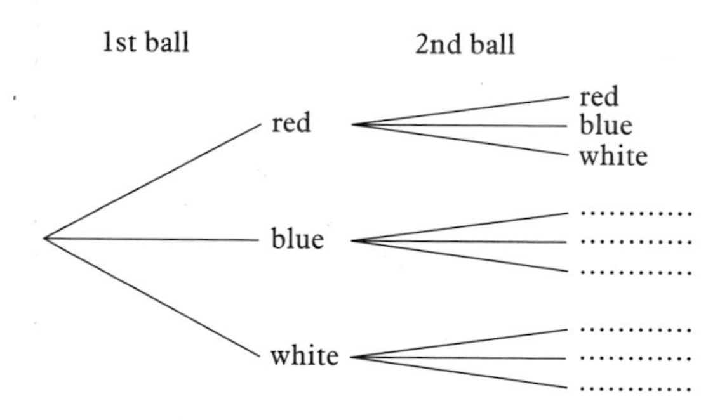

Find the probability that:
b both balls are red
c both balls are the same colour
d only one ball is red
e one is blue and one is white.
f at least one ball is red.

13 Using factorials

1 Factorial notation
Defining factorial notation
Using factorials to count the number of arrangements
Arrangements with restrictions
Using factorials to find probabilities

2 Selections or combinations
Defining selections or combinations

3 The binomial distribution
Rules for the binomial distribution
The probability of x successes in n trials

EXAM QUESTIONS

SUMMARY

ICT IN STATISTICS

TEST YOURSELF

1 Factorial notation

Noah counted his animals in two by two: monkeys, giraffes, zebras … *or* giraffes, monkeys, zebras *or* … *or* … *or*!!

Order is important in probability.

Look at these products: $4 \times 3 \times 2 \times 1$, $6 \times 5 \times 4 \times 3 \times 2 \times 1$ and $3 \times 2 \times 1$.
There is a quick way of writing this type of calculation.
If you want to multiply all the numbers starting with a number x and going down to 1, it is called **x factorial**. This is written **$x!$**

So 6 factorial is written 6!

$$\begin{aligned} 6! &= 6 \times 5 \times 4 \times 3 \times 2 \times 1 \\ &= 720 \end{aligned}$$

Two factorials can be multiplied together:

$$\begin{aligned} 6! \times 3! &= 6 \times 5 \times 4 \times 3 \times 2 \times 1 \times 3 \times 2 \times 1 \\ &= 4320 \end{aligned}$$

You can also divide by factorials:

$$\begin{aligned} \frac{5! \times 4!}{3!} &= \frac{5 \times 4 \times 3 \times 2 \times 1 \times 4 \times 3 \times 2 \times 1}{3 \times 2 \times 1} \\ &= \frac{2880}{6} \\ &= 480 \end{aligned}$$

Note that 6! $= 6 \times 5 \times 4 \times 3 \times 2 \times 1$

In the same way $x!$ $= x \times (x-1) \times (x-2) \times (x-3) \ldots \times 2 \times 1$

Exercise 13:1

1 Calculate:

a $7!$

b $8!$

c $3! \times 4!$

d $5! \times 2!$

e $\dfrac{6! \times 5!}{4!}$

f $\dfrac{4! \times 5!}{2!}$

There is a key on your calculator that works out factorials.
It is marked $n!$

Example You can use the calculator to find the value of:

a $12!$ **b** $0!$ **c** $\dfrac{10! \times 4! \times 3!}{8!}$

a $12! = 479\,001\,600$ Key in: [1] [2] [n!] [=]

b $0! - 1$ Key in: [0] [n!] [=]

$0!$ is a special case. It is always equal to 1.

c $\dfrac{10! \times 4! \times 3!}{8!} = 12\,960$ Key in: [1] [0] [n!] [×] [4] [n!] [×] [3] [n!] [÷] [8] [n!] [=]

---

2 Use a calculator to work out:

a $11! \times 3!$

b $9! \times 4!$

c $5! \times 3! \times 2!$

d $\dfrac{6! \times 7!}{2!}$

e $\dfrac{12! \times 6!}{9!}$

f $\dfrac{4! \times 8!}{7!}$

g $\dfrac{14! \times 5! \times 2!}{13!}$

h $\dfrac{4! \times 0! \times 7!}{3! \times 5!}$

i $\dfrac{6! \times 15!}{12!}$

• **3** Write the answer to each of these in terms of x:

a $\dfrac{10x}{4}$

b $\dfrac{7!x}{6!}$

c $\dfrac{10!x}{8!}$

d $\dfrac{x!}{(x-1)!}$

e $\dfrac{(x+1)!}{x!}$

f $\dfrac{(x+2)!}{x!}$

Flags are used to send messages while at sea. You can send many different messages with just three flags. You arrange the flags in different orders for each message.
Here are three flags:

These are the different ways that the flags can be arranged.

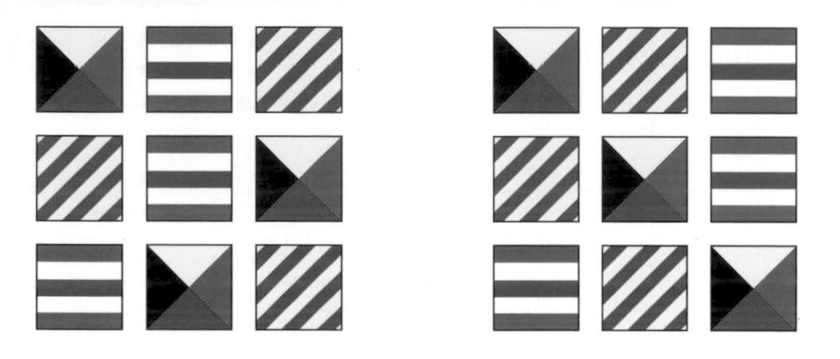

So the number of arrangements = 6

It is often easier to calculate the number of arrangements rather than draw them.

Number of ways of choosing the **first** flag = **3**

This leaves two flags.

Number of ways of choosing the **second** flag = 2

This leaves one flag.

Number of ways of choosing the **third** flag = 1

*This leaves **no** flags.*

The total number of arrangements is found by multiplying these numbers together.

The total number of arrangements = $3 \times 2 \times 1 = 3! = 6$

Example A bag contains 1 black, 1 red, 1 yellow, 1 purple and 1 orange counter. Find the number of ways that these five counters can be taken from the bag, one at a time.

Number of ways of choosing the **first** counter $= 5$

This leaves four counters.

Number of ways of choosing the **second** counter $= 4$

This leaves three counters.

Number of ways of choosing the **third** counter $= 3$

This leaves two counters.

Number of ways of choosing the **fourth** counter $= 2$

This leaves one counter.

Number of ways of choosing the **fifth** counter $= 1$

*This leaves **no** counters.*

The total number of arrangements $= 5 \times 4 \times 3 \times 2 \times 1$
$= 5!$
$= 120$

In the same way, $n!$ is the number of ways of placing n **different objects** in a line.

Exercise 13:2

1 Find the number of ways that six children, Luke, Sara, George, Christina, Jason and Rachel can stand in a line outside the canteen waiting for lunch.

2 Jonty puts 12 different French books on a bookshelf. Find the number of ways that the 12 books can be placed on the shelf.

3 Hannah is playing a game of Scrabble. She takes the letters T, W, J, K, Z, Y and A from the bag. Find the number of ways that she can place these on her letter rack.

4 There are five different coins in my pocket, a 50p, a 10p, a 20p, a 1p and a 2 p. Find the number of ways that I can take them out of my pocket one at a time.

5 Lisa always has a packed lunch for school. Every day, she has four different pieces of fruit. Today she has an apple, a pear, a plum and a banana.
She eats one piece of fruit completely before she eats the next.
Find the number of different ways that she can eat her fruit.

Sometimes the number of arrangements can be restricted.

Example **1**

These six picture cards are to bc placed in a line.

a Find the number of arrangements which can be made if the octopus is to be placed on the far right.

b Find the number of arrangements which can be made if the octopus is to be placed on the far right and the whale is to be placed on the far left.

a The arrangements must follow this form:

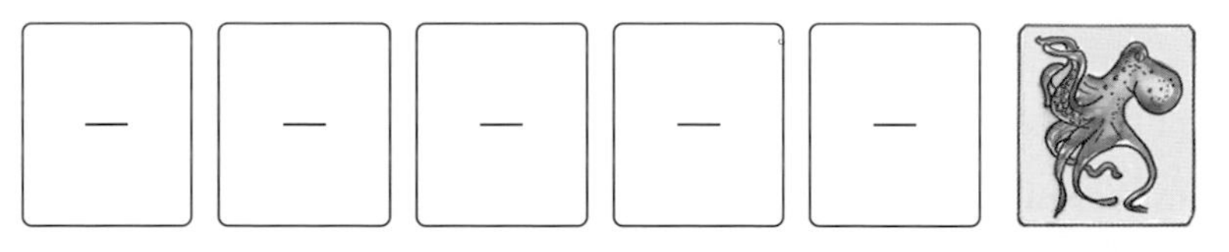

There is only **one** way of choosing the card on the far right.
The remaining five spaces on the left can be filled in $5 \times 4 \times 3 \times 2 \times 1 = 5!$ ways.

So the number of arrangements $= 5! \times 1$
$= 120$

b The arrangements must follow this form:

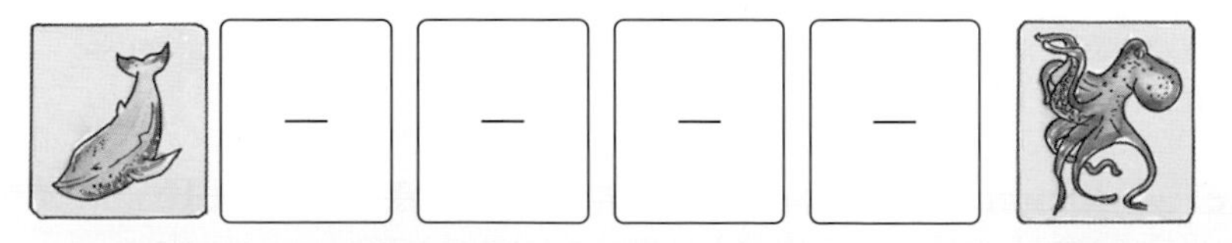

There is only **one** way of choosing both the right-hand card and the left-hand card.
The four middle spaces can be filled with the remaining cards in $4 \times 3 \times 2 \times 1 = 4!$ ways.

So the number of arrangements $= 1 \times 4! \times 1$
$= 24$

6 Katherine takes the 7, 8, 9, 10, jack, queen, king and ace of hearts from a pack of playing cards. She places these cards in a line on the table. Find the number of arrangements if:

a the jack must be placed first in the line

b the ace must be placed first in the line and the queen placed last in the line.

7 The letters from the word RANDOM are to be placed in a line. Find the number of arrangements if the letter A must always be placed last.

8 Yesterday, the school netball team won the regional finals. They are all to shake hands with the headteacher. Write down the number of ways the seven girls can shake hands with the headteacher, if the captain must go first.

9 A French number plate has six numbers. The numbers used on Pierre's car are 9, 1, 3, 7, 5 and 2. If the registration number is even, find the number of different possible number plates for Pierre's car.

Example Find the number of arrangements of the letters from the word **GERMAN**, if there must be a vowel at each end of the arrangements.

There are two vowels, so the arrangements must be:

E					A

or

A					E

Number of arrangements
$= 1 \times 4! \times 1$
$= 24$

Number of arrangements
$= 1 \times 4! \times 1$
$= 24$

Total number of arrangements $= 24 + 24$
$= 48$

---

10 Find the number of arrangements of the letters from the word HISTORY, if there must be a vowel at each end of the arrangement.

11 There are 10 children in a classroom. They are asked to stand in a line, with either the tallest first and the smallest last, or with the smallest first and the tallest last. Find the number of possible arrangements.

You can use factorials to help you work out probabilities.

Example Shaun has five counters in a box. They are of five different colours: red, grey, white, yellow and green. He picks out the counters one at a time. Shaun wants to find the probability that the first counter is green.

Shaun needs to work out the total number of possible outcomes.

The total number of arrangements of the five counters $= 5!$
The number of arrangements with the green counter first $= 1 \times 4!$

The probability that the first counter is green $= \dfrac{1 \times 4!}{5!}$

$= \dfrac{1}{5}$

Exercise 13:3

1 Seven different coloured coaches are in a line outside a hotel. Find the probability that the first coach is green and the second coach is white.

2 Jez is trying to see which numbers he can create by placing five numbered cards in a line. This is one of his arrangements:

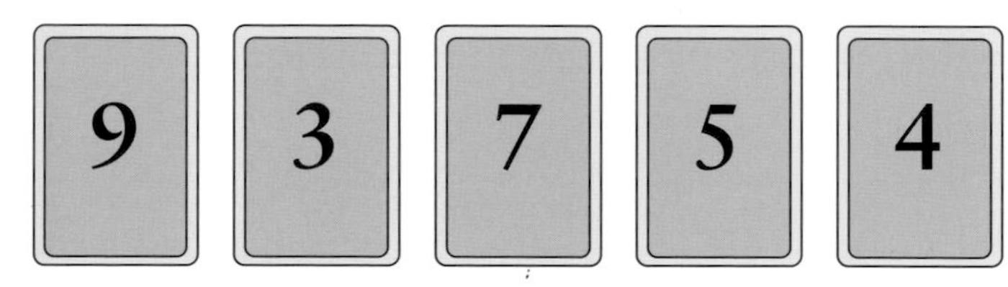

a Find the probability that the number formed in this way is divisible by five.

b Find the probability that the number which is made from all five of these cards is greater than 70 000.

3 The letters from the word MACHINE are placed at random in a line.

a Find the probability that the first letter is an M.

b Find the probability that the last letter is either an A or an E.

4 Jennifer is making a necklace. She has 10 different coloured beads. She makes the necklace by threading beads on a cord. Find the probability that her first chosen bead is red, the second chosen bead is green and the third bead is white.

2 Selections or Combinations

There are eight people in the school chess club. Rani has to choose three people to take part in a competition. Rani wonders how many different ways she could choose the team.

Selection

A **selection** or a **combination** is a group of objects where order is not important.

Example 1 Jon has four text books in his bag; a French book, a History book, an English book and a Geography book. He selects two books from his bag.

These are the possible selections:

French, History	History, English	French, Geography
French, English	History, Geography	English, Geography

The first selection is a French book and a History book.
The order in which Jon picks these two books does not matter. He can pick out the French book first and then the History book or the other way round. He will still end up with the same two books. There are six possible selections.

You can find the total number of selections using factorials instead.

You use the formula: ${}^nC_x = \frac{n!}{(n-x)!x!}$

${}^nC_x$ gives the number of selections of x items from a total of n items.
Jon is selecting 2 books from 4 books so $x = 2$ and $n = 4$.

$${}^4C_2 = \frac{4!}{(4-2)!2!} = \frac{4!}{2! \times 2!} = 6 \text{ selections}$$

Both methods give the same answer.

Example 2 Rani has to choose 3 people to take part in the chess competition from a total of 8 people.
The order of the 3 people does not matter.

The total number of ways $= {}^8C_3$

$$= \frac{8!}{(8-3)! \times 3!}$$

Key in:

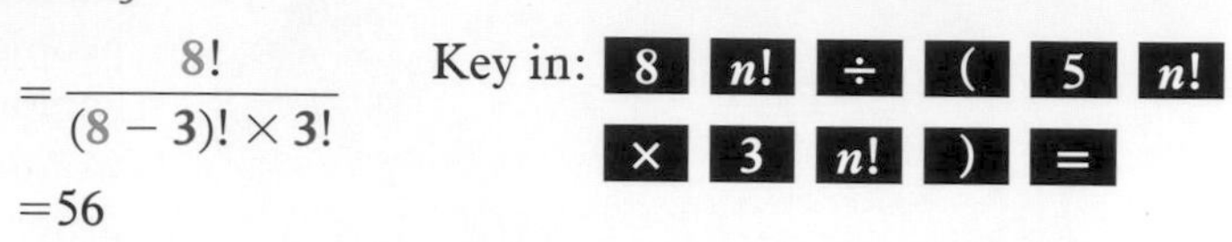

$= 56$

Rani can choose the 3 people in 56 different ways.

It would take a long time to write down all the possible ways.
Using the formula is quicker.

Exercise 13:4

1 There are 6 different tins of soup in the cupboard. James chooses 4 of these tins. Find the total number of selections that he can make.

2 Rebecca has a bag of 10 different chocolates. She gives 5 of them to her friend Lisa. Find the number of selections of 5 chocolates that Lisa can make from the bag of 10 chocolates.

3 Matthew has a set of 12 different coloured crayons. He chooses 6 crayons from his set. Find the number of different combinations that he can make.

4 Tim chooses a tennis team of 4 from 8 players.
Find the number of different selections that he can make.

5 From a group of 16 children, a team of 4 is chosen to take part in a competition. Find the number of different teams that can be made.

6 Fifteen different cakes are displayed in a shop window. Shona chooses 5 of these for her friends. Find the number of different selections that she can make.

7 Kris is going on holiday. He has 6 different shirts. He decides that he only needs to take 3. Find the number of different selections that he can make.

3 The binomial distribution

Jim has four rather unusual coins. He has found that when he tosses each of the coins the probability of a head showing is always $\frac{2}{5}$. He tosses the four coins together. Jim wants to know the probability that there will be three heads and one tail.

The **3 heads** and **1 tail** can be arranged in **four** different ways.

$P(\mathbf{H, H, H, T}) = \frac{2}{5} \times \frac{2}{5} \times \frac{2}{5} \times \frac{3}{5} = (\frac{2}{5})^3 \times \frac{3}{5}$

$P(\mathbf{H, H, T, H}) = \frac{2}{5} \times \frac{2}{5} \times \frac{3}{5} \times \frac{2}{5} = (\frac{2}{5})^3 \times \frac{3}{5}$

$P(\mathbf{H, T, H, H}) = \frac{2}{5} \times \frac{3}{5} \times \frac{2}{5} \times \frac{2}{5} = (\frac{2}{5})^3 \times \frac{3}{5}$

$P(\mathbf{T, H, H, H}) = \frac{3}{5} \times \frac{2}{5} \times \frac{2}{5} \times \frac{2}{5} = (\frac{2}{5})^3 \times \frac{3}{5}$

The total probability for **3 heads** and **1 tail** $= 4 \times (\frac{2}{5})^3 \times \frac{3}{5}$

$= \frac{96}{625}$

You can work out the answer using the formula for ${}^nC_x$.

${}^4C_3$ is the number of selections of **3 heads** from 4 coins, so the number of different arrangements is ${}^4C_3$.
The probability of the three heads is $(\frac{2}{5})^3$
The probability of the one tail is $1 - \frac{2}{5} = \frac{3}{5}$

The probability of **3 heads** and **1 tail** $= {}^4C_3\,(\frac{2}{5})^3 \times (\frac{3}{5})$

$= 4 \times (\frac{2}{5})^3 \times (\frac{3}{5})$

$= \frac{96}{625}$

Exercise 13:5

1 For Jim's coins, find the probability of obtaining
   a two heads and two tails
   b one head and three tails.

2 Remya has five more unusual coins. The probability of a head showing on each coin is always $\frac{1}{4}$. The five coins are tossed together. Find the probability of obtaining:
   a three heads and two tails
   b one head and four tails.

The binomial distribution is a special probability model.
It is used to find the probability of an outcome in a number of trials.
A probability experiment for a binomial distribution must have these characteristics:

- Each trial has two possible outcomes; success or failure. When tossing Jim's coins, getting a head was success and not getting a head was failure.
- The two outcomes must be independent.
- There must be a fixed number of trials, n. For Jim this was 4.
- It must have a constant probability. This is called p. For Jim $p = 0.4$

The probability of x successes in n trials is ${}^nC_x\, p^x(1-p)^{n-x}$

Example A marksman firing at a target hits the bull's eye once in every five shots. If he fires 10 times, find the probability that he will hit the bull's eye:
a six times **b** three times.

$n = 10$ (This is the fixed number of trials.)
$p = 0.2$ (Hitting the bull's eye is a success.)
$1 - p = 0.8$ (This is the probability of failing.)

a $x = 6$ P(of hitting target 6 times) $= {}^{10}C_6\, 0.2^6\, 0.8^4$
$= 0.00551$ (3 s.f.)

b $x = 3$ P(of hitting target 3 times) $= {}^{10}C_3\, 0.2^3\, 0.8^7$
$= 0.201$ (3 s.f.)

3

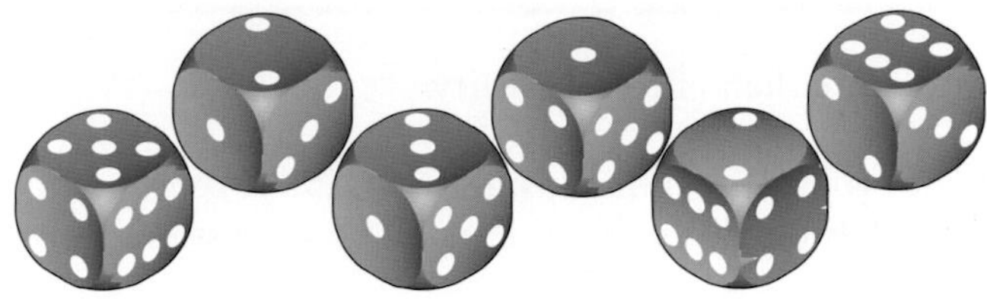

Six fair dice are rolled together. Find the probability of obtaining:

a five sixes **b** three sixes.

Give the answers to three significant figures.

4 The probability that a pen is faulty is $\frac{1}{10}$. The pens are sold in packs of eight. Find the probability, to three significant figures, that there are exactly:

a five faulty pens **b** two faulty pens.

5 Jean either walks to school or goes by bus. The probability that she walks to school on a summer morning is 0.7 For a school week of five days, find the probability that:

a she walks to school only once

b she walks to school exactly three times.

6 In a pack of lettuce seeds there are 20 seeds. The probability of a seed germinating is 0.8 Find the probability of exactly 19 seeds germinating. Give the answer to three significant figures.

7 In any group of students, the probability that a student wears glasses is 0.3 Find the probability that, in a group of 10 students, exactly four wear glasses. Give the answer to four decimal places.

8 A fair £2 coin is tossed six times.
Find the probability of obtaining:

a exactly four heads

b exactly three heads.

Give the answers to 2 decimal places.

9 The probability that a target is hit is 0.2 Jon shoots at the target eight times. Find the probability that he hits the target exactly four times. Give the answer to three decimal places.

10 The probability of a new born baby being a boy is 0.52 Mr and Mrs Svenson are planning to have four children. Find the probability, to 3 significant figures, that they will have exactly three boys.

Sometimes you need to find more than one probability.

Example In a family, the probability of having a boy is 0.52 If there are five children, find the probability that there will be at least four boys.

First you need to list the outcomes that give at least four boys.

Favourable outcomes are: **4 boys** and 1 girl $x = 4$
5 boys and 0 girls $x = 5$

Now find the probability for each of these outcomes

$P(\text{4 boys, 1 girl}) = {}^5C_4\, 0.52^4\, 0.48^1$
$= 0.17548$ (5 s.f.)

$P(\text{5 boys, 0 girls}) = {}^5C_5\, 0.52^5\, 0.48^0$
$= 0.03802$ (5 s.f.)

Add the probabilities together:

$P(\text{at least 4 boys}) = 0.17548 + 0.03802$
$= 0.214$ (3 s.f.)

11 The probability that a bulb is faulty is 0.01 Gillian has bought eight bulbs. Find the probability that at least two bulbs are faulty. Give the answer to four decimal places.

12 In a punnet of strawberries, the probability that a strawberry is mouldy is 0.12 Find the probability that, in a punnet of 12 strawberries, at most three are mouldy. Give the answer to four decimal places.

13 A fair coin is tossed 10 times. Find the probability of obtaining at least eight heads. Give the answer to four decimal places.

14 Sean uses this spinner five times. Find the probability that he gets more reds than greens.

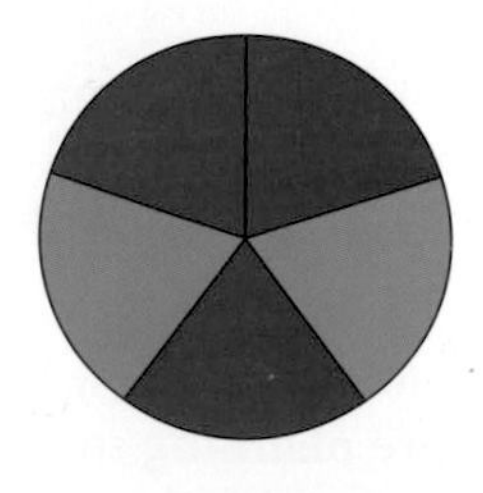

1

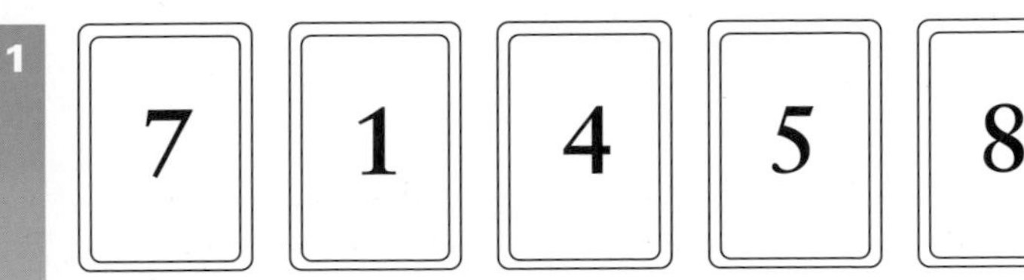

Jane arranges these cards to make five-digit numbers. Some possible numbers are: 45187, 81457, 17458 and so on. No card can be used more than once.

a Write down how many different numbers Jane can make. Give your answer as a factorial. *(1)*

b How many different numbers can Jane make if the number must end with 4? *(1)*

c How many different numbers can Jane make if each number starts and ends with an even digit? *(2)*

2 **a** At a mathematics competition pupils were asked to arrange three cards to make up three-digit numbers.

No card could be used twice in any number.

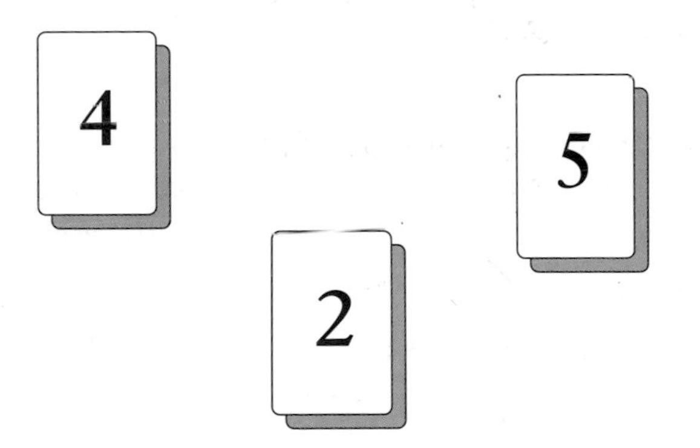

i How many different three-digit numbers can be formed with these three cards? *(1)*

ii What is the probability that a number formed is greater than 500? *(2)*

b Use factorial notation to work out how many six-digit numbers could be formed from six cards each showing a different number. *(2)*

SEG, 1996, Paper 1

3 A shop sells shirts in five different colours:

red white blue brown black

Rudi buys two shirts of different colours. How many different combinations of colours could he buy? *(2)*

4 A packet of geranium seeds contains 12 seeds.
The probability of a geranium seed germinating is 0.95
Use the binomial distribution to calculate the probability, to four decimal places, that:

a all 12 seeds germinate *(2)*

b exactly 11 seeds germinate *(2)*

c exactly 10 seeds germinate *(2)*

d 9 seeds or less germinate. *(2)*

SEG, 1997, Paper 3

5 Robin shoots arrows at a target. The probability that he will hit the target with any one shot is 0.6

a Use the binomial distribution to work out the probability that, in five shots, he will hit the target exactly three times. *(3)*

b Copy and complete the table to show the probability distribution of the number of hits in five shots. *(3)*

Number of hits in five shots	0	1	2	3	4	5
Probability	0.01	0.077	0.2304			

c There are two rounds in a competition.
In each round there are five shots.
The score is the sum of the number of hits in each round.
Using the table, calculate the probability that Robin will achieve a score of two in the competition. *(4)*

SEG, 1996, Paper 3

6 The probability that a girl is left-handed is 0.2
A group of 10 girls is selected at random.

a Why is the binomial distribution a suitable model for calculating the probability of the number of left-handed girls in the group? *(2)*

b **i** Calculate the probability that there are two left-handed girls in the group. *(2)*

ii Calculate the probability that there is one left-handed girl in the group. *(2)*

iii Calculate the probability that there are more than two left-handed girls in the group. *(3)*

SEG, 1998, Paper 3

7 Eight different models of cars are to be placed on a shelf in a row.

a Find the number of ways the eight models can be arranged. *(2)*

b There is only room for five cars on the shelf. Find the number of ways any five of the eight models can be arranged. *(2)*

c How many selections of five models can be made from the eight models? *(2)*

8 At a greyhound race meeting, each race has six dogs numbered 1, 2, 3, 4, 5 and 6. They can be placed in any of the six lettered starting traps.

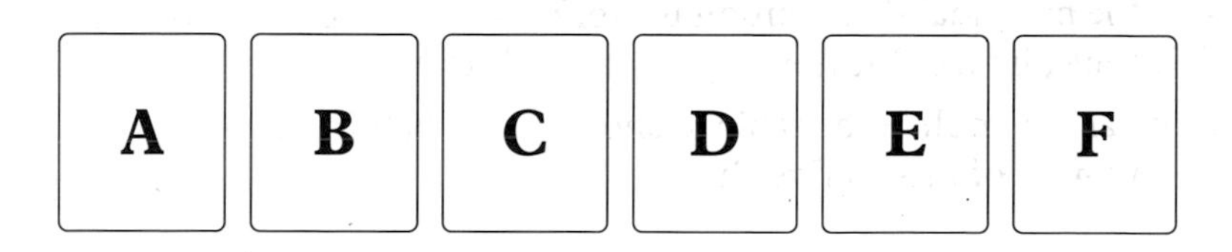

a In how many ways can the six dogs in one race be arranged in the traps? *(2)*

b In how many of these arrangements will dog 4 be in trap **B**? *(2)*

Due to faulty trap doors only four of the six dogs could run in the last race.

c Find the number of different selections of four dogs which could be made for this race. *(2)*

SEG, 1999, Paper 3

9 Rhian has 12 different pots to display on a shelf.
She puts them in a row.

a Find the number of ways the 12 pots can be arranged. *(2)*

b Only one of the pots has been decorated. Find the number of ways the 12 pots can be arranged if the decorated pot must be displayed on the left of the display. *(2)*

c Rhian has to select 5 of these pots for a display at a parents evening. How many selections can she make? *(2)*

d Find the probability that the decorated pot is included in the 5 pots. *(3)*

10 Leo rolls 4 unbiased ordinary dice.

a Use the binomial distribution to work out the probability that he will get exactly 3 sixes. Give your answers to 3 significant figures. *(2)*

b Copy and complete the table to show the probability distribution of the number of sixes. *(3)*

Number of sixes	0	1	2	3	4
Probability	0.482				

c Leo rolls the set of 4 dice twice.
Find the probability he gets a total of:
(1) 2 sixes in the two goes *(2)*
(2) 5 sixes in the two goes *(2)*

11 Jan buys a special offer pack of 6 light bulbs. The probability that a light bulb is damaged is 0.03 Use the binomial distribution to calculate the probability, to 3 significant figures, that:

a all the light bulbs are damaged. Give your answer in standard form. *(2)*

b exactly two light bulbs are damaged *(2)*

c less than two light bulbs are damaged *(2)*

d not more than two light bulbs are damaged *(2)*

e Explain your answers to parts **c** and **d**. *(1)*

Factorial notation $x! = x \times (x-1) \times (x-2) \ldots \times 3 \times 2 \times 1$

Example $5! = 5 \times 4 \times 3 \times 2 \times 1$
$= 120$ Key in: 5 $n!$ =

Arrangements $n!$ is the number of arrangements of n different objects when they are placed in a line.

Example Six different books are placed on a shelf in a bookcase. Find the number of arrangements.

Number of arrangements $= 6!$
$= 720$

If there are restrictions, these are considered first.

Selections *or* **combinations** A selection or combination is a group of objects where order is not important.
${}^nC_x$ gives the number of selections of x **items** from n **items.**

$${}^nC_x = \frac{n!}{(n-x)!x!}$$

Example There are 10 different coloured counters in a jar. Find the number of selections of **3** counters which can be made.

$${}^{10}C_3 = \frac{10!}{(10-3)!3!}$$
$$= 120$$

Binomial distribution The probability of x successes in n trials is ${}^nC_x\, p^x\,(1-p)^{n-x}$

Example The probability of a marksman hitting a target is always 0.7 He fires 10 shots. Find the probability that he hits the target **6** times.

The probability of 6 successes in 10 trials is ${}^{10}C_6\, 0.7^6\, 0.3^4 = 0.2001$ (4 d.p.)

Working with *Excel*

Calculating the probabilities for a binomial distribution

A marksman firing at a target has a probability of 0.2 of hitting the bull's-eye. He fires 10 times.

Use *Excel* to calculate the probability that he will hit the bull's-eye 0 times, once, twice, three times and so on up to ten times.

Enter the information into a new *Excel* document.

In Cell A1, type **Number of Successes**, in Cell B1, type **Number of Bull's-eyes** and in Cell C1, type **Probability**.

In Cell A2, type **0** and in Cell A3, type **1**.

These two values of 0 and 1 allow the pattern to be copied.

In Cell B2, type **0 Bull's-eyes** and in Cell B3, type **1 Bull's-eyes**.

*These two labels again allow the pattern to be copied. They must both say **Bull's-eyes**.*

	A	B	C	D
1	Number of Successes	Number of Bull's-eyes	Probability	
2	0	0 Bull's-eyes		
3	1	1 Bull's-eyes		
4	2	2 Bull's-eyes		
5	3	3 Bull's-eyes		
6	4	4 Bull's-eyes		
7	5	5 Bull's-eyes		
8	6	6 Bull's-eyes		
9	7	7 Bull's-eyes		
10	8	8 Bull's-eyes		
11	9	9 Bull's-eyes		
12	10	10 Bull's-eyes		
13				

Select Cells A2 and B3.

Place the cursor on the bottom right-hand corner of Cell B3 and drag down to Cell B12.

This copies both patterns.

Calculate the probabilities.

In Cell C2, insert the formula: **=BINOMDIST(A2,10,0.2,FALSE)** and press the **Enter** key.

Select Cell C2.

Place the cursor over the bottom right-hand corner of Cell C2 and drag down to Cell C12.

This copies the formula.

***10** is the number of trials.*

***0.2** is the constant probability.*

*In Cell A2, there is a **0**.*

*In Cell C2, the formula calculates the probability of **0** successes in **10** trials.*

*In Cell A3, there is a **1**.*

*In Cell C3, the formula calculates the probability of **1** success in **10** trials.*

- Select Cells B2 to C12.

- Use the Chart Wizard button.

- **Drawing a bar-chart.**
  From Chart type, choose Column.
  From Chart sub-type, choose the top left diagram.

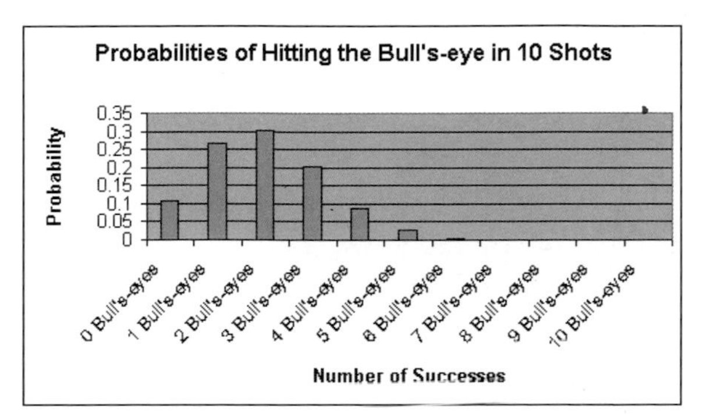

- **Labelling the bar-chart.**
  Click the Next> button until you get to the Chart Options window.
  In the Chart title box, type **Probabilities of Hitting the Bull's-eye in 10 Shots.**
  In the Category (X) axis box, type **Number of Successes.**
  In the Value (Y) axis box, type **Probability.**
  Click Finish.
  On the bar-chart, click on the Series 1 box and press Delete.
  Click on a blank area of the graph to select it before printing it.
  The graph should look like the one shown.
  It shows that the most likely number of bull's-eyes is 2.

Exercise 13:6

1 Prudeep uses this spinner eight times.
Use *Excel* to calculate the probability that Prudeep will get 0 reds, 1 red, 2 reds, and so on up to 8 reds.
Use a fraction instead of a decimal for the probability.

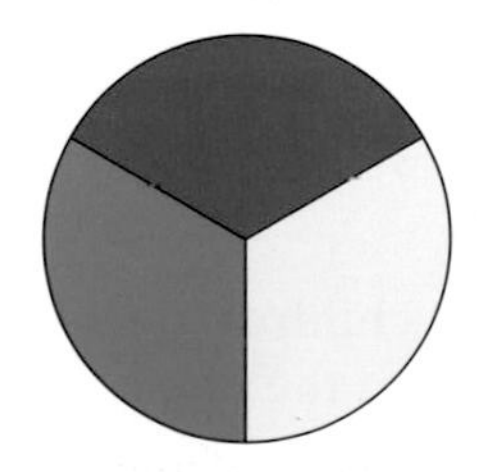

2 Christian tosses a coin 12 times.
Use *Excel* to calculate the probability that he gets 0 heads, 1 head, 2 heads and so on up to 12 heads.

1 Calculate:

a $10!$

b $8! \times 3!$

c $\dfrac{12! \times 7!}{8!}$

d $\dfrac{11! \times 4!}{8! \times 6!}$

e $\dfrac{6! \times 0!}{4!}$

f $\dfrac{4! \times 10!}{0! \times 7!}$

2 Find the number of ways that the letters from the word RHOMBUS can be arranged in a line.

3 Martin places the digits 7, 1, 3, 4, 5 and 9 in a line at random. Find the number of different orders of these digits if the resulting number must be divisible by 5.

4 A team of four children is to be selected from 10 children in the class. Find the probability that the oldest child and the youngest child will be part of the team.

5 Jonathan has eight different CDs. He has two heavy metal CDs, and six pop CDs. He is placing them in a single storage rack. He insists that the heavy metal CDs must go together at the top, or together at the bottom of the rack. Find the number of ways that he can place the eight CDs in the rack.

6 There are 20 different coloured sweets in a jar. Find the number of different selections of six sweets that can be taken from the jar.

7 The supermarket sells baked beans in six different varieties. Janet buys three different varieties. How many different combinations can she buy?

8 Five children are taking a music exam. The probability of each child passing the exam is 0.8 Find the probability of three children passing.

9 The probability that Paul is woken by his alarm clock is 0.3 Find, for a school week of five days, the probability that he will be woken by the alarm on at least three days.

14 Scatter graphs

1 Scatter diagrams

Gina and Greta are on a pebble beach. They are measuring the widths and lengths of pebbles.

They think that the wider pebbles might have longer lengths.
They say that this is obvious.

In maths this is called correlation.

Scatter graph

A **scatter graph** is a diagram that is used to see if there is a connection between two sets of data.
Here are some of Gina's results.

Length (mm)	42	28	50	24	37	52	22
Width (mm)	30	21	35	18	28	38	16

She plots the results in a graph.
Gina gets the co-ordinates from the table.
The first two points are (**42**, **30**) and (**28**, **21**).

The points almost lie in a straight line.
Gina needs to draw a line through her points.
First she has to work out the mean of the lengths and widths.

Lengths and widths of pebbles on the beach

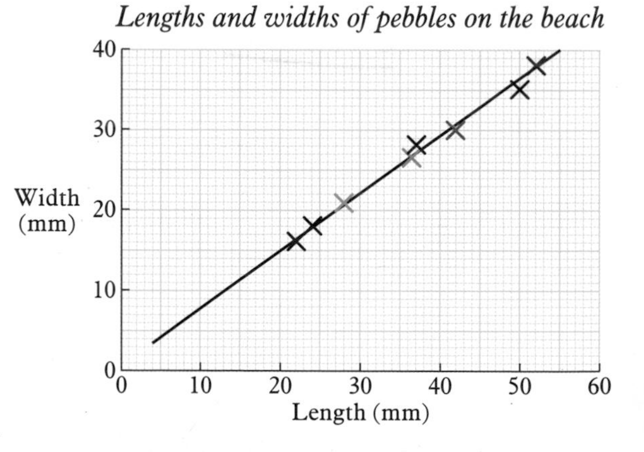

$$\text{Mean of the lengths} = \frac{42 + 28 + 50 + 24 + 37 + 52 + 22}{7} = \frac{255}{7} = 36.4 \text{ (3 s.f.)}$$

$$\text{Mean of the widths} = \frac{30 + 21 + 35 + 18 + 28 + 38 + 16}{7} = \frac{186}{7} = 26.6 \text{ (3 s.f.)}$$

Gina plots the mean point (**36.4, 26.6**).
She draws a line through the mean point and in the middle of the rest of the points.
This line is called the **line of best fit**.

The line of best fit is used to estimate other data values.

Example Estimate the width of a pebble of length **a** 30 mm **b** 10 mm.

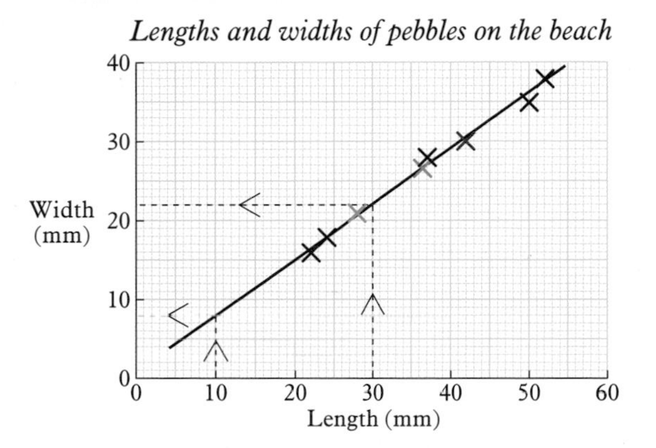

a Draw a vertical line from 30 on the length axis to the line of best fit then a horizontal line to the width axis.
Read off this value. An estimate for the width is 22 mm.

b An estimate for the width is 8 mm. This estimate is not as good as the estimate in part **a**.
This is because the length 10 mm is outside the range of the lengths in the table.

Interpolate **Interpolate** means to estimate a value between two values that you know.

The length 30 mm is between other lengths in the table.
This is an example of interpolation.

Extrapolate **Extrapolate** means to find a value by following a pattern and going outside the range of values that you know.

The length 10 mm is outside the rest of the lengths in the table.
This is an example of extrapolation.

Extrapolation is less reliable because the pattern may not continue outside the existing values.

Exercise 14:1

1 These are the English and History results for some pupils in Year 11.

English	32	39	31	33	39	32	45	26	40	32	34
History	34	43	27	34	32	37	48	25	37	30	36

a Find the mean English mark and the mean History mark.

b Copy the axes. Draw a scatter graph for the data.

c Plot the point you found in **a**.

d Draw the line of best fit.

e John scored 35 in English. Estimate his History mark.

f Ann scored 55 in History. Estimate her English mark.

g Which of your two estimates is less reliable? Give a reason.

History marks: 0, 10, 20, 30, 40, 50, 60

English marks: 0, 10, 20, 30, 40, 50

Correlation

Correlation is a measurement of how strongly connected two sets of data are.
There are different types of correlation.

Positive correlation

This scatter graph shows the weights and heights of people.
As the weight increases so does the height.
This is **positive correlation.**

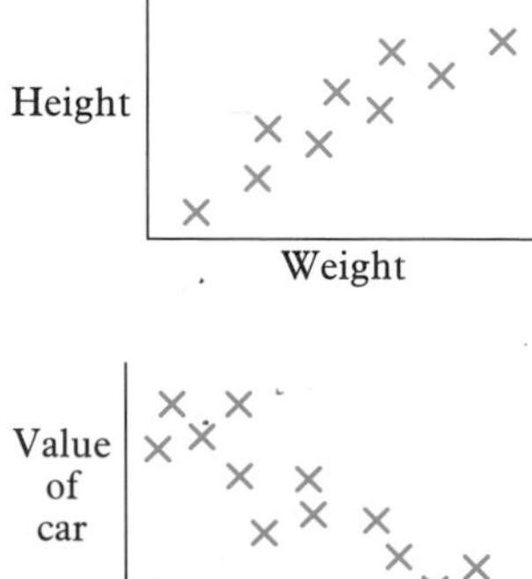

Negative correlation

This graph shows the values of cars and their ages. As age increases, value decreases.
This is **negative correlation.**

Zero correlation

This graph shows the height of some students against their Maths scores. There is no connection between these two things. There is **zero correlation** or **no correlation**.

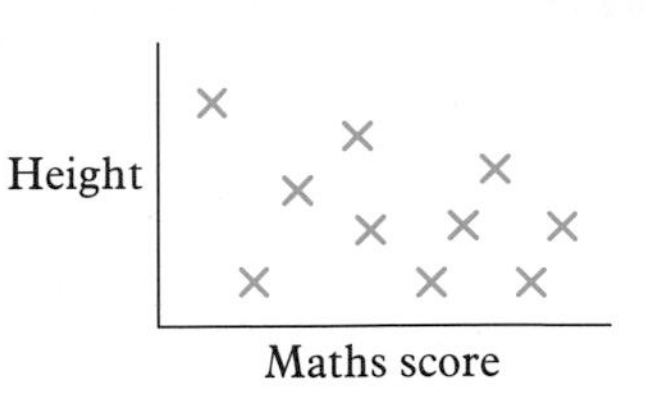

Correlation can be strong or weak.

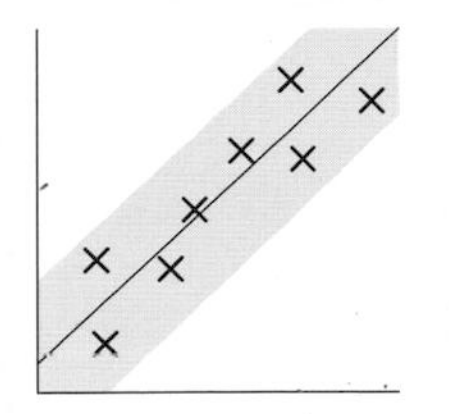

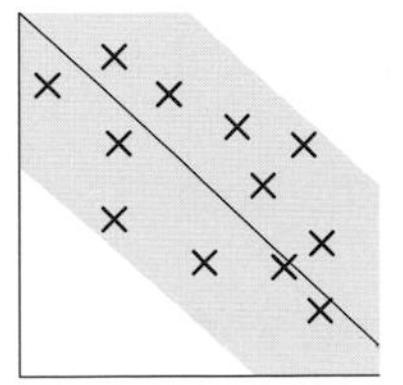

Strong correlation
The points all lie close to the line of best fit.

Weak correlation
The points are well spread out from the line of best fit but still follow the trend.

Two variables may show correlation but are not linked. Monthly sales of cameras and monthly sales of computers is an example of this.
Two variables may be linked and an increase in one *causes* an increase in the other. If this happens the two variables are said to have a **direct causal relationship**. An example of this is temperature and sales of ice-cream.

---

2 Barbara is doing a Physics experiment. She is measuring the resistance and the current in a circuit to see if they are related. She calculates the reciprocal of the current.

Resistance (ohms)	5	10	15	20	25	30	35	40
Reciprocal of current (amps^{-1})	0.13	0.26	0.36	0.20	0.59	0.77	1.0	1.1

a Find the mean resistance and the mean reciprocal current.

b Copy the axes.
Draw a scatter graph for the data.

c Draw the line of best fit.

d What type of correlation does your graph show?

e Estimate the current for a resistance of 23 ohms.

f Estimate the resistance for a current of 10 amps.

g Which of your two estimates is less reliable?
Give a reason for your choice.

Reciprocal of current (amps^{-1}): 0, 0.1, 0.2, 0.3, 0.4, 0.5, 0.6, 0.7, 0.8, 0.9, 1.0, 1.1

Resistance (ohms): 10, 20, 30, 40

3 The ages of Astra cars and their sale values are shown in the table.

Age (years)	4	3	5	2	3	4	3	5	6	2	7
Value (£1000s)	4.9	5.8	4.2	6.7	4.9	4.2	5.2	3.7	3.6	5.7	3.0

a Find the mean age and the mean value.

b Copy the axes.
Draw a scatter graph for the data.

c Draw the line of best fit.

d What type of correlation does your graph show?

e Estimate the age of a car worth £4500.

f Estimate the value of a one-year old car.

g Which of your two estimates is less reliable?
Give a reason for your choice.

h Do the two variables have a direct causal relationship?
Explain your answer.

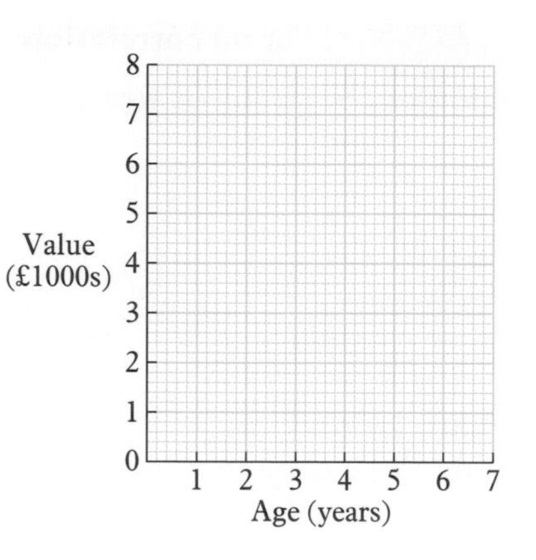

4 The scatter diagram shows the weights, in kg, and the heights, in cm, of 20 male members of the basketball club.

a Write down the weight of the heaviest member.

b Write down the height of the shortest member.

c One member is particularly heavy for his height.
Write down the height and weight of this member.

Weight (kg)
100
90
80
70
60
50
160 170 180 190 200 210
Height (cm)

There can be correlation between two variables which is not linear.
You need to draw a curve instead of a line of best fit when this occurs.

Tanya collects data on the height of a bouncing ball against time.

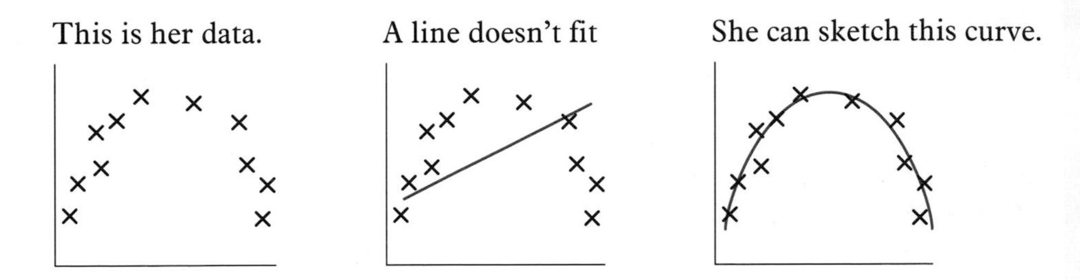

The curve best represents the correlation between the variables.

5 Plot each of these sets of data.
Draw either a straight line or a curve that best fits the data.

a

x	1	7	4	6	2	8	9	10	5
y	8	5	5	4	7	7	9	10	4

b

x	60	45	55	75	85	65	80	80	65
y	30	40	40	25	15	25	17	22	32

c

x	6	4	8	14	4	10	16	6	16	10	20	19
y	30	15	50	40	25	54	30	40	42	50	25	10

d

x	108	110	103	105	108	121	112	130	105	106	123	112
y	75	73	97	85	65	63	65	60	64	90	57	75

2 The equation of the line of best fit

Both sides of this bridge are at the same angle.
As you travel from left to right the left side has a positive gradient and the right side has a negative gradient. You need to remember about gradients to find the equation of the line of best fit.

The line of best fit is a straight line.
The equation of a straight line can be written in the form $y = mx + c$.
m is the gradient of the line.
c is the point where the line cuts the vertical axis.

Meg bought an antique table 10 years ago.
She has drawn a scatter graph comparing the value and the age in years.
She wants to find the equation of the line of best fit.
She needs to find the values of m and c.

The graph cuts the vertical axis at **200**.
The value of c is **200**.
To find the gradient Meg chooses two points on the line that are far apart.
She writes down the vertical and horizontal change between the two points.

$$\text{Gradient} = \frac{\text{vertical change}}{\text{horizontal change}} = \frac{960 - 350}{10 - 2}$$

$$= \frac{610}{8} = 76 \text{ (2 s.f.)}$$

The value of m is **76**.

The equation of the line of best fit is $V = 76A + 200$.

The values of m and c can sometimes have a meaning.
Here c is the original price that Meg paid for the table.
m is the appreciation in value of the table each year on average. This is how much more the table is worth each year on average.
The line of best fit can also be called the regression line of y on x.

Exercise 14:2

1 Find the equation of the line of best fit for each of these scatter graphs. Give your answers in the form $y = mx + c$.

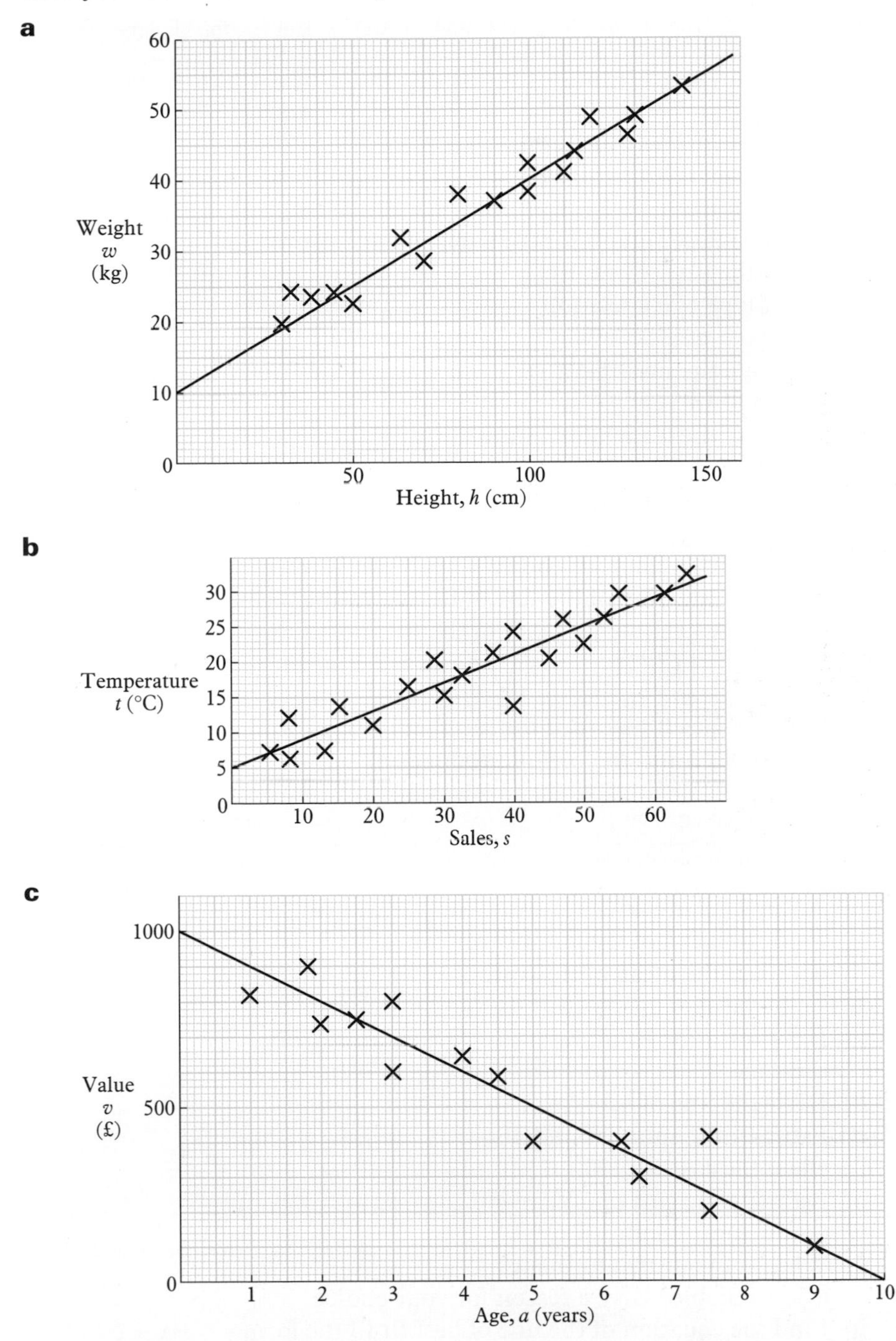

2 Zac has done an experiment in science.
He adds masses to the end of a spring and measures the new length L of the spring. These are his results.

M (g)	100	200	300	400	500	600	700	800
L (cm)	10.3	12.9	15.2	17.4	20.1	22.2	24.9	27

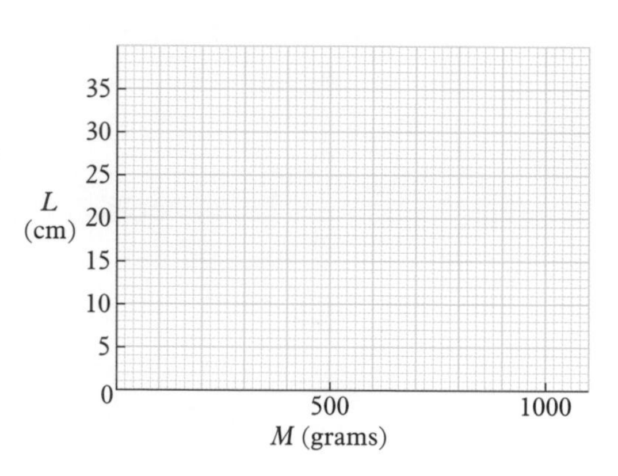

a Find the mean of the mass and the mean of the length.
b Copy the axes.
Draw a scatter graph for the data.
c Draw the line of best fit.
d What type of correlation does your graph show?
e Estimate the mass for a length of 18.5 cm.
f Estimate the length for a mass of 1 kg.
g Which of your two estimates is less reliable?
Give a reason for your choice.
h Find the equation of the line of best fit in the form $y = mx + c$
i Write down the significance of the value of:
(1) c (2) m.

3 A factory manager recorded the number of dolls produced on a production line and the cost of production.

Number of dolls	50	96	68	61	72	85	91	84	55
Cost (£)	170	314	226	202	231	280	291	278	185

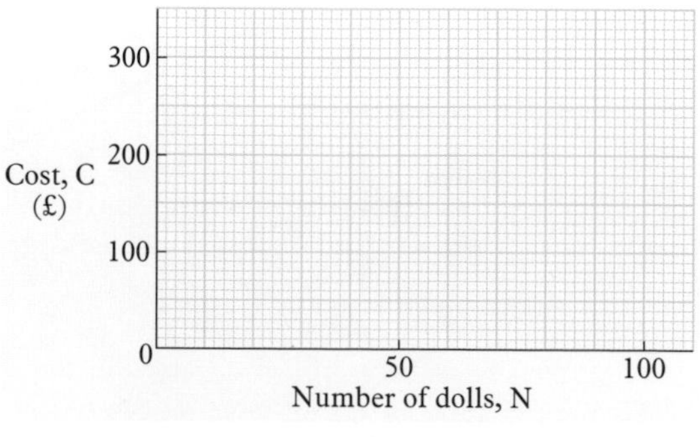

a Find the mean of the number of dolls and the mean of the cost.
b Copy the axes.
Draw a scatter graph for the data.
c Draw the line of best fit.
d Estimate the number of dolls when the cost is £100.
e Estimate the cost of producing 75 dolls.
f Which of your two estimates is less reliable? Give a reason for your choice.
g Find the equation of the line of best fit in the form $y = mx + c$.

The graph shows the relationship between the height and age of pupils.
The line of best fit goes through the two points (11, 127) and (14, 175).

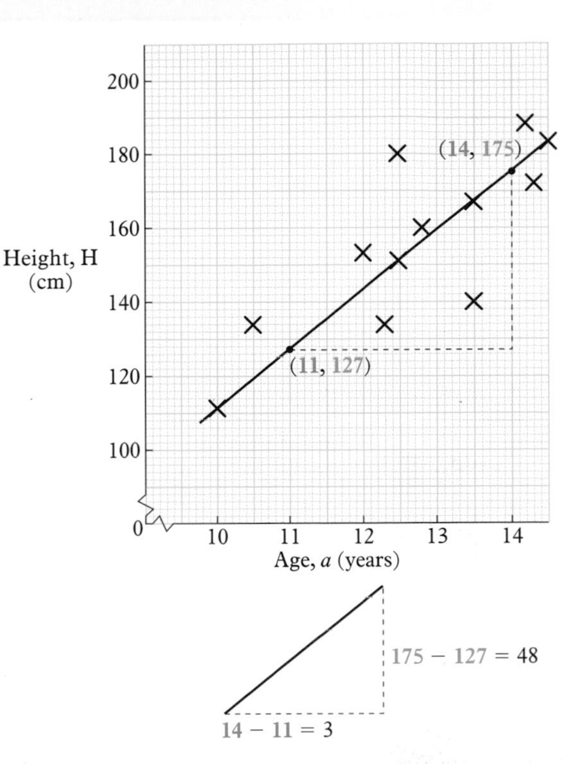

You cannot see where the line cuts the vertical axis.
You use the two points to find the equation of the line of best fit.

Draw the red triangle to find the gradient.

Gradient $= 48 \div 3 = 16$

The value of m is 16 so $\text{H} = 16a + c$.

You now use one of the two points to find the value of c.
Using the point (11, 127) you know that $\text{H} = 127$ when $a = 11$.
Substitute for a and H in $\text{H} = 16a + c$.

$$127 = 16 \times 11 + c$$
$$127 = 176 + c$$
$$127 - 176 = c$$
$$c = -49$$

The equation of the line of best fit is $\text{H} = 16a - 49$.

You can use the other point to check the equation.

Using the point (14, 175) you know that H should be 175 when $a = 14$.

When $a = 14$ $\quad\quad$ $\text{H} = 16 \times 14 - 49 = 224 - 49 = 175$ ✓

As the age, a, increases, you get a change in the value of H, the height.
a is called the **explanatory variable**.
H is called the **response variable**.

4 (1) Find the equation of the line of best fit for each of these scatter graphs. Give your answers in the form $y = mx + c$.
(2) Write down the explanatory and response variables.

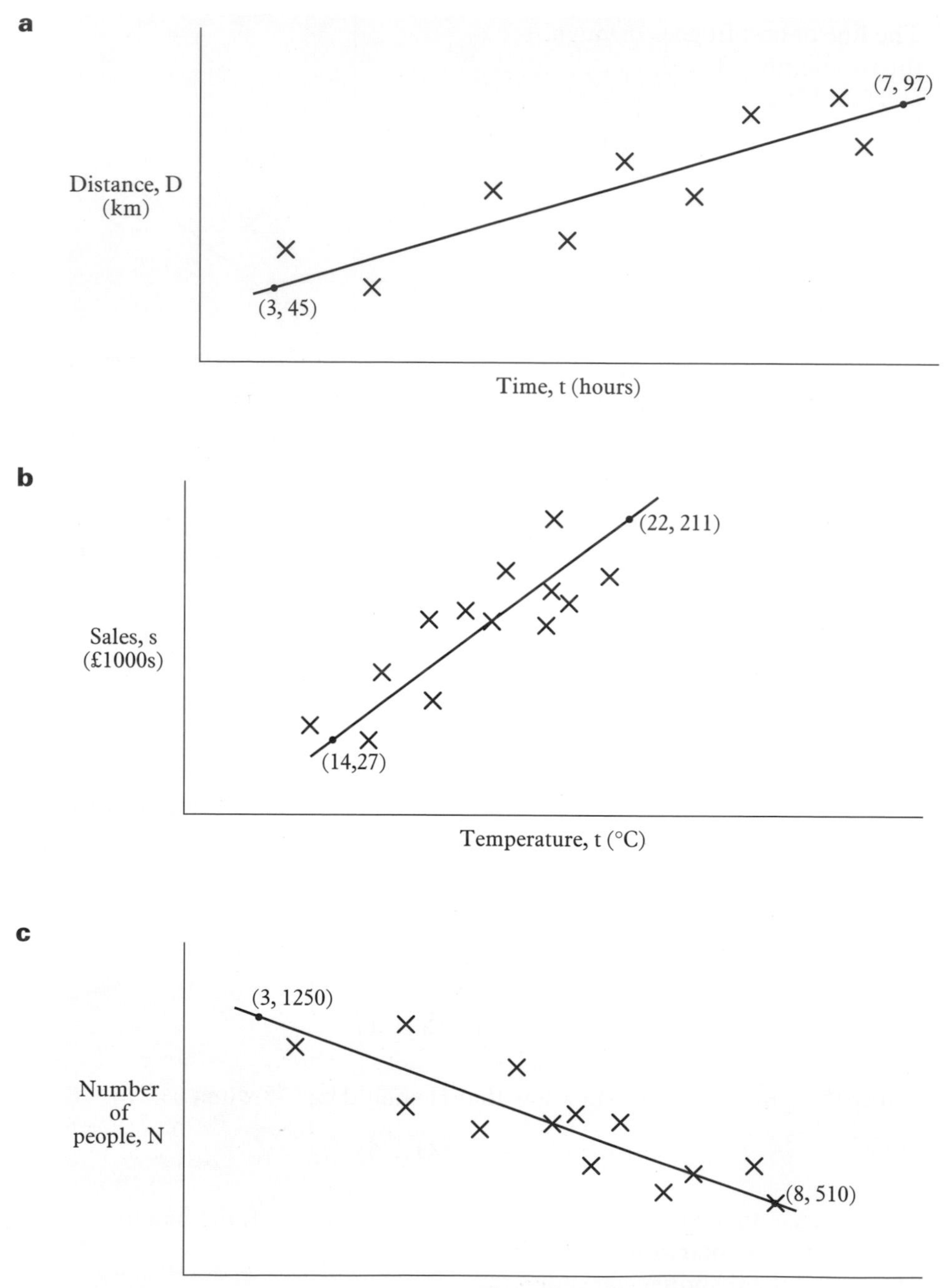

Not all variables are linked by linear equations.
Some may be linked by quadratic equations or other types of function.
Amazingly, you can still use straight line graphs to find the equations.

Example Tom is doing an experiment in science.
He thinks that the temperature of liquid is related to the square of the pressure it is under.
He has the following data.

Pressure (atm)	1	2	3	4	5
Temperature (°C)	20.2	23.1	26.4	35.8	45.2

Tom plots temperature against pressure.
He thinks the graph looks like a quadratic.

To check this, Tom plots temperature against (pressure)^2.
This gives an approximate straight line.
This shows that the temperature of the liquid *is* proportional to the square of the pressure it is under.

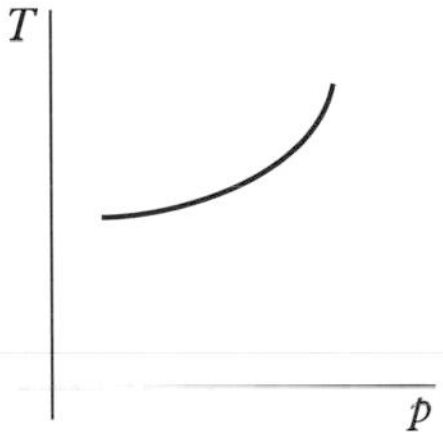

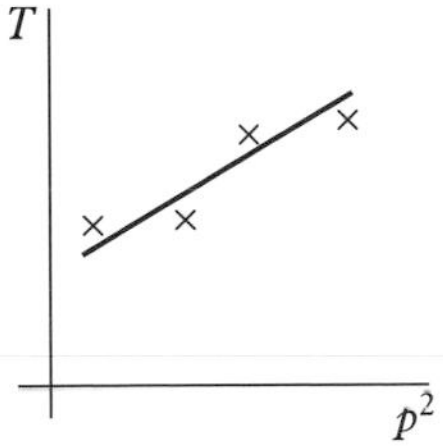

To find the equation, Tom finds the gradient and intercept from his straight line graph.
The gradient is **1.2** and the intercept is **18.5**
The equation is $T = 1.2p^2 + 18.5$

This can be done with any type of function.

For example, if you have two variables p and q and you suspect p is related to $\frac{1}{q}$, plot a graph of p against $\frac{1}{q}$
If this graph is a straight line, your suspicion is correct.
You can work out the gradient and write down the intercept from the *straight line* graph. From this you can write down the full equation.
If the gradient is m and the intercept is c the formula will be

$$p = m \times \frac{1}{q} + c = \frac{m}{q} + c$$

If $c = 0$ then p is proportional to $\frac{1}{q}$

Exercise 14:3

1 In each of the following sets of data, y is related to x^2.
For each one: (1) plot a graph of y against x^2
(2) find the gradient of the line
(3) write down the intercept with the y axis
(4) write down an equation connecting y and x^2.

a

x	1	2	3	4
y	7	16	31	52

b

x	0.2	0.4	0.6	0.8
y	1.5	3	5.5	9

2 In each of the following sets of data, y is related to $\sqrt{x}$
For each one: (1) plot a graph of y against $\sqrt{x}$
(2) find the gradient of the line
(3) write down the intercept with the y axis
(4) write down an equation connecting y and $\sqrt{x}$.

a

x	0.6	1.0	1.4	1.8
y	3.3	3.9	4.4	5.0

b

x	1	2	3	4
y	2.2	3.2	3.7	4.2

3 Rebecca has some results from a Physics experiment.

She suspects that y is related to $\frac{1}{x}$.

These are Rebecca's results:

x	1	2	3	4	5
y	4.9	3.9	3.6	3.2	3.5

a Plot a graph of y against $\frac{1}{x}$.

b Find the gradient of the line.

c Write down the intercept with the y axis.

d Write an equation connecting y and $\frac{1}{x}$.

3 Spearman's coefficient of rank correlation

These boys and girls are ranked in order of height.
The first person in each line is the shortest.

Ranking Putting data in order is called **ranking**.

Rank The **rank** of an item of data is its position in the ranking.

Tie A **tie** is when two or more items are given the same ranking.

Example These are the scores of six pupils in a test.

Jim 8 Sally 9 Pete 8 Tara 2 Viv 5 Eve 10

They are ranked in order starting with the highest score.

Eve 10 Sally 9 Jim 8 Pete 8 Viv 5 Tara 2

Eve is 1st, Sally 2nd, Jim and Pete have tied for 3rd place and so on.

This can be shown in a table:

	Jim	Sally	Pete	Tara	Viv	Eve
Mark	8	9	8	2	5	10
Ranking	3=	2	3=	6	5	1

There is no 4th place as two people have tied for third.

Sometimes they are said to be ranked $3\frac{1}{2}$th instead of both being ranked 3. $\left(\frac{3+4}{2}\right) = 3\frac{1}{2}$

If three pupils tie for 3rd place, you use $\left(\frac{3+4+5}{3}\right)$ as the ranking and so on.

Exercise 14:4

1 Two judges have ranked eight exhibits in an art competition.
These are their rankings.

Exhibit	A	B	C	D	E	F	G	H
Judge 1	4	6	2	3	1	8	7	5
Judge 2	1	5	8	5	2	3	7	4

a Which exhibit got the same ranking from both judges?

b Write down the letters of the exhibits which tied for the same place in the rankings of the second judge.

You can compare two sets of ranking using Spearman's coefficient of rank correlation.

You use the formula $\rho = 1 - \dfrac{6\Sigma d^2}{n(n^2 - 1)}$

d is the difference between the two rankings of one item of data.
n is the number of items of data. ρ is Spearman's coefficient of rank correlation.

To work out the value of ρ for the results of the two judges you add another two rows to the table. The first row is for the value of d and the second for d^2.
You use the values $5\frac{1}{2}$ for the tied ranking.

Exhibit	A	B	C	D	E	F	G	H
Judge 1	4	6	2	3	1	8	7	5
Judge 2	1	$5\frac{1}{2}$	8	$5\frac{1}{2}$	2	3	7	4
d	3	$\frac{1}{2}$	6	$2\frac{1}{2}$	1	5	0	1
d^2	9	0.25	36	6.25	1	25	0	1

Σd^2 means add all the d^2 values, so $\Sigma d^2 = 9 + 0.25 + 36 + 6.25 + 1 + 25 + 0 + 1$
$= 78.5$

$n = 8$ so: $\rho = 1 - \dfrac{6 \times 78.5}{8(64 - 1)} = 1 - \dfrac{471}{504} = 1 - 0.935 = 0.065$

The value of ρ will always be between -1 and $+1$.

-1		0		1
ranking in reverse order strong negative correlation	weak negative correlation	no correlation	weak positive correlation	same ranking strong positive correlation

If the value of ρ is close to 0 there is almost no correlation.
If the value of ρ is close to 1 or -1 there is strong positive or strong negative correlation respectively.

A value of 0.06 for ρ means there is almost no correlation between the marks of the two judges.

2 Which of these correlation coefficients shows the least correlation?

0.45 −0.8 0.07 −0.1 0.9 −1

3 Two judges had to rank seven ice skaters. These are the rankings.

Ice skater	A	B	C	D	E	F	G
Judge 1	5	3	7	1	4	2	6
Judge 2	4	3	6	2	5	1	7

a Use the table to calculate Spearman's coefficient of rank correlation.
b What does your answer suggest about the level of agreement between the two judges?

4 These are the scores for eight pupils in their Maths and Science exams.

Pupil	A	B	C	D	E	F	G	H
Maths	75%	48%	59%	72%	61%	67%	71%	42%
Science	69%	55%	50%	68%	60%	67%	64%	31%

a The table shows the ranking of the pupils in each exam. The highest mark is ranked 1. Copy and complete the table.

Pupil	A	B	C	D	E	F	G	H
Maths	1	...	...	2	...	...	...	...
Science	1	6	...	...	...	...	...	...

b Use the table to calculate Spearman's coefficient of rank correlation.
c Use your answer to **b** to comment on the results of the pupils in the two exams.

5 Write the value of Spearman's coefficient of rank correlation that best matches each of these graphs.
Choose from: 0.8, 0.3, −0.4, −0.9 and 0.03

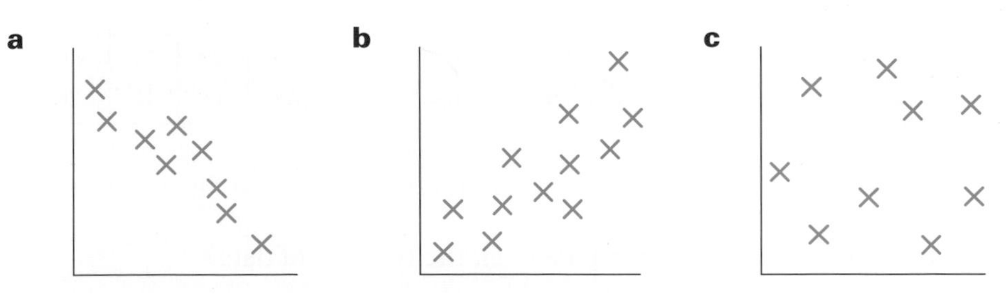

6 Eight people entered a competition to make the best homemade jam. These are the marks awarded to the 8 jams by two judges.

Jam	A	B	C	D	E	F	G	H
Judge 1	34	51	60	44	15	29	36	34
Judge 2	29	56	71	33	48	62	60	59

a Rank the two sets of marks.
b Calculate Spearman's coefficient of rank correlation for the two sets of marks.
c Use your answer to part **b** to compare the marks of the two judges.

7 The table shows the number of people attending the cinema and the outside temperature for each of seven days in a school holiday.

Day	Mon	Tues	Wed	Thurs	Fri	Sat	Sun
Number of people	74	91	107	45	50	66	88
Temperature (°F)	60	55	57	83	79	61	62

a Calculate Spearman's coefficient of rank correlation for the two sets of data.
b What is the relationship between the two sets of data? Explain your answer.

8 A hospital investigated the level of lung damage of patients and the number of years they had smoked. This is the data for 10 of these patients.

Patient	A	B	C	D	E	F	G	H	I	J
Lung damage (%)	45	31	50	49	32	30	22	15	31	48
Number of years	21	12	30	27	28	9	11	11	15	26

a Calculate Spearman's coefficient of rank correlation for the two sets of data.
b What is the relationship between the two sets of data? Explain your answer.

1 Mr Bean often travels by taxi and has to keep details of the journey in order to complete his claim form at the end of the week. Details for journeys made during a week are:

Distance travelled (miles)	2	7	$8\frac{1}{2}$	11	6	3	$4\frac{1}{2}$
Cost (£)	3.00	5.40	6.10	7.40	5.00	3.20	4.20

a On graph paper, plot a scatter graph of the above data. *(2)*

b Calculate the mean cost and mean distance and use these to draw the line of best fit on your scatter diagram. *(4)*

c Does the diagram show positive, zero or negative correlation? *(1)*

d **i** Estimate the cost of an 8 mile journey by taxi. *(1)*

ii Mr Bean had only £7 to pay for his taxi.
He had to make a journey of 12 miles.
How far short of his destination would he be at the completion of his taxi journey? *(2)*

SEG, 1999, Specimen Papers

2 A student carried out an experiment to investigate the effectiveness of a fertiliser.
Five plants of the same height were chosen and a different amount of fertiliser was given to each plant. The student plotted the results of the experiment on a scatter diagram and drew the line of best fit.

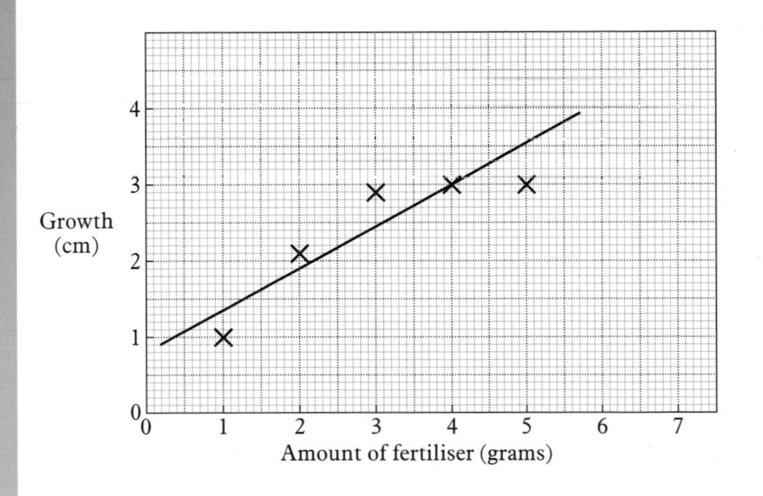

a Comment on the line of best fit. *(2)*

b Explain why it would be inadvisable to use this line of best fit to predict the growth for a 6 gram dose of fertiliser. *(2)*

NEAB, 1998, Paper 2

3 The table shows the number of deaths per thousand of the population for England and Wales over a 35-year period.

Year	0	5	10	15	20	25	30	35
Deaths per thousand	23.5	20.7	19.4	19.6	18.5	18.2	15.3	13.7

a Calculate the mean number of deaths per thousand. *(2)*

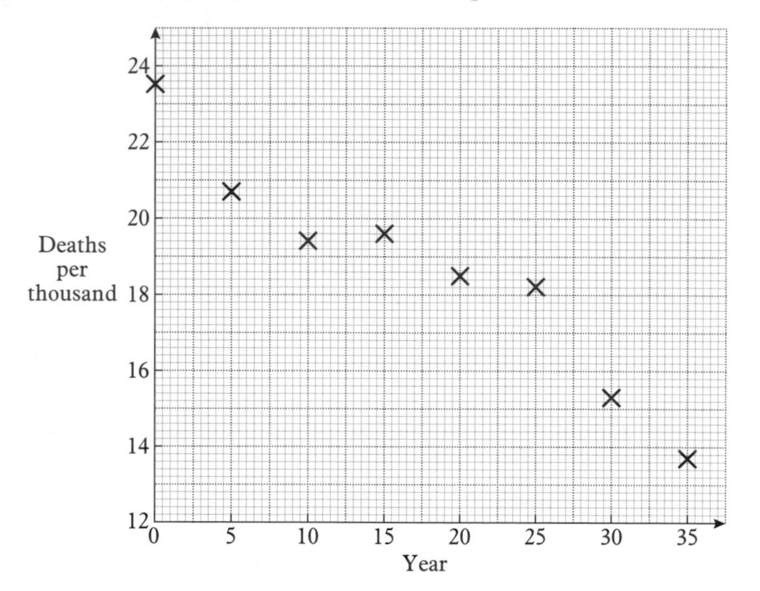

The mean of the years is 17.5

b Copy this graph on graph paper. Draw a line of best fit on the scatter graph. *(2)*

c Use your line of best fit to estimate the number of deaths per thousand of the population for the 27th year. *(1)*

d Calculate the gradient of your line of best fit. *(2)*

e What does the gradient of this line of best fit represent? *(1)*

SEG, 1999, Paper 3

4 For twelve consecutive months a factory manager recorded the number of items produced by the factory and the total cost of their production.

Number of items (x)	18	36	45	22	69	72	13	33	60	79	10	53
Production cost (£y)	37	54	63	42	84	91	33	49	79	98	32	70

a On graph paper, draw a scatter diagram for the data. *(3)*

b Find the mean value of x and of y. *(2)*

c Use your results from **b**, and the fact that the line of best fit for the data passes through the point (70, 88), to draw this line on the graph. *(2)*

d Estimate, from your line, the production cost for
   i 65 items *(2)*
   ii 84 items. *(2)*
e Which of your forecasts in **d** is the more reliable?
   Justify your choice. *(2)*
f Find, from your graph, the gradient of the line. *(2)*
g Describe briefly, in the context of the question, what the gradient measures. *(1)*

NEAB, Paper 3, 1998

5 Mrs Madan is a wine expert. At a wine tasting evening she is asked to taste the wines of a producer taken from each of ten different years and place them in order of quality. She regards 1983 as the best quality and ranks it 1.

Year	1990	1989	1988	1987	1986	1985	1984	1983	1982	1981
Rank age of wine	10	9	8	7	6	5	4	3	2	1
Rank quality	8	3	7	6	2	9	5	1	4	10

a Calculate Spearman's coefficient of rank correlation between the age and the quality of the wine. The formula for calculating Spearman's rank correlation coefficient is

$$\rho = 1 - \frac{6\Sigma d^2}{n(n^2 - 1)}$$ *(4)*

b On the basis of your answer to **a** comment on the statement 'wine improves with age'. *(1)*
c A similar tasting between the age and quality of beer resulted in a correlation coefficient of -1. What does this suggest about the age and quality of beer? *(1)*

SEG, 1999, Specimen Papers

6 The table below shows the weather recorded on Monday 14 March 1997 in ten Scottish towns.

WEATHER
Last night's reports for 24 hours to 6 p.m.

	Sunshine hours	Maximum temperature (°F)	Weather (day)	Rank sunshine hours	Rank temperature
Aberdeen	2.7	36	snow		
Aviemore	1.8	32	cloudy		
Edinburgh	0.2	33	cloudy		
Eskdalemuir	0.0	34	cloudy		
Glasgow	0.2	39	cloudy		
Kinloss	2.6	37	bright		
Lerwick	3.4	35	snow		
Leuchars	5.7	38	bright		
Tiree	1.2	40	bright p.m.		
Wick	6.4	37	snow a.m.		

Source: *The Guardian*, 15 March 1997

a Copy the table and rank each town in relation to hours of sunshine and maximum temperature °F. *(3)*

b Using the formula $1 - \dfrac{6\Sigma d^2}{n(n^2 - 1)}$ calculate Spearman's rank correlation coefficient for these data. *(4)*

c Comment on the correlation shown. *(2)*

d Suppose the equivalent rank correlation coefficients for Saturday 12 March and Sunday 13 March were, respectively, +0.07 and −0.95.
What would these two values suggest about the relationship between daily hours of sunshine and maximum temperature? Comment for the:

i Saturday value *(2)*

ii Sunday value. *(3)*

NEAB, 1998, Paper 3

7 The table shows the percentage of the total lottery ticket sales and the National Lottery grant for each of the ten different regions.

Region	Percentage of total lottery sales	Lottery grants (£m)	Ranks		d	d^2
			Sales	Grants		
London	22.3	125.0				
Midlands	15.7	11.7				
North West	11.7	5.7				
The South	10.4	17.5				
Yorkshire	9.5	5.1				
Scotland	8.8	5.9				
Wales	7.4	4.3				
The East	6.5	12.6				
North East	5.3	2.4				
N. Ireland	2.4	0.5				

a Copy the table and complete the ranking columns for each set of data. *(2)*

The formula for calculating Spearman's rank correlation coefficient is:

$$\rho = 1 - \frac{6\Sigma d^2}{n(n^2 - 1)}$$

b Calculate this rank coefficient for the above data. *(3)*

c Use your value of Spearman's rank correlation coefficient to comment about the way money is distributed. *(1)*

d By what amount could the National Lottery grant for London be reduced without changing the value of Spearman's rank correlation coefficient? *(2)*

SEG, 1997, Paper 2

8 The table shows the number of points achieved, and the number of goals conceded, by eight netball teams.

Team	A	B	C	D	E	F	G	H
Points	29	28	27	25	24	22	23	21
Goals conceded	15	12	13	16	19	14	21	17

a Use the table to calculate Spearman's coefficient of rank correlation between the number of points achieved and the number of goals conceded.

$$\text{Spearman's coefficient} = 1 - \frac{6\Sigma d^2}{n(n^2 - 1)}$$ *(6)*

Spearman's coefficient of rank correlation between the number of points achieved and the number of goals **scored** is 0.45.

b Which is the better relationship to use for predicting the points obtained, the relationship between:

i the number of goals conceded and the number of points achieved, or

ii the number of goals scored and the number of points achieved?

Explain why you chose this answer. *(1)*

SEG, 1998, Paper 2

9 A teacher believes that pupils can type faster if it is warmer. To test her theory she measures the times taken by the two groups of pupils to type 100 words. One group works in a temperature of 18 °C, the other group in a temperature of 24 °C.

a What is the explanatory variable? *(1)*

b What is the response variable? *(1)*

c Give **two** factors the teacher should consider when placing the pupils in the two groups. *(2)*

10 An article in a magazine states that

'FLOWERS IN A VASE WILL STAY FRESH LONGER IF THEY ARE FED ASPIRIN'

To test this theory one vase of flowers is given 5 mg of aspirin and another vase of flowers is not given any aspirin.

a What is the explanatory variable? *(1)*

b What is the response variable? *(1)*

c What else would need to be done to obtain reliable data from this simple statistical experiment? *(1)*

Scatter graph — A **scatter graph** is a diagram that is used to see if there is a connection between two sets of data.

Line of best fit — The mean values of both variables are found. These are the co-ordinates of the mean point.
Draw a line through the mean point and in the middle of the rest of the points. This line is called the **line of best fit**.
It can also be called the regression line of y on x.

Interpolate — **Interpolate** means to estimate a value between two values that you know.

Extrapolate — **Extrapolate** means to find a value by following a pattern and going outside the range of values that you know.

Extrapolation is less reliable because the pattern may not continue outside the existing values.

The line of best fit is a straight line.
The equation of a straight line can be written in the form $y = mx + c$.
m is the gradient of the line.
c is the point where the line cuts the vertical axis.
Straight line graphs can still be used to find the equations when variables are linked by other types of functions.

Ranking — Putting data in order is called **ranking**.

Rank — The **rank** of an item of data is its position in the ranking.

Tie — A **tie** is when two or more items are given the same rank.

You can compare two sets of ranking using Spearman's coefficient of rank correlation.

You use the formula $\rho = 1 - \dfrac{6\Sigma d^2}{n(n^2 - 1)}$

d is the difference between the two rankings of one item of data.
n is the number of items of data. ρ is Spearman's coefficient of rank correlation.

The value of ρ can lie between -1 and $+1$.

-1		0		1
ranking in reverse order strong negative correlation	weak negative correlation	no correlation	weak positive correlation	same ranking strong positive correlation

If the value of ρ is close to 0 there is almost no correlation. If the value of ρ is close to 1 or -1 there is strong positive or strong negative correlation respectively.

Working with *Excel*

Drawing scatter diagrams

Gina and Greta are on a pebble beach. They measure the widths and lengths of pebbles. This is their data

Length (mm)	42	28	50	24	37	52	22
Width (mm)	30	21	35	18	28	38	16

Use *Excel* to:

a draw a **scatter diagram** and the **line of best fit**
b calculate the **equation** of the line of best fit.

 **Enter the information on pebble sizes into a new *Excel* document.**

In Cell A1, type Length (mm) and in Cell B1, type Width (mm).
In Cell A2, type 42, in Cell A3, type 28 and so on.
In Cell B2, type 30, in Cell B3, type 21 and so on.

	A	B	C
1	Length (mm)	Width (mm)	
2	42	30	
3	28	21	
4	50	35	
5	24	18	
6	37	28	
7	52	38	
8	22	16	
9			
10	Mean	Mean	
11	36.4285714	26.5714286	
12			
13			

 **Calculate the mean.**

In Cell A10, type Mean.
In Cell A11, insert the formula: =**AVERAGE(A2:A8)**
Remember to press the Enter key.
Cell A11 displays the mean of the lengths.

Repeat this to find the mean of the widths.

Select Cells A1 to B8.

Use the Chart Wizard button.

Draw a scatter diagram.

From Chart type, choose XY (Scatter).
From Chart sub-type, choose the top left diagram.

Label the scatter diagram

Click the Next > button until you get to the Chart Options window.
In the Chart title box, type Pebble Sizes.
In the Value (X) axis box, type Length (mm).
In the Value (Y) axis box, type Width (mm).
Click Finish.
On the graph, click on the Width (mm) box and press Delete.

Add the line of best fit to the scatter diagram.
Click on a blank area of the scatter diagram to select it.
On the Toolbar, click **Chart**.
On the drop-down menu, choose **Add Trendline....**
In the **Add Trendline** window, click on the **Type** tab.
From **Trend/Regression type**, choose the top left diagram.
Click **OK**.

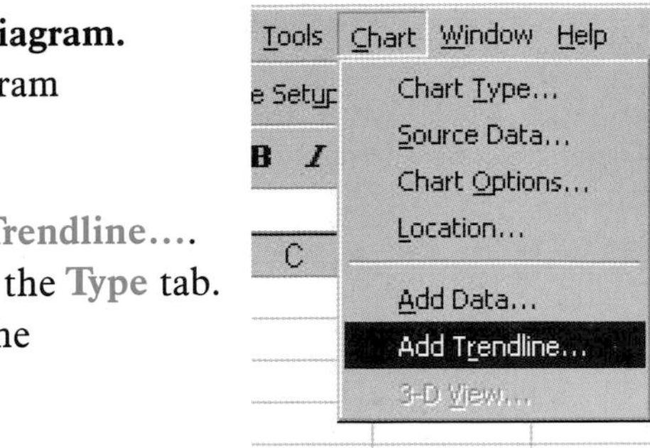

Calculate the equation of the line of best fit.
In Cell A13, type **Gradient**.
In Cell A14, insert the formula: **=SLOPE(B2:B8,A2:A8)**
Cell A14 displays the gradient of the line.

In Cell B13, type **Value of y-intercept**.
In Cell B14, insert the formula: **=INTERCEPT(B2:B8,A2:A8)**
Cell B14 displays the value of the y-intercept.

The equation of the line is: $y = 0.693x + 1.313$

Click on a blank area of the graph to select it before printing it.

Exercise 14:5

1 These are the marks of two judges for 10 entries in a cookery competition.

Entry	A	B	C	D	E	F	G	H	I	J
1st Judge	63	84	31	47	59	72	80	93	38	86
2nd Judge	51	77	36	45	60	64	72	75	43	85

Use *Excel* to:

a draw a scatter diagram and the line of best fit
b calculate the equation of the line of best fit.

2 Ten pupils sat a Maths test and an English test. These are their results.

Maths	14	21	39	18	49	30	42	33	28	6
English	21	30	44	25	47	36	37	34	27	14

Use *Excel* to:

a draw a scatter diagram and the line of best fit
b calculate the equation of the line of best fit.

1 The table shows the retail price and the sale price of some video recorders in Ted's Discount Store.

Retail price (£)	160	225	280	340	360	420	460	540	600
Sale price	115	170	210	260	265	310	350	405	460

a Find the mean of the retail price and the mean of the sale price.

b Copy the axes. Draw a scatter graph for the data.

c Draw the line of best fit.

d What type of correlation does your graph show?

e Estimate the sale price for a video with a retail price of £700.

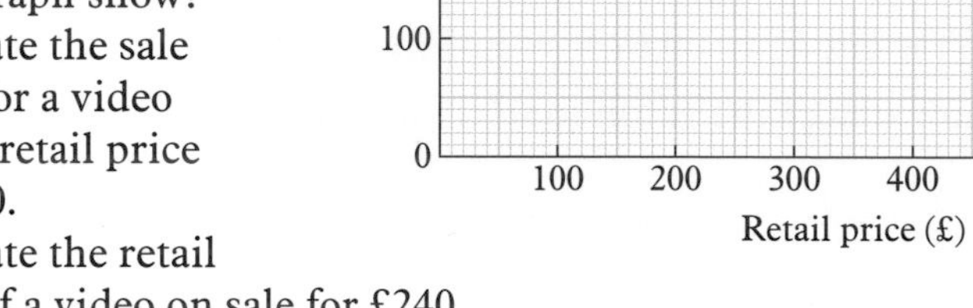

f Estimate the retail price of a video on sale for £240.

g Which of your two estimates is less reliable? Give a reason for your choice.

h Find the equation of the line of best fit.

2 The table shows the results of six students for their two Maths papers.

Student	A	B	C	D	E	F
Paper 1	62	45	73	59	60	78
Paper 2	53	47	68	61	54	65

a Calculate Spearman's coefficient of rank correlation for the two papers.

b Comment on the results of the students on the two papers.

15 Coursework

STEP 1	**Choose a topic**
STEP 2	**Form the hypotheses**
STEP 3	**Collect the data**
STEP 4	**Analyse the data**
STEP 5	**Interpret the results**
STEP 6	**Form conclusions and state the limitations**

Coursework

A census takes place once in every ten years.

A questionnaire is sent to every household in the country.

The purpose of this is to collect and analyse data on many different topics.

Coursework This is an extended piece of work or a project.
There are six stages in a piece of coursework.

Think about each stage before you begin your coursework task.

STEP 1 Choose a topic

This is very difficult, but it is very important.
Your topic must interest you.
Your topic must allow you to demonstrate the statistical skills that you have learnt.

Here is a list of possible coursework titles.

a The relationship between the heights of boys and their shoe sizes.
b The age of parents at the birth of their first child.
c Pupils' journeys to school.
d The viewing habits of pupils in the school.
e The weather in popular holiday destinations.

In this chapter, 'Pupils' journeys to school' and 'The weather in popular holiday destinations' are investigated.

STEP 2 Form the hypotheses

To form hypotheses, you must ask questions.

Example 1 *Pupils' journeys to school*

Here is a list of possible questions:

a How long does it take you to travel to school?
b How far away from school do you live?
c How do you travel to school?

From these questions, hypotheses can be formed.
Here is a list of possible hypotheses:

a The average journey time to school for each pupil is 30 minutes.
b The further away from school that a pupil lives, the longer their journey takes.
c Most students travel to school by bus.

Example 2 *The weather in popular holiday destinations*

Here is a list of possible questions:

a Are popular holiday destinations always sunny?
b How much rain is there?
c What is the temperature?

From these questions, hypotheses can be formed.
Here is a list of possible hypotheses:

a The mean amount of sunshine exceeds 8 hours a day in the months of July and August.
b The mean amount of rain per day is very small in the months of July and August.
c The monthly average temperature throughout the year is higher than where I live.

Exercise 15:1

Here are the three other coursework titles from before:

a The relationship between the heights of boys and their shoe sizes.
b The age of parents at the birth of their first child.
c The viewing habits of pupils in the school.

1 Write down three questions for each of the coursework titles.

2 Write down three hypotheses for each of the coursework titles.

STEP 3 Collect the data

a Decide on the population.

b Think carefully about the size of the sample.
The sample must be large enough to represent the population, but small enough to be manageable.
It should be at least 50.
The larger the sample, the more accurate your results are likely to be.

c Choose your method of sampling.
Samples can be chosen by different methods:

Random sampling	Quota sampling
Systematic sampling	Cluster sampling
Stratified sampling	

d Choose the members of the sample and collect the data.

The aim is to choose the sample without **bias**, so that the results will apply to the whole population.

Now consider two important questions.

Is the data to be found in published statistics? This is called secondary data. A list of possible sources can be found below.
Look at the section ***'Collecting secondary data'***.

Is the data to be collected by you? This is primary data.
Look at the section ***'Collecting primary data'***.

Collecting secondary data

Methods	**a** Newspapers **b** Books **c** Computerised data bases **d** The Internet **e** Published statistics in the library
Advantages	There is a large quantity of data. The data is easy to find.
Disadvantages	The data may not be exactly what you require. The accuracy of the data may not be known.

Collecting primary data

Methods

a Design an experiment. It may be possible to program a machine to take results at regular intervals. This is called data logging.
b Take measurements.
c Write a questionnaire.

A questionnaire is a set of questions on a given topic.

1 Questions should not be biased.
2 Questions can give a choice of possible answers.
3 Questions should not upset people or embarrass them.
4 Questions should be clear, short and easily understood.
5 Questions should be relevant.
6 Don't ask questions that allow people to give many different answers (open questions).
7 Questions should be in a sensible order.

Are you going to ask the questions, or are the members of your sample filling in the questionnaire?

Conduct a pilot survey to check the questions work.
Rewrite the questions if necessary.
Good questionnaires need to be written carefully.

Advantages

You can collect the data that you want.
You know the accuracy of your data.
You know how you have collected the data.

Disadvantages

It will take you a long time.
The questionnaires may not be filled in accurately.
Some questionnaires may not be returned at all.

Whichever type of data you have collected, you now have some information to analyse.

Check

a Your population is decided.
b Your sample is chosen.
c Your data is collected.
d The results are recorded clearly.
e The results are recorded accurately.
f The results are recorded in a suitable form.

Example 1 Pupils' journeys to school

Population — The population is all the pupils in Year 7 and all the pupils in Year 11.

Method of sampling — Random sampling.
Use alphabetical year lists and a table of random numbers.

Size of sample — There are 200 pupils in Year 7. A sample of 50 is manageable and representative. There are 198 pupils in Year 11. A sample of 50 is manageable and representative. You would need to use a stratified sample if the sizes of the year groups were very different.

Collect primary data — Ask each pupil in the sample for the information required.

Results — Make a clear table of the required information.

Name	Year	Transport	Journey time	Distance from School

Example 2 The weather in popular holiday destinations

Population — The population is all of the pupils in years 7–9.

Method of sampling — Obtain, from a systematic sample, the pupils holiday destinations for the previous year.

Size of sample — Take a large sample of more than 50 students. From this sample the six most popular destinations can be found.

Collect secondary data — Use published statistics to find the information required.

Results — For each day in the months of July and August of the previous summer, make a clear table showing:

Date	Number of hours of sunshine	Amount of rain	Daily average temperature

For each month of the previous two years, make a clear table showing the monthly average temperature:

Year	Month	Average monthly temperature	
		Where I live	Holiday destination

Exercise 15:2

For the questions and hypotheses below, write down how you would collect the data. Think carefully about each of the following:

Population
Method of sampling
Size of sample
Collecting the data
Recording the results

1 Consider 'The age of parents at the birth of their first child'.

a *Question*
'What is the age of the mother at the birth of her first child?'
Hypothesis
'The average age of a mother at the birth of her first child is 24 years.'

b *Question*
'Has the age of a mother at the birth of her first child changed in the last fifty years?'
Hypothesis
'The average age of a mother at the birth of her first child has increased in the last fifty years.'

2 Consider 'The viewing habits of pupils in the school'.

a *Question*
'On which day do pupils watch most television?'
Hypothesis
'Pupils watch most television on a Sunday.'

b *Question*
'What is the pupils' favourite television programme?'
Hypothesis
'The favourite television programme is not the same for Year 7 and Year 11 pupils.'

STEP 4 Analyse the data

a Organise your data into clear frequency tables.
The data can be ungrouped.
The data can be grouped.
Do not use too many or too few groups.
The aim is to show the important facts about your data.

b Analyse your data.
Some diagrams and calculations are more suited to discrete data.
Other diagrams and calculations are more suited to continuous data.

Now consider these important questions.

Is the data discrete data?
Look at the section ***'Discrete data'***.

Is the data continuous?
Look at the section ***'Continuous data'***.

Is the data in pairs?
Look at the section ***'Pairs of data'***.

Discrete data

Diagrams	Pictograms	Vertical line graphs
	Bar-charts	Box plots
	Stem and leaf	Cumulative frequency step polygons
	Pie-charts	Time series
Calculations	Mean	Range
	Median	Interquartile range
	Mode	Standard deviation

Continuous data

Diagrams	Histograms Frequency polygons Cumulative frequency polygons Stem and leaf Box plots Time series
Calculations	Mean Median Mode Range Interquartile range Standard deviation

Pairs of Data

Diagram	Scatter graph
Calculations	Equation of the line of best fit Spearman's rank correlation coefficient

Whichever type of data you have collected, you now have some results.

Check

a The tables are clear.

b The diagrams are labelled clearly.
The diagrams have titles.
There is a wide variety of diagrams.
They are all appropriate.
Remember that there is no need to use every type.

c The calculations are clearly written out.
They are appropriate and relevant.

Example 1   Pupils' journeys to school

Question — 'How long does it take you to travel to school?'

Hypothesis — 'The average journey time to school is 30 minutes.'

Population — Pupils in Year 7.
Pupils in Year 11.

Sample — 50 pupils from each year, chosen at random.

Collect primary data — Ask each pupil in the sample for the information required.

Results — Make a clear table of the required information.

Year 7		Year 11	
Name	Journey time	Name	Journey time

Analyse the data — This is continuous data. The results for each year could be arranged into grouped frequency tables.

Diagrams — For the ungrouped data, appropriate diagrams are:
Stem and leaf   Box plots

For the grouped frequency data, appropriate diagrams are:
Histograms   Frequency polygons
Cumulative frequency polygons

Calculations — Appropriate calculations are:
Mean   Range
Median   Interquartile range
Mode   Standard deviation

Use the ungrouped data for the calculations.
Information is lost when you group data.
However, if the sample is very large then you may have to group the data before doing the calculations to make it manageable.

Example 2 The weather in popular holiday destinations

Question	'Is the temperature high?'
Hypothesis	'The monthly average temperature throughout the year is higher than where I live.'
Population	All of the popular holiday destinations.
Sample	The first six most popular holiday destinations.
Collect secondary data	Use published statistics to find the information required.
Results	Make a clear table of the required information.
Analyse the data	Use an ungrouped table showing the average monthly temperature for the last two years.

Year	Month	Average monthly temperature	
		Where I live	Holiday destination

Diagrams	One time series graph showing all seven places. This allows monthly temperatures to be compared.
Calculations	It is inappropriate to calculate measures of spread or location, as each month is being compared.

Exercise 15:3

For the hypotheses in questions **1** and **2**, write down how you would analyse the data.
Think carefully about each of the following:
Are the tables organised?
Is the data, discrete, continuous or in pairs?
What are the appropriate diagrams and calculations?

1 Consider 'The age of parents at the birth of their first child'.

Hypotheses

a 'The average age of a mother at the birth of her first child is 24.'
b 'The average age of a mother at the birth of her first child has increased in the last fifty years.'

2 Consider 'The viewing habits of pupils in the school'.

Hypotheses

a 'Pupils watch most television on a Sunday.'

b 'The favourite television programme is not the same for Year 7 and Year 11 pupils.'

STEP 5 Interpret the results

a Think about your coursework title.
Look at all of the original hypotheses.

b Consider your analysis of the data.
Think about both the diagrams and the calculations.
Discuss the ways in which your data, diagrams and calculations support each of your hypotheses.

STEP 6 Form conclusions and state the limitations

a Very briefly summarise your results.
State a representative figure, if this is appropriate.
State a measure of spread, if this is appropriate.

b Look at each hypothesis.
Does your data support each hypothesis?
State clearly whether you accept or reject each of your original hypotheses.
It is important to only make statements and form conclusions that you have shown in your project.

c Discuss the limitations of your project.
The sample size may be too small.
The method of sampling may be inappropriate.
The data collected may be biased.
The questionnaire may be poorly written and may have collected inappropriate data.

d Make suggestions for further investigations.

At the end of the report, include all sources of secondary data and copies of questionnaires.

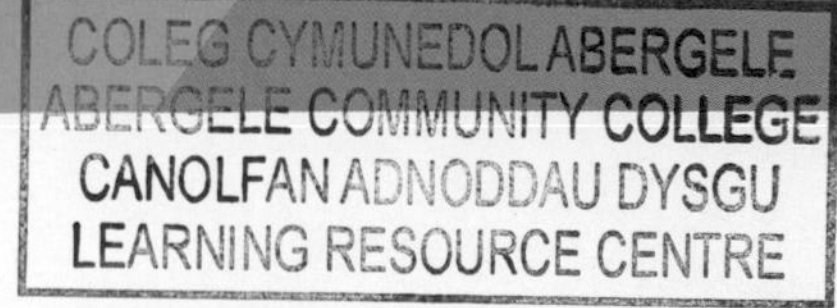

Sources of published statistics

The following journals are examples of the sources of statistics that can be found in any large Reference Library.

Key Data

Annual Abstract of Statistics

Guide to Official Statistics

Monthly Digest of Statistics

Social Trends

Census of Population Reports

Economic Trends

Family Expenditure Survey

Regional Trends

Europe in Figures

CHAPTER 1

Exercise 1:1

1 **a** Qualitative **b** Quantitative **c** Quantitative **d** Qualitative
2 **a** Quantitative **b** Quantitative **c** Qualitative **d** Quantitative **e** Qualitative
3 There are many questions which could be asked here. Some examples are given.
a (1) How many people live in your city?
(2) What are the heights of the three highest mountains in Scotland?
(3) How many countries are there in the EU?
(4) What are the lengths and widths of a sample of ten pebbles from the beach?
b (1) What is the name of your favourite holiday resort?
(2) How do most of your class travel to school?
(3) What is the main agricultural crop in each of the countries in the EU?
(4) Which state in North America has the most sunshine?
4 **a** Quantitative **b** Qualitative **c** Quantitative **d** Quantitative
5 **a** Secondary **b** Primary **c** Primary **d** Primary **e** Secondary **f** Secondary **g** Primary

Exercise 1:2

1 **a** Biased sample. Pupils in the library are likely to read more books.
b Biased sample. There are too few pupils, and they are chosen from one year group only.
c Biased sample. The time between 2pm and 3pm is not going to indicate traffic flows throughout the rest of the day.
d Biased sample. The interviewer is only going to ask customers in the shop, who are probably already interested in sport.
2 **a** 1.53 m, female 1.44 m, male 1.21 m, female 1.23 m, female 1.43 m, male
b 1.21 m, female 1.23 m, female 1.43 m, male 1.25 m, female 1.00 m, female
c 1.53 m, female 1.25 m, female 1.2 m, female 1.44 m, male 1.23 m, female
d This could generate any of the thirty children.

Exercise 1:3

1 **a** A. Coombes, A. Dowle, D. Jones, B. Cook. **b** J. Crabtree, B. Clifford, J. Darnton, A. Wardle.
2 **a** J. Crabtree, A. Smith, C. Patel, D. Jones, C. Dorris. **b** J. Reeves, A. Dowle, A. Stuart, A. Wardle, L. Torvalds.
3 **a** Random **b** Convenience **c** Random **d** Systematic **e** Convenience

Exercise 1:4

1 **a** 13 boys **b** 17 girls
2 19, 9, 12, 8, 2
3 207
4 36
5 0, 1, 5, 4, 1, 1
6 129, 111, 91, 169

Exercise 1:5

1 Systematic sampling **2** Cluster sampling **3** Quota sampling **4** Random sampling
5 Cluster sampling **6** Stratified sampling **7** Stratified sampling
8 When the effect of new drugs is being tested, two groups of patients are studied. The first group is given the new drug. The second group is given a 'dummy' drug called a placebo. The 'placebo effect' is when the second group, thinking it is receiving the drug, improves.

Exercise 1:6

1 **a** Most passengers will be flying to other countries.
b The whole of the local population.
c A systematic or random sample from the electoral register.
2 **a** Workers may be rushing to complete the cars before the weekend or before the shift ends.
b All of the cars produced by the production line.
c A systematic sample from the production line.
3 **a** He is excluding pupils who do not travel on the bus.
b All the pupils in the school.
c Stratified sampling.
4 **a** Only some people belong to golf club. **b** All people who live in Polly's village. **c** Stratified sampling.
5 Julie splits the seedlings into 3 lots of 100. One group is the control, 1 group is treated with one of the fertilisers and the remaining group with the other fertiliser.
6 **a** Have equal numbers of fields randomly chosen in one particular area. One group to receive fertiliser only, the other to act as a control. Fields in different parts of the country need to be considered.
b Rainfall, sunshine, temperature, past use of field …
7 720
8 250

Exercise 1:7

1 **a** Not biased.
b Biased. The first sentence makes you think that your school coat must have a designer label.
c Biased. The first sentence makes you think that you are not normal if you do not enjoy pizza.
d Biased. The first sentence makes you think that eating fruit is important.
e Not biased.

Exercise 1:8

1 c **2** b or d **3** a **4** b or d

Exercise 1:9

1 a c f
2 a. Some people are very embarrassed by their weight.
c. Some people find age embarrassing.
f. A person earning very little might find it embarrassing to answer this.
3 24%

Questions

1 **a** Questioning or measuring or counting *(1)*
b Secondary data *(1)*
c Out of date or not exactly what is required *(1)*
2 **a** A census involves the whole population. Sampling involves only a part of the population. *(1)*
b The bulbs are tested until they fail. *(1)*
c Saves time, it is cheaper, or any other suitable reason. *(1)*
3 **a** Put the numbers 1–30 in a hat. Pick 5 from these. *(2)*
b Randomly choose a number from 1 to 5. Take every 6th number after this. *(2)*
c Systematic and an argument for choosing this. *(2)*
4 There are a number of suitable questions. Two examples could be
(1) Where did you take your holiday last year? Tick the appropriate box.
☐ In Britain ☐ Elsewhere in Europe ☐ In the rest of the world
(2) What kind of accommodation did you take? Tick the appropriate box.
☐ Self catering ☐ Hotel ☐ Camping *(4)*
5 **a** Method 2. All groups are represented in the electoral register. *(2)*
b Put all the names in a hat and randomly pick 20 or give each pupil a number from 1 to 100 and use random tables to select a sample of size 20 *(2)*
6 **a** 0.4 or $\frac{2}{5}$ *(1)*
b Short people are less likely to respond. The 'non-response' category is much too large. The question is very vague. *(2)*
c What is your height in cm? Tick the appropriate box.
☐ <120 ☐ 120–139 ☐ 140–159 ☐ 160–179 ☐ $\geqslant 180$
(Answers must include at least three non-overlapping boxes and no gaps.) *(2)*
7 Question A: It is an ambiguous and personal question. What is your age? Tick the appropriate box.
☐ Under 10 ☐ 10–19 ☐ 20–29 ☐ 30–39 ☐ 40–49 ☐ Over 49 *(3)*
Question B: It is a leading question. Do you think that the new supermarket is successful?
☐ Yes ☐ No ☐ Don't know *(3)*
8 **a** **i** Sample is 59, 68, 62, 35 Mean = 56 *(2)*
ii Sample is 54, 57, 59, 56 Mean = 56.5 *(3)*
iii Sample is 60, 61, 42, 41 Mean = 51 *(3)*
b Sample 3. This reflects the ages of boys and girls in the sample. *(2)*
9 **a** This will give a poor cross-section *(1)*
b Stratified sample according to age/electoral register of people in Medford and choose a random sample/choose a more central location. *(1)*
c **i** Unreliable question as it is too personal and too open. *(1)*
ii The question is useful but it is too vague. *(1)*
iii The question is too open. Should include only relevant sports and tick boxes. *(1)*
d There are many further questions which could be included. These could include: opening times, type of sport, frequency of use, facilities required. *(2)*
10 **a** Quick/cheaper/easier *(1)*
b Sample needs to represent the floors and the male/female population. *(1)*
c 1st floor – 20; 2nd floor – 10; 3rd floor – 20 *(3)*
d 4 *(2)*
e Choose a random number from 1 to 5. Using this as the starting point, select every 5th person after this. *(2)*

11 a Easy to obtain or equivalent; bias or equivalent; random sample or equivalent; does not reflect proportion of students in each year or equivalent.

b i 800 students so take groups of 3 digits. Ignore any numbers >800 and also any repeated numbers. Continue until sample of 10 students obtained

ii Assign numbers to each of 800 students. Take every 80th student.

c Reflects proportion of students in each year.

d |————————————|

very poor excellent

Respondent to mark the above scale

☐	☐	☐	☐	☐	☐
very poor	poor	fair	good	very good	excellent

Test Yourself

1 a Quantitative **b** Qualitative **c** Qualitative **d** Quantitative

2 a Primary data **b** Secondary data

3 a Team players are more likely to enjoy sports than a random sample from the school.

b People visiting the showrooms are more likely to be interested in the new model of the car.

4 22, 13, 5, 10 **5** 800 **6 a** Systematic sampling **b** Quota sampling

CHAPTER 2

Exercise 2:1

1 2, 5, 15, 24, 36, 17, 0. **2** 0, 2, 10, 23, 32, 28, 3.

3 a 6 **b** 9 **c** 26 **d** 26.9%

4 a Football **b** 203 **c**

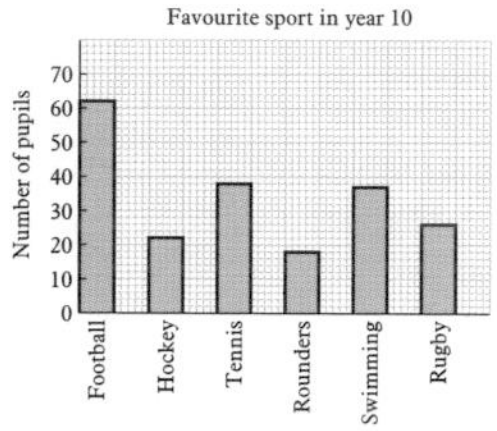

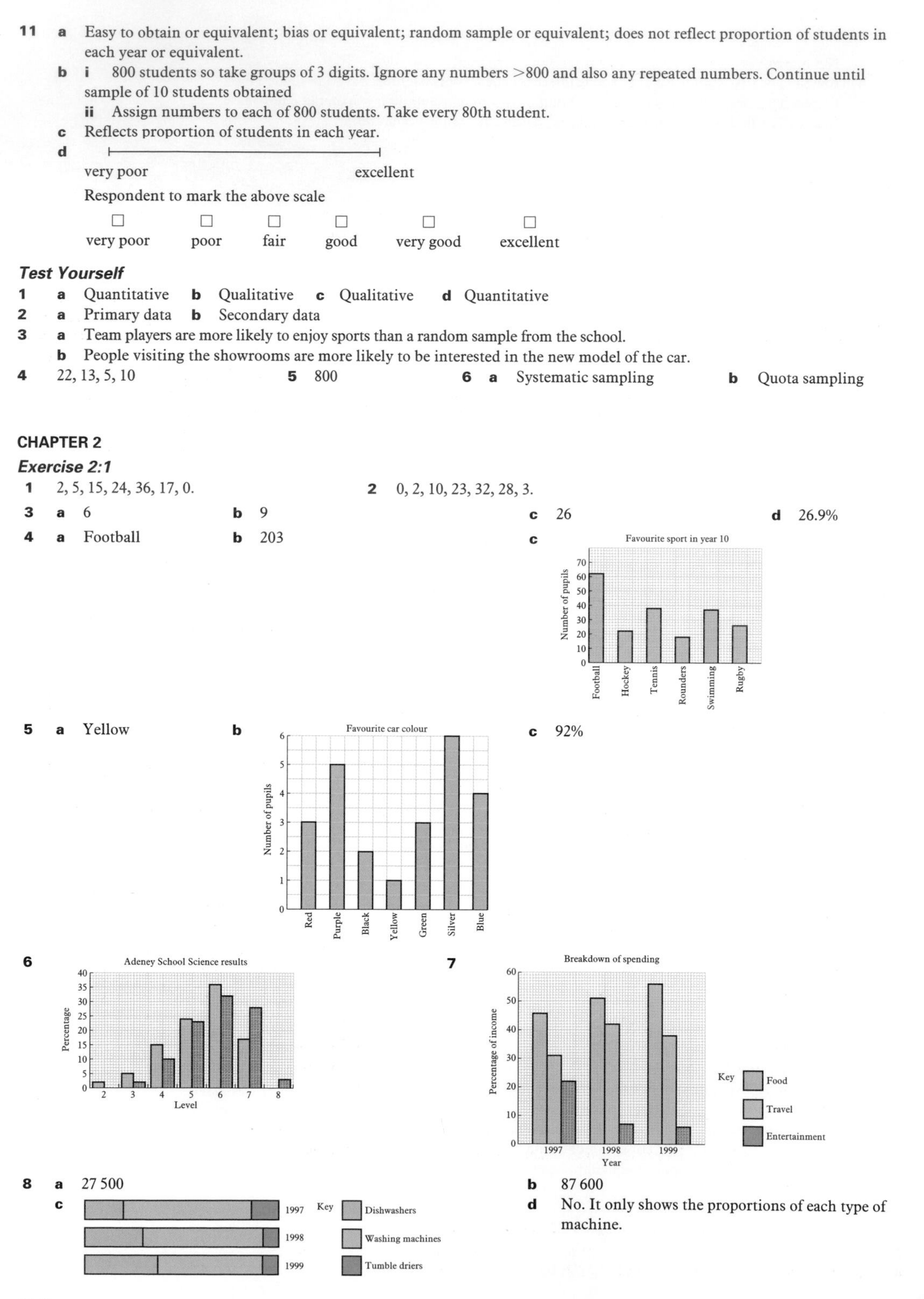

5 a Yellow **b**

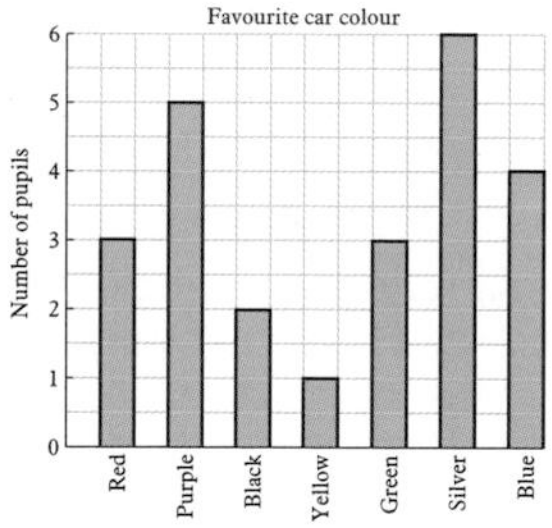

c 92%

6

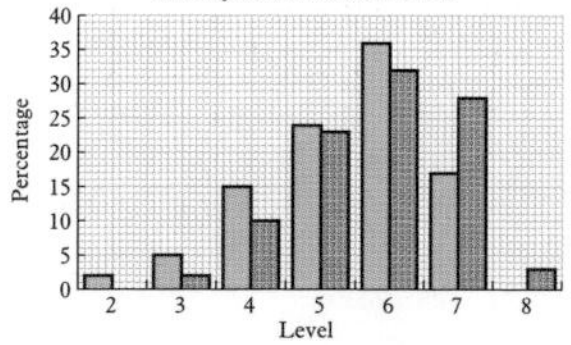

7

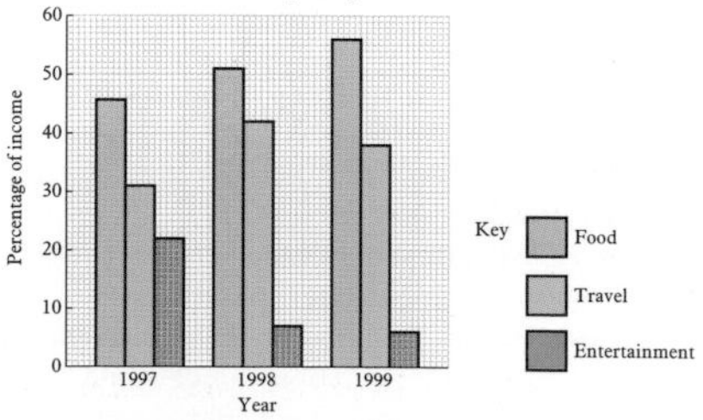

8 a 27 500 **b** 87 600

c

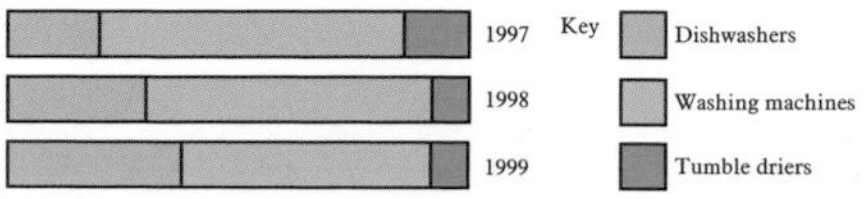

d No. It only shows the proportions of each type of machine.

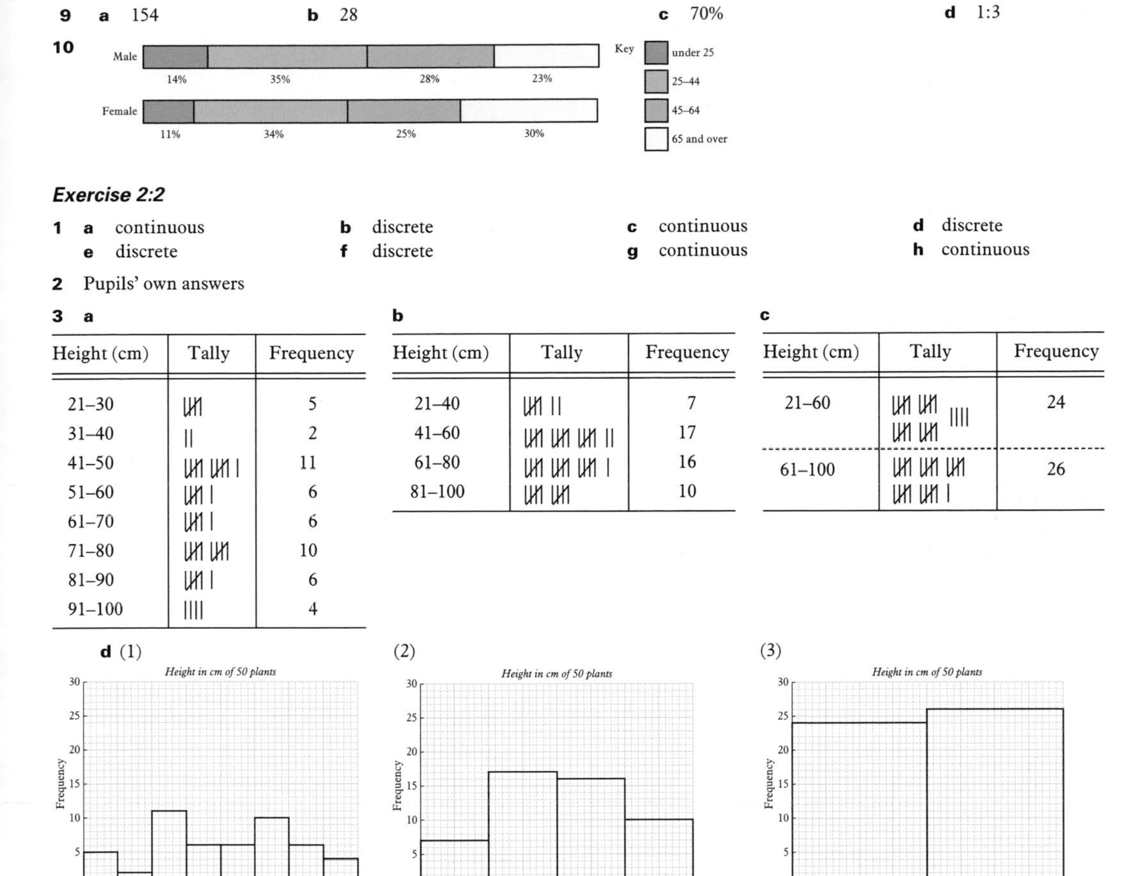

9 **a** 154 **b** 28 **c** 70% **d** 1:3

10

Exercise 2:2

1 **a** continuous **b** discrete **c** continuous **d** discrete
e discrete **f** discrete **g** continuous **h** continuous

2 Pupils' own answers

3 **a**

Height (cm)	Tally	Frequency
21–30	卌	5
31–40	||	2
41–50	卌 卌 |	11
51–60	卌 |	6
61–70	卌 |	6
71–80	卌 卌	10
81–90	卌 |	6
91–100	||||	4

b

Height (cm)	Tally	Frequency
21–40	卌 ||	7
41–60	卌 卌 卌 ||	17
61–80	卌 卌 卌 |	16
81–100	卌 卌	10

c

Height (cm)	Tally	Frequency
21–60	卌 卌 卌 卌 ||||	24
61–100	卌 卌 卌 卌 卌 |	26

d (1) (2) (3)

e The first bar-chart has appropriate number classes and shows the shape of the data well. The remaining two bar-charts lose detail about the shape of the data since there are too few groups.

4 **a** The groups overlap
b 10 and 51 are not included

5 **a** (1)

Total score	Tally	Frequency
$1 \leqslant$ marks $\leqslant 20$	|	1
$21 \leqslant$ marks $\leqslant 40$	卌 卌 |||	13
$41 \leqslant$ marks $\leqslant 60$	卌 卌 卌 ||	17
$61 \leqslant$ marks $\leqslant 80$	卌 卌 |	11
$81 \leqslant$ marks $\leqslant 100$	卌 |	6
$101 \leqslant$ marks $\leqslant 120$		0
$121 \leqslant$ marks $\leqslant 140$	|	1
$141 \leqslant$ marks $\leqslant 160$	|	1

(2)

Total score	Tally	Frequency
$1 \leqslant$ marks $\leqslant 20$	|	1
$21 \leqslant$ marks $\leqslant 40$	卌 卌 |||	13
$41 \leqslant$ marks $\leqslant 60$	卌 卌 卌 ||	17
$61 \leqslant$ marks $\leqslant 80$	卌 卌 |	11
$81 \leqslant$ marks $\leqslant 100$	卌 |	6
marks $\geqslant 101$	||	2

b The second tally-table is the more suitable since there are so few values after 100.

Exercise 2:3

1 46.5 g to 47.5 g **2** 47.5 min to 48.5 min

3 **a** yes **b** no **c** yes **d** yes **e** no **f** yes

4 **a** 5.5 to 6.5 **b** 8.25 to 8.35 **c** 35 to 45 **d** 16.5 to 17.5

ANSWERS

Exercise 2:4

1 Tower Hamlets **2** City **3** Camden, Islington, Newham **4** −6.3 to −7.8

Exercise 2:5

1 **a** Vertical **b** Line **c** Line **d** Vertical

2 **a** 6 **b** 47 **c** 8 **d** Number of households is 'discrete data'

3 **a** 17°C
b 01.00 to 04.00 and 15.00 to 24.00
c ≃12:15 and 19:15
d ≃14.5° No actual reading taken at this time.

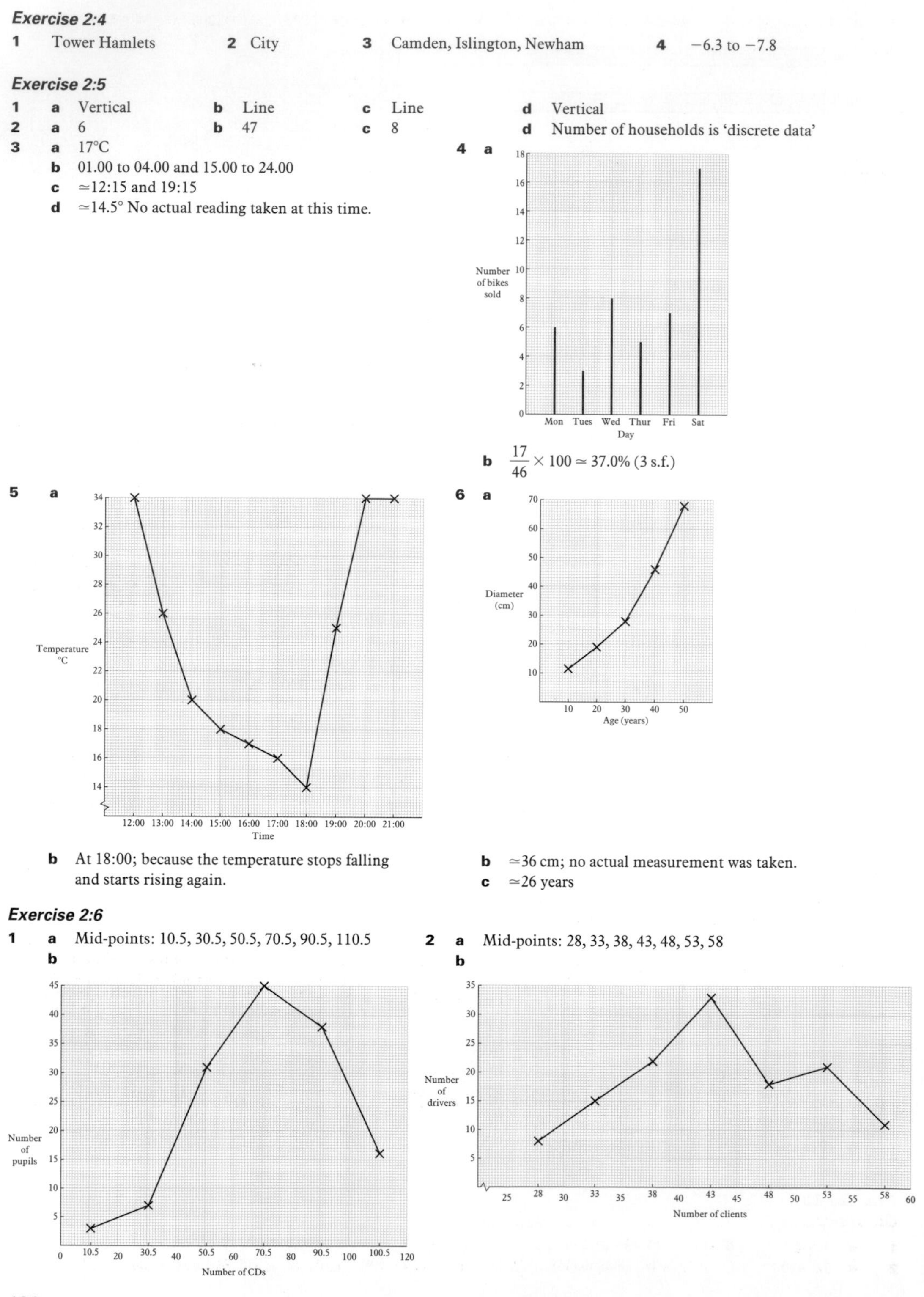

4 **a**
b $\frac{17}{46} \times 100 \simeq 37.0\%$ (3 s.f.)

5 **a**
b At 18:00; because the temperature stops falling and starts rising again.

6 **a**
b ≃36 cm; no actual measurement was taken.
c ≃26 years

Exercise 2:6

1 **a** Mid-points: 10.5, 30.5, 50.5, 70.5, 90.5, 110.5
b

2 **a** Mid-points: 28, 33, 38, 43, 48, 53, 58
b

3 **a** Mid-points: 50.5, 150.5, 250.5, 350.5, 450.5, 550.5

b

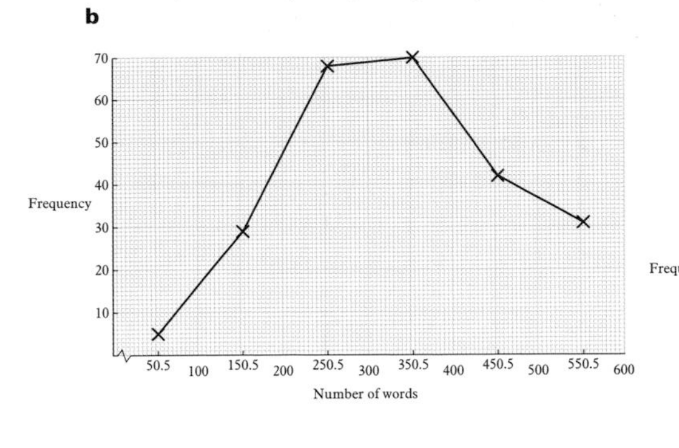

4 **a** Mid-points: 60.5, 80.5, 100.5, 120.5, 140.5, 160.5, 180.5

b

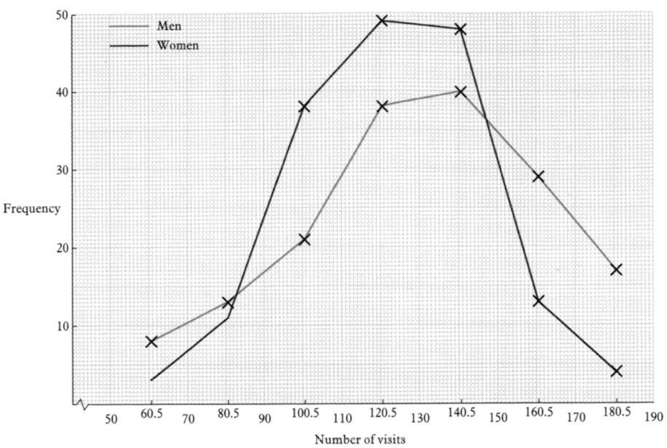

c Less women than men at the lower and higher number of visits. More women than men in the middle number of visits.

Exercise 2:7

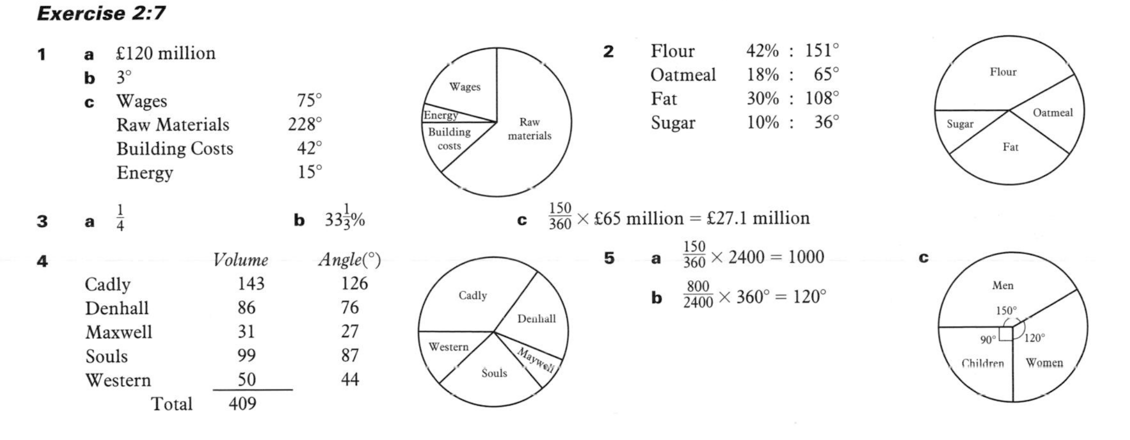

1 **a** £120 million

b 3°

c

Wages	75°
Raw Materials	228°
Building Costs	42°
Energy	15°

2

Flour	42%	151°
Oatmeal	18%	65°
Fat	30%	108°
Sugar	10%	36°

3 **a** $\frac{1}{4}$ **b** $33\frac{1}{3}\%$ **c** $\frac{150}{360} \times$ £65 million = £27.1 million

4

	Volume	*Angle*(°)
Cadly	143	126
Denhall	86	76
Maxwell	31	27
Souls	99	87
Western	50	44
Total	409	

5 **a** $\frac{150}{360} \times 2400 = 1000$

b $\frac{800}{2400} \times 360° = 120°$

c

Exercise 2:8

1 5.4 cm (2 s.f.) **2** 11.3 cm (3 s.f.)

3 Jones' total expenditure: £84 Williams' total expenditure £145

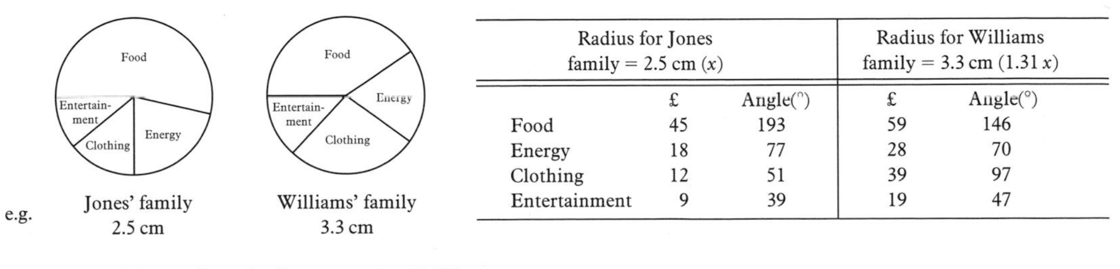

e.g. Jones' family 2.5 cm Williams' family 3.3 cm

	Radius for Jones family = 2.5 cm (x)		Radius for Williams family = 3.3 cm (1.31 x)	
	£	Angle(°)	£	Angle(°)
Food	45	193	59	146
Energy	18	77	28	70
Clothing	12	51	39	97
Entertainment	9	39	19	47

4 2.7 cm, 3.2 cm, 3.8 cm (2 s.f.) **5** 24 000

Questions

1 **a** **i** 100 *(1)* **ii** 70 *(1)* **iii** $4\frac{1}{2}$ *(2)* **b** 25 *(2)*

2 **a** 32 *(1)* **b** 8:12 or 4:6 or equivalent. *(1)* **c** 36.8% *(2)* **d** 0 *(1)* **e** 170 + 114 = 284 *(2)*

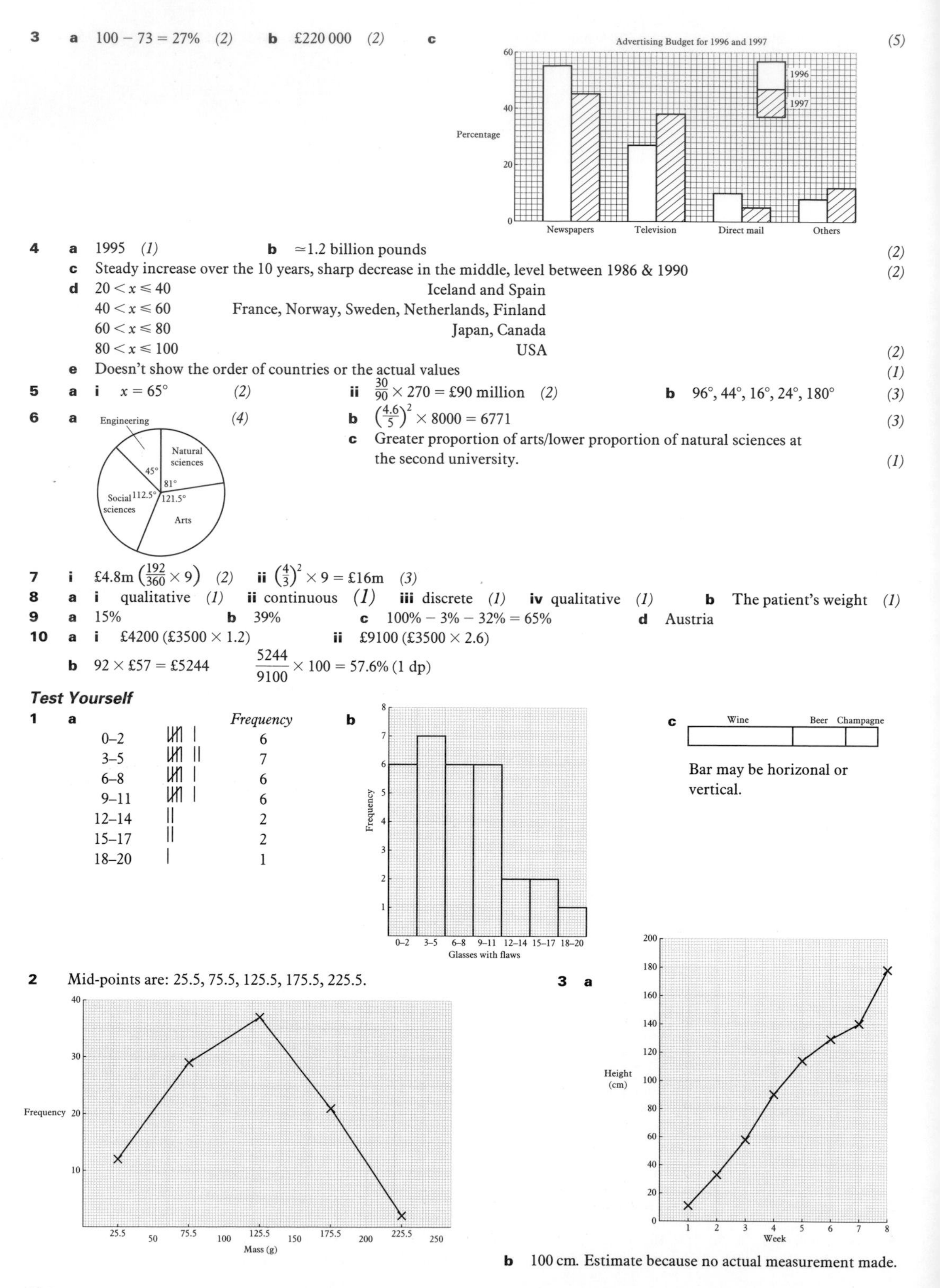

3 **a** $100 - 73 = 27\%$ *(2)* **b** £220 000 *(2)* **c** *(5)*

4 **a** 1995 *(1)* **b** $\simeq$1.2 billion pounds *(2)*

c Steady increase over the 10 years, sharp decrease in the middle, level between 1986 & 1990 *(2)*

d

$20 < x \leqslant 40$	Iceland and Spain
$40 < x \leqslant 60$	France, Norway, Sweden, Netherlands, Finland
$60 < x \leqslant 80$	Japan, Canada
$80 < x \leqslant 100$	USA

(2)

e Doesn't show the order of countries or the actual values *(1)*

5 **a** **i** $x = 65°$ *(2)* **ii** $\frac{30}{90} \times 270 = £90$ million *(2)* **b** 96°, 44°, 16°, 24°, 180° *(3)*

6 **a** *(4)* **b** $\left(\frac{4.6}{5}\right)^2 \times 8000 = 6771$ *(3)*

c Greater proportion of arts/lower proportion of natural sciences at the second university. *(1)*

7 **i** £4.8m $\left(\frac{192}{360} \times 9\right)$ *(2)* **ii** $\left(\frac{4}{3}\right)^2 \times 9 = £16$m *(3)*

8 **a** **i** qualitative *(1)* **ii** continuous *(1)* **iii** discrete *(1)* **iv** qualitative *(1)* **b** The patient's weight *(1)*

9 **a** 15% **b** 39% **c** $100\% - 3\% - 32\% = 65\%$ **d** Austria

10 **a** **i** £4200 (£3500 × 1.2) **ii** £9100 (£3500 × 2.6)

b $92 \times £57 = £5244$ $\frac{5244}{9100} \times 100 = 57.6\%$ (1 dp)

Test Yourself

1 **a**

		Frequency
0–2	𝍸 I	6
3–5	𝍸 II	7
6–8	𝍸 I	6
9–11	𝍸 I	6
12–14	II	2
15–17	II	2
18–20	I	1

b

c Bar may be horizonal or vertical.

2 Mid-points are: 25.5, 75.5, 125.5, 175.5, 225.5.

3 **a**

b 100 cm. Estimate because no actual measurement made.

4 **a** $\frac{1}{3}$ **b** $\frac{1}{4}$ of £120 = £30 **c** Radius of Cara's pie chart is 2 cm

Area of Cara's pie chart $= \pi \times 2^2 = 12.57\text{ cm}^2$ (4 s.f.)

Area for £1 $= 12.57 \div 120 = 0.1047\text{ cm}^2$

Area for £200 $= 0.1047 \times 200 = 20.94\text{ cm}^2$

Radius for Tom's pie chart is r

so $\pi r^2 = 20.94$

$r^2 = 20.24 \div \pi$

$r = 2.6$ cm (1 d.p.)

CHAPTER 3

Exercise 3:1

1 **a** 16 **b** 16 **c** 19.1 **d** The median as it is not affected by the extreme values.

2 **a** 8 **b** 8 **c** 8 **d** Any of these values is representative of this group. The distribution is fairly symmetrical and there are no extreme values which distort the data. The high and low values are evenly distributed.

Exercise 3:2

1 **a** 3 **b** 2 **c** 2.1 **2** **a** 4 **b** 5 **c** 4.8

3 **a** 4 **b** 4 **c** 4.25

Exercise 3:3

1 **a** 11–15 **b** 12.7 **c** 13.2 **2** **a** 7–9 **b** 8.0 **c** 7.4

3 **a** 9–11 **b** 10.8 **c** 10 **4** **a** 8–12 **b** 11.9 **c** 11.7

Exercise 3:4

1 **a** 161–180 g **b** 165 g **c** 166 g **2** **a** $110 \leqslant x < 120$ **b** 129.5 cm **c** 127.5 cm

Exercise 3:5

1 117.5 cm **2** 119 cm **3** 2.875 bottles **4** 30 runs

5 **a** £17 000 **b** The mean does not represent the wages fairly. At each type of work there will be a different number of people employed.

6 £14 531.25

Exercise 3:6

1 38.8% **2** 73.7% **3** 11 kg **4** £20.56 **5** £1.14 **6** £7.75 **7** **a** 60 **b** 47.6

Exercise 3:7

1 **a** −2, −9, 6, 9, 8, −5, 0 **b** 1 **c** 51

2 122 **3** 499 **4** 717 **5** 250 **6** 401 **7** 824 **8** 2005 **9** 2400

10 **a** 4 **b** 6.48 **c** 10 **d** 6.14 **e** 15.1 **f** 28.0 **11** 4.99% (3 sf) **12** 33.7% (3 sf)

13 **a** 8.00% (3 sf) **b** 8.47% (3 sf) **c** the second share group

14 **a** Brompton 7.65% Cornwall 9.63% **b** Cornwall

Exercise 3:8

1 **a**

Length in cm	Frequency
5	1
6	1
7	6
8	2
9	6
10	5
11	4
12	4
13	3
14	6
15	6
16	4
17	1
18	1

b 11.54 cm

c 11.54 cm

d Tricia could take a larger sample.

2 **a** 8.57 words **b** 8.57 words **3** **a** $\frac{3}{60} = \frac{1}{20}$ **b** $\frac{1}{20}$

4 **a** $\frac{25}{40} = \frac{5}{8}$ **b** $\frac{5}{8}$ **5** **a** $\frac{20}{100} = \frac{1}{5}$ **b** $\frac{1}{5}$

Exercise 3:9

1 **a, b**

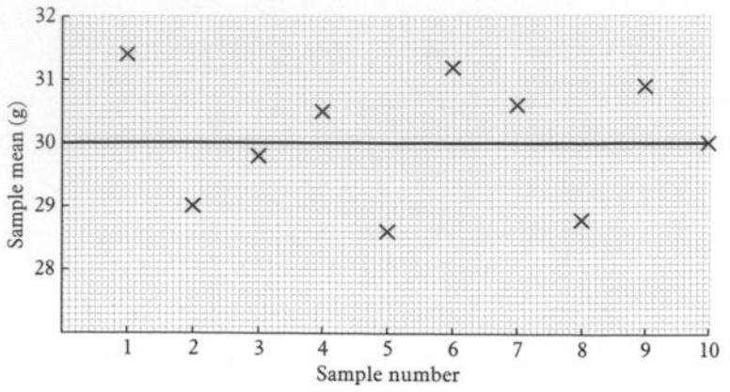

c Yes. The points are in a band about the target line, although this band is quite wide.

d The weights of the packets are increasing. There could be a fault in the machine.

2 Both machines A and B seem to be working properly but Machine B seems to be more consistent than Machine A. Machine A has a wider variability of sample means. Machine C seems to have developed a fault as the trend of the sample means is downward. Machine D consistently produces overfull bags as the sample means are falling above the target value.

3 **a**

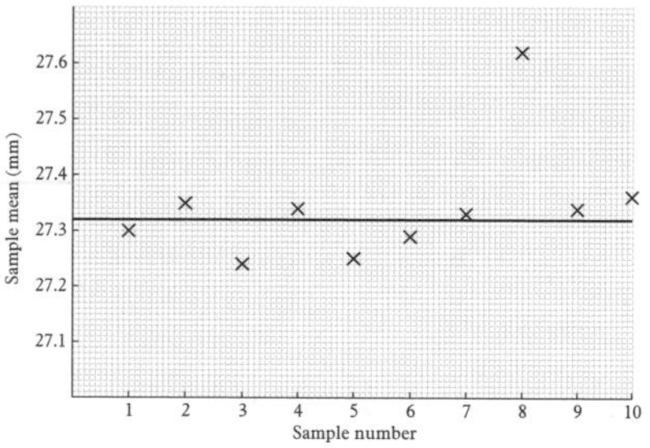

b Sample 8, because this sample does not follow the trend and so warns of a fault in the machine.

4 **a**

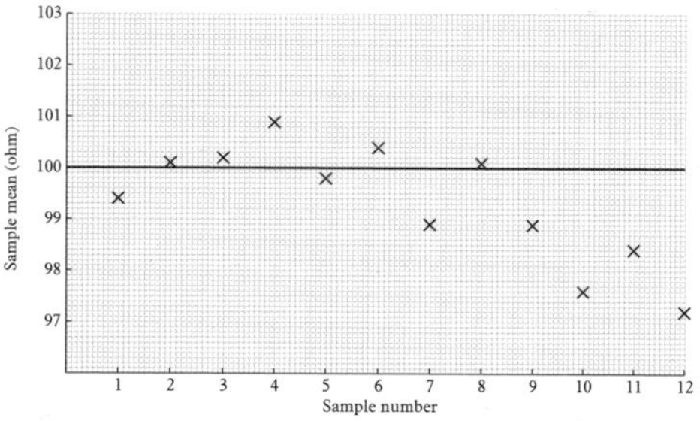

b No because the sample mean trend is downwards.

Questions

1 **a** 2, 5, 3, 5, 8, 7 *(2)* **b** 48.1 g *(3)* **c** Larger sample/more information/larger period of time. *(1)*

2 The mode would not have been affected. *(1)* He had never failed to catch fish before. *(1)*

3 **a** The median

£7500 would distort the mean/The median ignores extreme values/£3900 is the mode but it is too low to represent the data. *(2)*

b **i** Discrete data *(1)* **ii** Continuous data. *(1)*

4 **a** *(4)*

Score	Frequency
0	2
1	7
2	4
3	2
4	3
5	2

b 1 *(1)*

c 2 *(1)*

5 **a** Mean *(1)*

b $\frac{600}{8} = 75$ kg *(1)*

6 **a** *(4)*

Number of days absent	Frequency
0	10
1	7
2	4
3	2
4	1
5	4
6	1
7	1

b 0 days *(1)*

c 1.93 (2 d.p.) *(2)*

d May not be 30 pupils in each class.
Different age groups/classes may have different absence rates. *(2)*

7 **a** 27.3 greenfly *(3)* **b** 20 greenfly *(4)* **c** 24.5 *(1)*

8 **a** *(2)*

Wage (in £)	Frequency
85	7
90	5
120	4
160	2
220	1
335	1

b Mode *(1)*
c Median *(1)*
d 10 *(1)*
e 20 *(2)*
f 10 *(2)*
g 12 *(2)*
h The form has not got a name on it. *(1)*
It enables information to be obtained on a sensitive subject. Only the person filling in the questionnaire knows why they are ticking the 'Yes' box.

9 **a**

x	1	3	4	6	8
f	2	4	6	5	3

b 4 **c** $(1 \times 2 + 3 \times 4 + 4 \times 6 + 6 \times 5 + 8 \times 3) \div 20 = 4.6$

Test Yourself

1 Mean of 21 pupils $= \dfrac{(20 \times 125) + 144}{(20 + 1)} = \dfrac{2644}{21} = 126 \text{ cm}$

2 Average price of ticket $= \dfrac{(3 \times 8) + (15 \times 5) + (25 \times 7)}{(3 + 5 + 7)} = £18.27$

3 **a** 4 musicians **b** Median is 11th item of data = 4 musicians

c Mean pop-group size $= \dfrac{(2 \times 4) + (3 \times 5) + (4 \times 7) + (5 \times 2) + (6 \times 3)}{(4 + 5 + 7 + 2 + 3)} = 3.8$

4 **a** Modal group is 5–9 cars. **b** Estimate of median number of cars $= 10 + (\frac{5}{4} \times 1) = 11.25$

c Estimate of mean number of cars $= \dfrac{(2 \times 1) + (7 \times 5) + (12 \times 4) + (17 \times 3)}{(1 + 5 + 4 + 3)} = 10.5$

5 **a** Modal group is $8 \leqslant x < 10$ hours.

b Estimate of mean number of hours $= \dfrac{(3 \times 5) + (15 \times 7) + (18 \times 9) + (11 \times 7) + (13 \times 5)}{(3 + 15 + 18 + 7 + 5)} = 8.83$ hours

c Estimate of median number of hours $= 8 + (\frac{2}{18} \times 6.5) = 8.7$ hours

CHAPTER 4

Exercise 4:1

1 **a** 19 **b** 3.2 **c** 756
2 **a** £37 **b** £154
3 **a** 9 **b** 6 **c** 3
4 **a** 17.5 **b** 9.5 **c** 4.75
5 **a** 4.0 **b** 2.35 **c** 1.175
6 **a** 62.5 **b** 18 **c** 9
7 **a** 17 **b** 2
8 **a** Red Glory give higher weights on average (medians). Red Glory weights are also more consistent (because of lower interquartile range).
b Sweet Giant. Although less weight on average, the high value of the interquartile range indicates the chances of a large weight more likely.
9 Machine A produces lengths that are closer to 80 cm on average, but the lengths are less consistent than machine B. Machine A could produce lengths less than 80 cm but this is less likely with machine B. Get rid of machine A.

Exercise 4:2

1 **a**

Number of aces	Cumulative frequency
0	14
up to 1	40
up to 2	77
up to 3	88
up to 4	97
up to 5	100

b (1) 77
(2) 88
(3) 12

2 **a**

Number of visits	Cumulative frequency
less than 3	21
less than 6	40
less than 9	54
less than 12	61
less than 15	65

b 65
c 54
d 25

3 **a**

Money in £	Cumulative frequency
< 100	7
< 120	17
< 140	22
< 160	36
< 180	55
< 200	60

b 17
c 24

4 Median = 2 Interquartile range = 1

5 **a**

Number of GCSEs	Cumulative frequency
up to 1	5
up to 2	12
up to 3	23
up to 4	39
up to 5	60
up to 6	88
up to 7	107
up to 8	115
up to 9	120

b (1) 39
(2) 81
(3) 115
c 5
d 3
e Girls passed more GCSEs on average but the number of GCSEs for girls was less consistent than those for boys.

Exercise 4:3

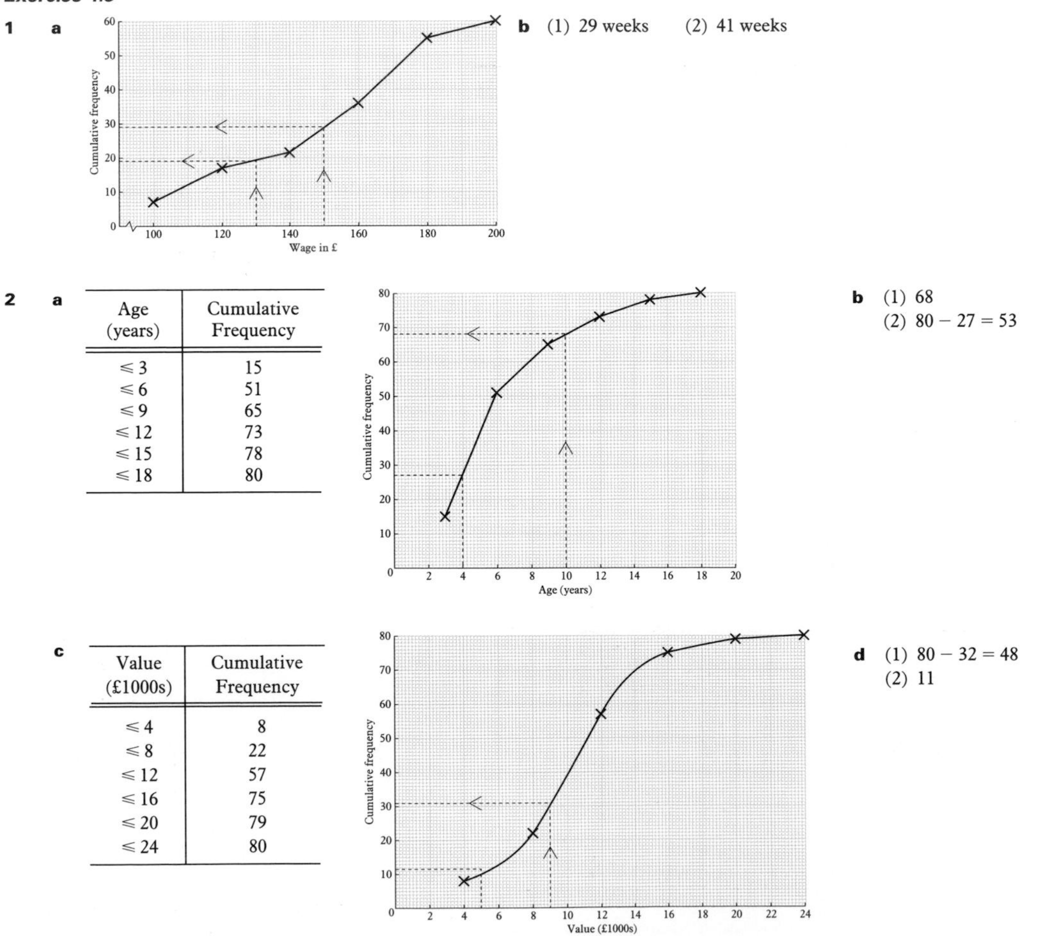

1 **a** (graph)

b (1) 29 weeks (2) 41 weeks

2 **a**

Age (years)	Cumulative Frequency
$\leqslant 3$	15
$\leqslant 6$	51
$\leqslant 9$	65
$\leqslant 12$	73
$\leqslant 15$	78
$\leqslant 18$	80

b (1) 68
(2) $80 - 27 = 53$

c

Value (£1000s)	Cumulative Frequency
$\leqslant 4$	8
$\leqslant 8$	22
$\leqslant 12$	57
$\leqslant 16$	75
$\leqslant 20$	79
$\leqslant 24$	80

d (1) $80 - 32 = 48$
(2) 11

3 **a**

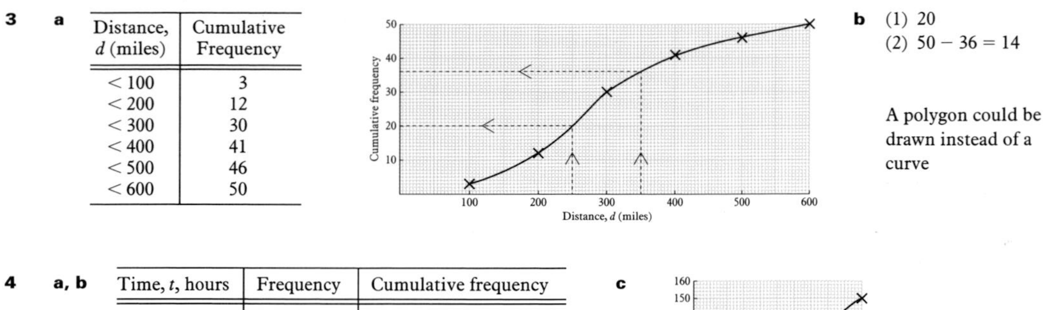

Distance, d (miles)	Cumulative Frequency
< 100	3
< 200	12
< 300	30
< 400	41
< 500	46
< 600	50

b (1) 20
(2) $50 - 36 = 14$

A polygon could be drawn instead of a curve

4 **a, b**

Time, t, hours	Frequency	Cumulative frequency
$0 \leqslant t < 2$	20	20
$2 \leqslant t < 4$	32	52
$4 \leqslant t < 6$	57	109
$6 \leqslant t < 8$	25	134
$8 \leqslant t < 10$	16	150

c

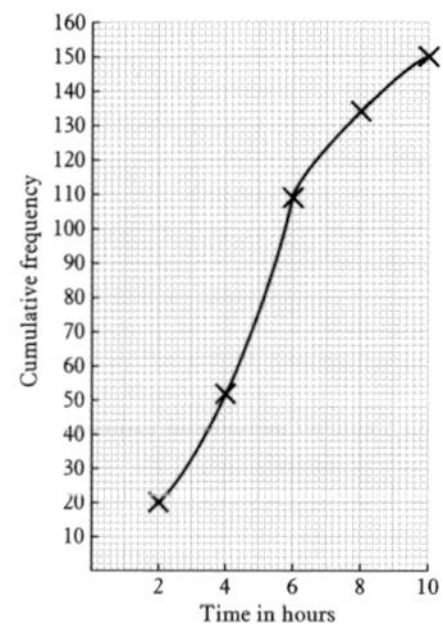

Points may be joined with a curve or with straight lines.

5 **a, b**

Distance, m, miles	Frequency	Cumulative frequency
$0 \leqslant m < 5$	2	2
$5 \leqslant m < 10$	6	8
$10 \leqslant m < 15$	13	21
$15 \leqslant m < 20$	15	36
$20 \leqslant m < 25$	18	54
$25 \leqslant m < 30$	11	65

c

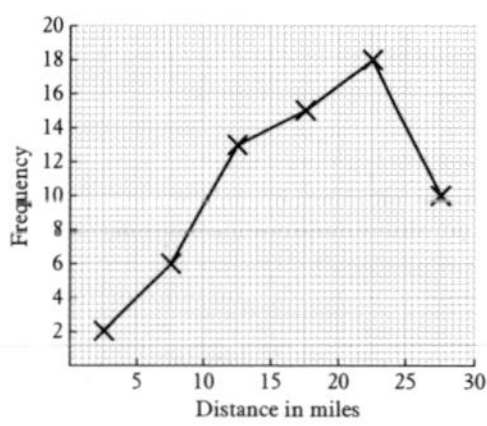

6

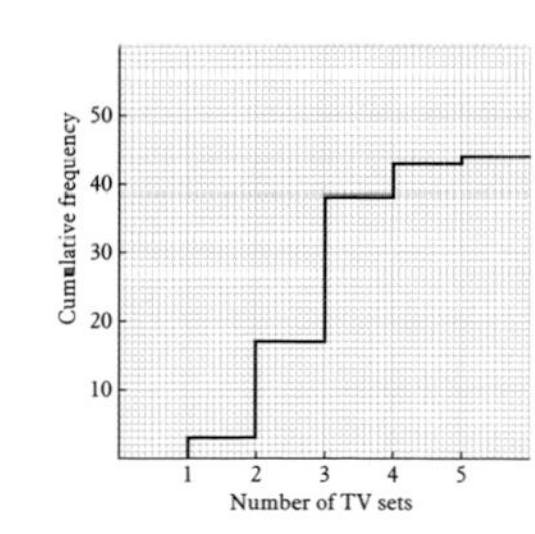

7 **a**

Number of eggs	Frequency	Number of eggs	Cumulative frequency
1	1	$\leqslant 1$	1
2	3	$\leqslant 2$	4
3	3	$\leqslant 3$	7
4	5	$\leqslant 4$	12
5	5	$\leqslant 5$	17
6	8	$\leqslant 6$	25
7	6	$\leqslant 7$	31
8	5	$\leqslant 8$	36
9	4	$\leqslant 9$	40

b

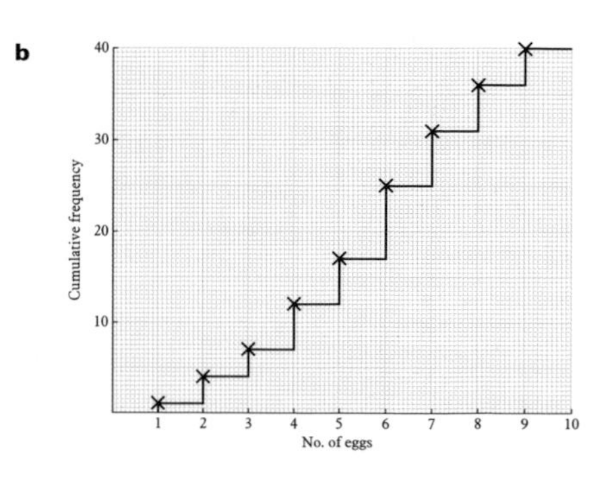

Exercise 4:4

1 a

Time (t) (mins)	Cumulative Frequency
⩽ 1	5
⩽ 2	18
⩽ 3	41
⩽ 4	77

Time (t) (mins)	Cumulative Frequency
⩽ 5	123
⩽ 6	159
⩽ 7	189
⩽ 8	200

b $4\frac{1}{2}$ mins
c 3.2 mins, 5.7 minutes
d 2.5 mins

2 a

Payment (p)	Cumulative Frequency
⩽ 50	7
⩽ 100	22
⩽ 150	43
⩽ 200	72

Payment (p)	Cumulative Frequency
⩽ 250	104
⩽ 300	133
⩽ 350	153
⩽ 400	160

b £215
c £145, £275
d £130

3 a 4.1 mins **b** 6.7 mins

4 a £125 **b** £290 **c** $128 - 32 = 96$

5 a Cumulative frequencies 2, 7, 32, 44, 46, 47, 49, 50
b (1) 3 items (2) 4 items
c (1) 2 items (2) 5 items
d $5 - 2 = 3$ items
e It includes 80% of all the data.

6 a Cumulative frequencies 11, 46, 104, 164, 185, 195, 198, 200
b (1) 3 people (2) 5 people **c** (1) 3 people (2) 5 people

7 a The median for the country cats is higher showing that they live longer on average. The city cats interpercentile range (of 12 years) is greater than the country cats interpercentile range (of 9 years) so city cats ages are more spread out.
b The ranges would probably be the same. It is highly likely that in both groups you have one early death and one long life.

Exercise 4:5

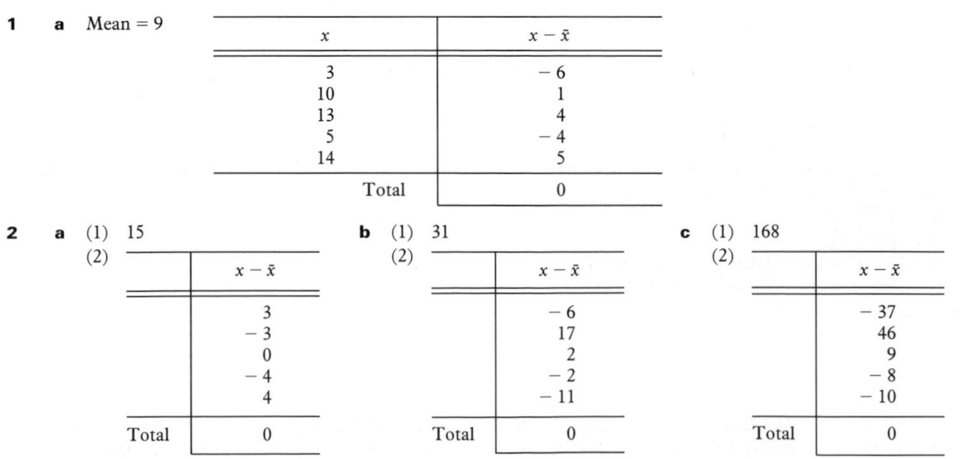

1 a Mean = 9

x	$x - \bar{x}$
3	−6
10	1
13	4
5	−4
14	5
Total	0

2 a (1) 15
(2)

	$x - \bar{x}$
	3
	−3
	0
	−4
	4
Total	0

b (1) 31
(2)

	$x - \bar{x}$
	−6
	17
	2
	−2
	−11
Total	0

c (1) 168
(2)

	$x - \bar{x}$
	−37
	46
	9
	−8
	−10
Total	0

3 It always comes to 0.

4 **a** 4 **b** 3 **c** $\frac{3}{8}$ **d** 2.5 **e** 1.756 **f** $1\frac{3}{4}$

5 **a** 2.8 **b** 2 **c** 3.9 **d** 3.5 **e** 3.4 **f** 1.67 (3 sf)

Exercise 4:6

1 **a** Mean = 32, standard deviation = 5.15 **b** Mean = 18, standard deviation = 5.73
c Mean = 56, standard deviation = 16.8 **d** Mean = 44, standard deviation = 2.28
e Mean = 91, standard deviation = 4.14 **f** Mean = 7, standard deviation = 5.10

2 **a** £111 **b** £45.80
c James' school raised less money on average, but the amounts raised were more consistent.

3 **a** Mean = 32, standard deviation = 5.15 **b** Mean = 18, standard deviation = 5.73
c Mean = 56, standard deviation = 16.8 **d** Mean = 44, standard deviation = 2.28
e Mean = 91, standard deviation = 4.14 **f** Mean = 7, standard deviation = 5.10

4 **a** Martin mean = 82.7, standard deviation = 5.93 Jane mean = 80, standard deviation = 9.35
b Martin takes more strokes on average but is more consistent than Jane.

5 **a** Mean = 35.14, standard deviation = 3.00, median = 35 **b** Mean = 39.14, standard deviation = 3.00 median = 39
c Mean and median have increased by 4, standard deviation has stayed the same.

6 **a** Mean = 14.71, standard deviation = 2.49 **b** Mean = 24.71, standard deviation = 2.49
c Mean = 17.71, standard deviation = 2.49 **d** Mean = 12.71, standard deviation = 2.49

7 **a** Mean = 8.17, standard deviation = 4.45 median = 7.5 **b** Mean = 24.5, standard deviation = 13.35 median = 22.5
c The mean, median and the standard deviation have been multiplied by 3.

8 **a** Mean = 2.3, standard deviation = 0.65 **b** Mean = 4.6, standard deviation = 1.30
c Mean = 23, standard deviation = 6.52 **d** Mean = 11.5, standard deviation = 3.26

9 Mean and standard deviation are multiplied by 1.175
Mean = £88.65, standard deviation = £43.70

Exercise 4:7

1 **a**

x	f	$f \times x$	x^2	$f \times x^2$
3	24	72	9	216
7	15	105	49	735
8	21	168	64	1344

b Mean = 5.75, standard deviation = 2.28

2 **a** Mean = 11.25, standard deviation = 1.02 **b** Mean = 3.51, standard deviation = 0.99

3 **c** (1) $\Sigma x = 149.4$ (2) $\bar{x} = 14.94$ (3) $\Sigma x^2 = 2255.8$ (4) $s = 1.54$
e (1) $\Sigma x = 196$ (2) $\bar{x} = 15.1$ (3) $\Sigma x^2 = 2985$ (4) $s = 1.53$

4 **a** Mean = 2.95, standard deviation = 0.30 **b** Mean = 23.74, standard deviation = 1.54

5 **b** Mid values are: 7.5, 12.5, 17.5, 22.5, 27.5 **c** 17.61 **d** 6.12

6 **a** Mid values are: 7 500, 12 500, 17 500, 22 500, 32 500
Mean = £17 895, standard deviation = £7259
b The second garage sells cheaper cars on average and also has a narrower price range than the first garage.

7 **a** 4 + 7 + 14 + 18 + 17 + 13 + 2 = 75
b Mid values are: 30, 50, 70, 90, 110, 130, 150 (£1000s)
Mean = £92 400, standard deviation = £29 568
c Lornton has the cheapest houses on average, with Tanley the most expensive. Tanley also has the widest price range and Cordworth the narrowest price range.

8 **a** Frequency 5, 8, 5, 12, 6, 2
b Mid values: 1, 3, 5, 7, 9, 11
Mean = 5.63 mins, standard deviation = 2.87 mins

9 **a** (1) 2.35 (3 sf) (2) 1.87 (3 sf) (3) 2.54 (3 sf)
b They are different samples.

Questions

1 **a** **i** £15 000 *(1)* **ii** £12 000 *(1)* **iii** £22 000 *(1)* **iv** £10 000 *(1)* **v** £20 000 *(1)*
b **i** £15 000 *(1)* **ii** £12 000 *(1)* **iii** £22 000 *(1)* **iv** £10 000 *(1)* **v** £22 000 *(1)*

2 **a** Cumulative frequency. 14, 42, 67, 85, 94, 98, 100, 100 *(2)*

b *(2)*

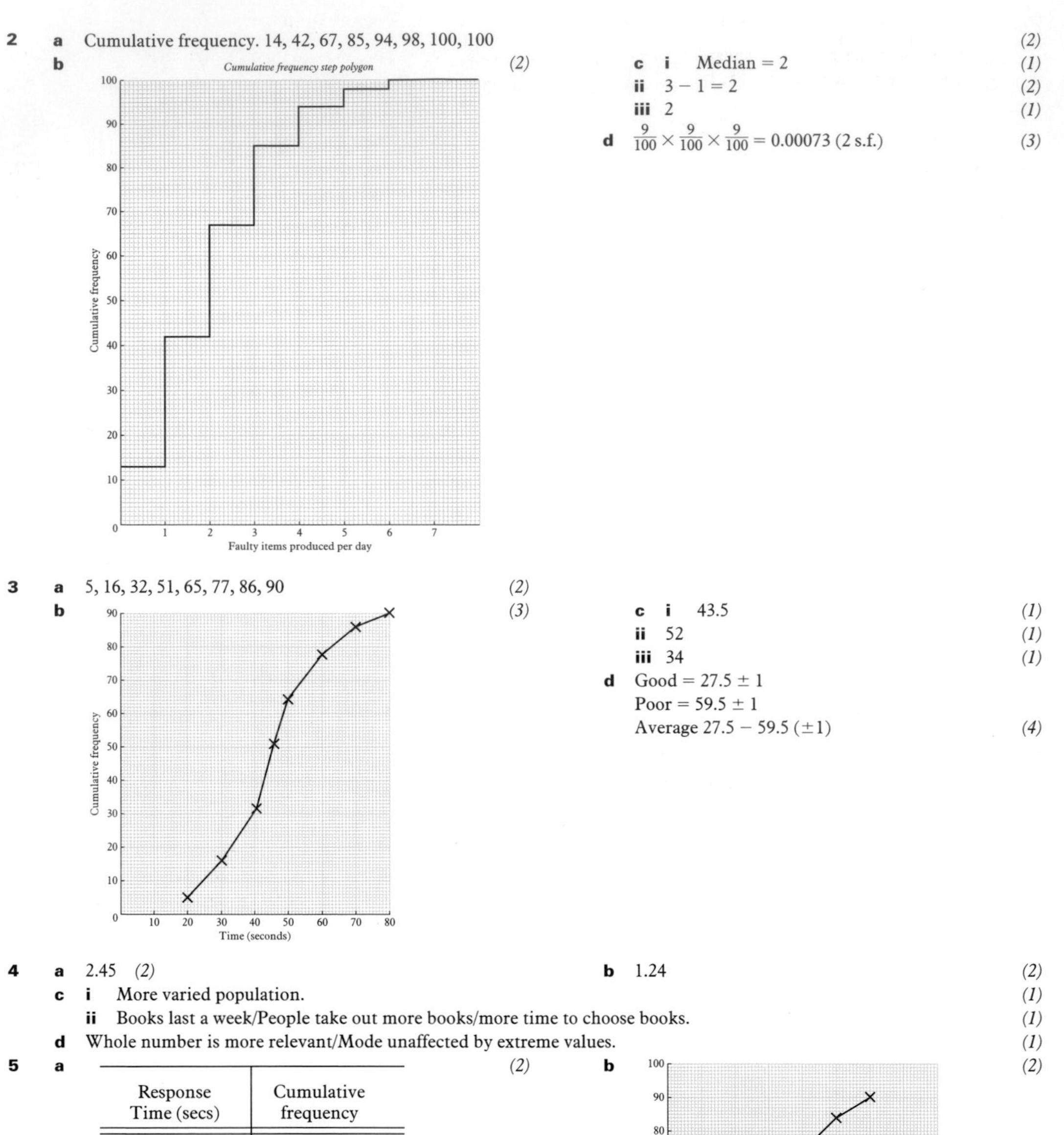

c **i** Median = 2 *(1)*

ii 3 − 1 = 2 *(2)*

iii 2 *(1)*

d $\frac{9}{100} \times \frac{9}{100} \times \frac{9}{100} = 0.00073$ (2 s.f.) *(3)*

3 **a** 5, 16, 32, 51, 65, 77, 86, 90 *(2)*

b *(3)*

c **i** 43.5 *(1)*

ii 52 *(1)*

iii 34 *(1)*

d Good = 27.5 ± 1

Poor = 59.5 ± 1

Average 27.5 − 59.5 (±1) *(4)*

4 **a** 2.45 *(2)* **b** 1.24 *(2)*

c **i** More varied population. *(1)*

ii Books last a week/People take out more books/more time to choose books. *(1)*

d Whole number is more relevant/Mode unaffected by extreme values. *(1)*

5 **a** *(2)*

Response Time (secs)	Cumulative frequency
0–10	10
0–20	28
0–30	44
0–40	74
0–50	84
0–60	90

b *(2)*

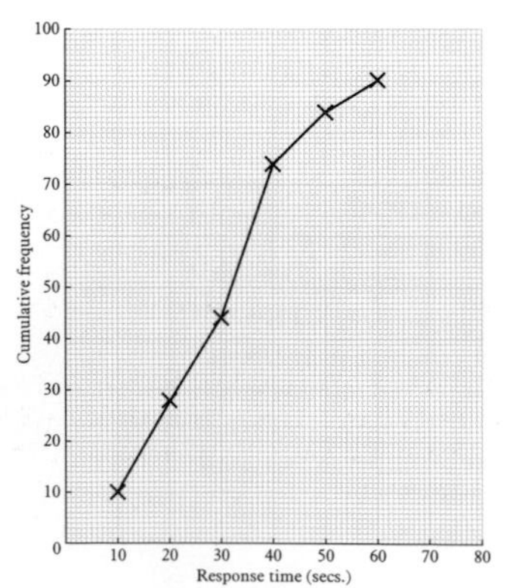

c LQ = 17 UQ = 38 Interquartile range = 21 *(2)*

d Middle 50% lie within a narrower range.
Values are closer together/not so spread out. *(2)*

6 **a** Mean = 10.6875, standard deviation = 1.12 (3 s.f.) *(3)* **b** $13\frac{1}{2} - 8\frac{1}{2} = 5$ kg *(2)*

c **i** will not change *(1)* **ii** Decreases to 1.11 (3 s.f.) *(1)* **iii** will not change. *(1)*

7 **a** **i** $\bar{x} = 1.5$ **ii** standard deviation = 0.98 *(6)* **b** $\frac{s}{\bar{x}} = 0.65$ and 0.36 *(2)*

c Decrease in mean rate or standard deviations approximately equal **but** relative variation higher over time. *(2)*

8 Cumulative frequency 0, 14, 47, 85, 115, 132, 143, 150

a *(4)*

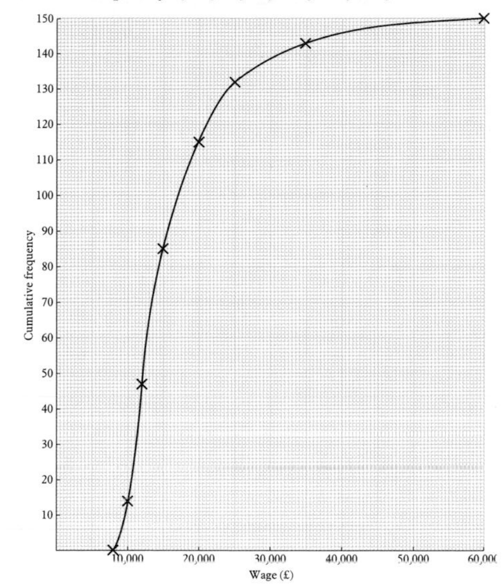

b $\frac{11}{150} = 7.3\%$ *(2)*

c £14 500 *(1)*

d **i** £16 750 *(3)*

ii Men's wages are more spread out *(1)*

iii 80% of data is used *(1)*

e **i** Increases by £300 *(1)*

ii None *(1)*

Test Yourself

1 23, 39, 41, 42.5, 44, 49, 50, 55, 58, 62, 63.5, 65, 74, 76

a $63.5 - 42.5 = 21$ **b** $21 \div 2 = 10.5$

2 **a** (1) 22 (2) 19 **b** (1) 19 (2) 23

3 Cumulative frequencies 15, 37, 67, 81, 91, 100

a

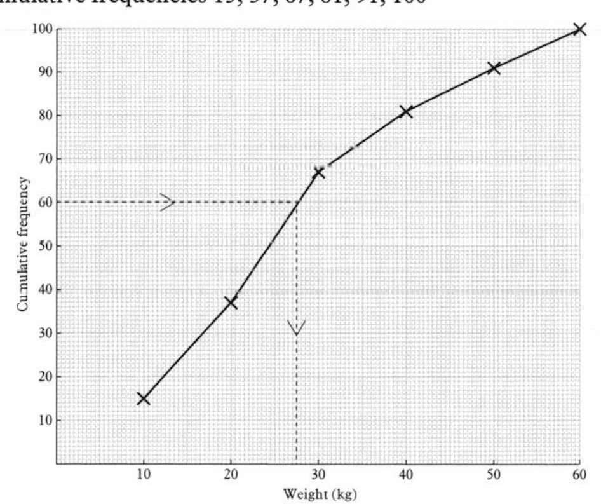

b 74

c 24.5 kg

d upper quartile = 36

lower quartile = 15

interquartile range = 36 − 15 = 21 kg

e 27.5 kg

4 Variance = 213.2, standard deviation = 14.6

5 $\sqrt{\frac{\Sigma x^2}{40} - \left(\frac{\Sigma x}{40}\right)^2} = \sqrt{\frac{6381}{40} - \left(\frac{387}{40}\right)^2} = 8.12$ (3 s.f.)

6 Mid points are: 2.5, 7.5, 12.5, 17.5

a Mean = 9.11 mins, standard deviation = 4.82 mins

b Patients waited longer on average to see Dr Job but these waiting times were more consistent than the waiting times for Dr Thompson.

CHAPTER 5

Exercise 5:1

1 **a** 21 **b** 5 **c** 76 **d** 59

2 **a** 5 **b** 15

3 **a** 6 **b** 10 **c** 3

4

	Ice cream	Drinks	Total
Chocolates	158	207	365
Toffees	99	43	142
Total	257	250	507

5

	France	Spain	Germany	Total
Car/ferry	15	8	5	28
Plane	3	6	3	12
Total	18	14	8	40

6

	Volvo	Renault	Ford	Total
Grey	8	9	14	31
Red	4	11	8	23
Blue	0	3	3	6
Total	12	23	25	60

7 **a**

	Car	Coach	Train	Total
Student	20	45	37	102
Tourist	12	1	60	73
Retailer	30	2	53	85
Total	62	48	150	260

b 11 more tourists travelled by car than by coach.

8 **a**

	Under 21s	21–45	Over 45s	Total
Satellite	48	23	19	90
Terrestrial	28	21	11	60
Cable	21	43	86	150
Total	97	87	116	300

b 150 visitors prefer cable.
c 44 visitors prefer either satellite or terrestrial.

Exercise 5:2

1 *Waiting time in minutes*

Stem	Leaf
1	1 2 4 8 9 9
2	4 8
3	3 6

Where 2|4 means 24 minutes

2 *Temperatures (°C)*

Stem	Leaf
1	8 9
2	4 5 6 6 7 8 8
3	0 2 3

Where 2|4 means 24°C

3 *Time (minutes)*

Stem	Leaf
6	1 3 8
7	2 4 7 8 9
8	3 4 8
9	9

Where 7|2 means 7.2 minutes

4 *Length (cm)*

Stem	Leaf
14	7 7 7 8
15	1 2 3 5 8 9
16	4 5 6 6 7
17	1 2 8
18	8 9 9

Where 15|1 means 15.1 cm

5 *Marks scored*

Stem	Leaf
10	1 2 3 4 8
11	2 3 8 8 9 9
12	2 4 4 4 5
13	1 6 9
14	0

Where 12|4 means 124 marks

6 *Time (seconds)*

Stem	Leaf
19	2 4 5 6 7 8
20	1 3 5 6 6 7
21	2 3
22	0 1

Where 20|1 means 20.1 seconds

7 *Height (cm)*

Stem	Leaf
14	3 5
15	2 4 6 6 7
16	4 5 8 8 8 9 9
17	3 4 5 7
18	2 3

Where 16|4 means 164 cm

Exercise 5:3

1 *Time (minutes)*

Stem	Leaf
3	0 1 2 3 4
3	6 7 8 9
4	0 1
4	6 8
5	4 4
5	5

Where 4|6 means 46 minutes

2 *Runs scored*

Stem	Leaf
12	0 1
12	6 7 8
13	0 1 2 4
13	6 7 7 8 8 9 9

Where 12|6 means 126 runs

3 *Time (minutes)*

Stem	Leaf
3	1 2 3 4
3	5 6 8
4	0 1 2
4	6 6 7 7 7
5	
5	9

Where 4|6 means 4.6 minutes

4 *Time (minutes)*

Stem	Leaf
7	1 2 3
7	5 6 8 8 9 9
8	1 4
8	6
9	2 4
9	5 6 7 7 7 8

Where 7|5 means 7.5 minutes

5 *Earnings (£/week)*

Stem	Leaf
2	
2	8 9
3	2 4 4
3	5 6 7 8
4	1 3
4	8

Where 3|5 means £35 per week

Exercise 5:4

1 **a** Mode = 23°C **b** Median = 19°C

2 **a** Mode = 57 marks **b** Interquartile range = 18 marks

3 **a** Median = 2.5 miles **b** Semi-interquartile range = 0.7 miles

Exercise 5:5

1 **a** Median = 26°C, mode = 23°C **b** Median = 22°C, mode = 22°C

c Median, because it more accurately reflects the generally higher temperatures in France.

2 **a** Mode = 35 lessons, median = 27 lessons **b** Mode = 34 lessons, median = 28 lessons

c There is no difference in the number of lessons taken by male and female students.

3 **a** *Time in hours*

Beth		Josie
5 4 3	2	
2	3	6 8 8
5 3	4	1 3 5 6
8 7 5 3	5	2 4 5

5|5 means 5.5 hrs

b Beth : Median = 4.4 hours, semi-interquartile range = 1.5 hrs
Josie : Median = 4.4 hours, semi-interquartile range = 0.7 hrs

c They both spend the same time on average at the zoo. Josie's visits are less variable in length than Beth's visits.

4 **a**

Females		Males
9 3	36	
9	37	
6 3 3 0 0	38	
9 9 6 4 4 3	39	1 2 6 7
7	40	1 2 2 3 5 5
	41	1 3 5 7
	42	
	43	
	44	
	45	2

where 36|9 means 36.9 hrs

b (1) Median time = 40.3
Interquartile range = 41.3 − 39.7 = 1.6
(2) Median time = 38.6
Interquartile range = 39.6 − 38.0 = 1.6

c The median time is higher for the males. Average hours worked by men is longer. The interquartile range is the same for both males and females.

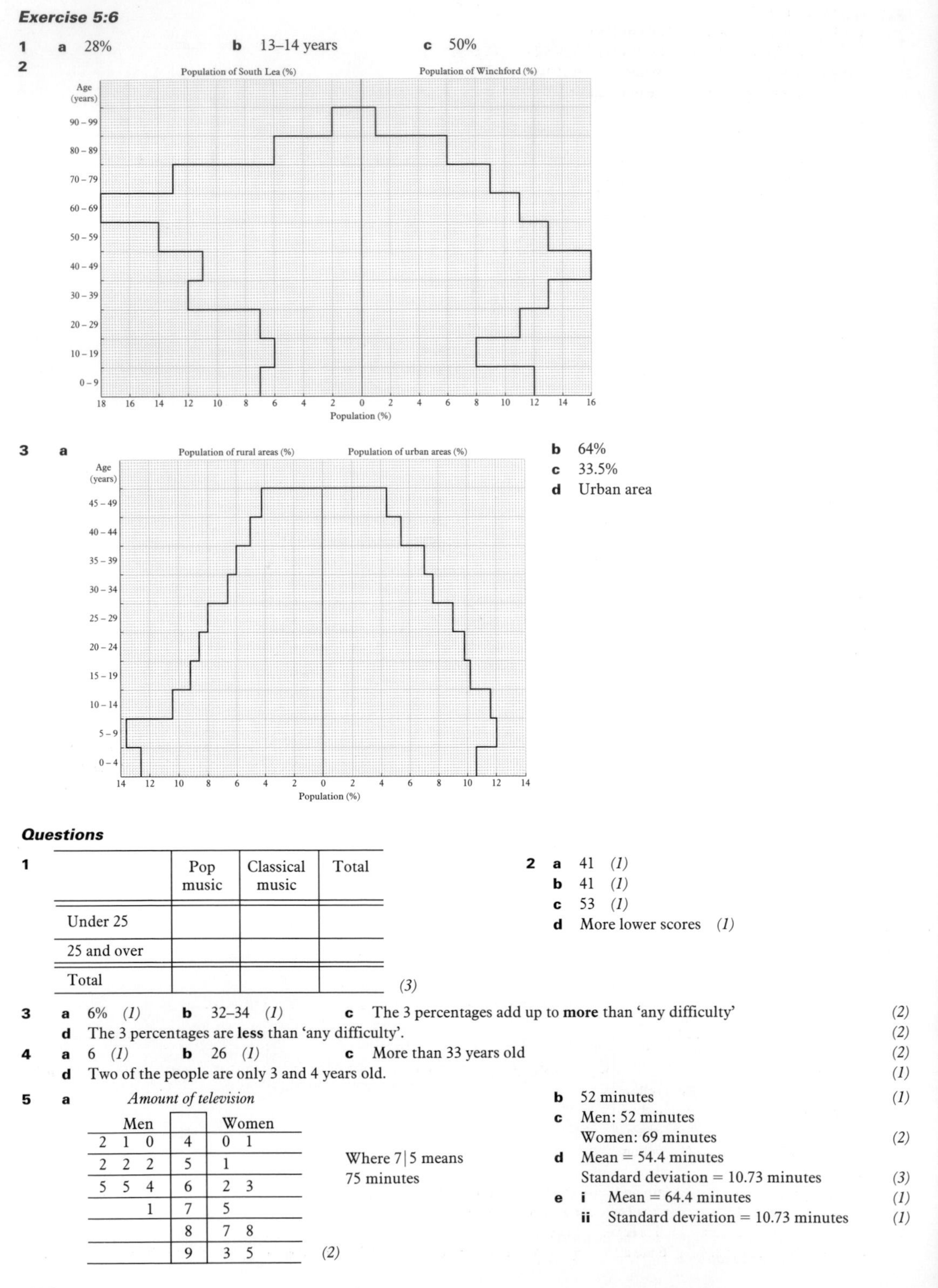

Exercise 5:6

1 **a** 28% **b** 13–14 years **c** 50%

2

3 **a**

b 64%
c 33.5%
d Urban area

Questions

1

	Pop music	Classical music	Total
Under 25			
25 and over			
Total			

(3)

2 **a** 41 *(1)*
b 41 *(1)*
c 53 *(1)*
d More lower scores *(1)*

3 **a** 6% *(1)* **b** 32–34 *(1)* **c** The 3 percentages add up to **more** than 'any difficulty' *(2)*
d The 3 percentages are **less** than 'any difficulty'. *(2)*

4 **a** 6 *(1)* **b** 26 *(1)* **c** More than 33 years old *(2)*
d Two of the people are only 3 and 4 years old. *(1)*

5 **a** *Amount of television*

Men		Women
2 1 0	4	0 1
2 2 2	5	1
5 5 4	6	2 3
1	7	5
	8	7 8
	9	3 5

Where 7|5 means 75 minutes *(2)*

b 52 minutes *(1)*
c Men: 52 minutes
Women: 69 minutes *(2)*
d Mean = 54.4 minutes
Standard deviation = 10.73 minutes *(3)*
e **i** Mean = 64.4 minutes *(1)*
ii Standard deviation = 10.73 minutes *(1)*

6 **a** **i** About 17% *(1)** **ii** About 15% *(1)*
b **i** Lower % of young people in 1991 *(2)*
ii In 1991 population is fairly evenly spread up to age of 70 years *(1)*
c Females – about 13% Males – about 10%
There is a higher percentage of females aged 60–90 years in 1991 *(1)*

7 **a** *(2)*

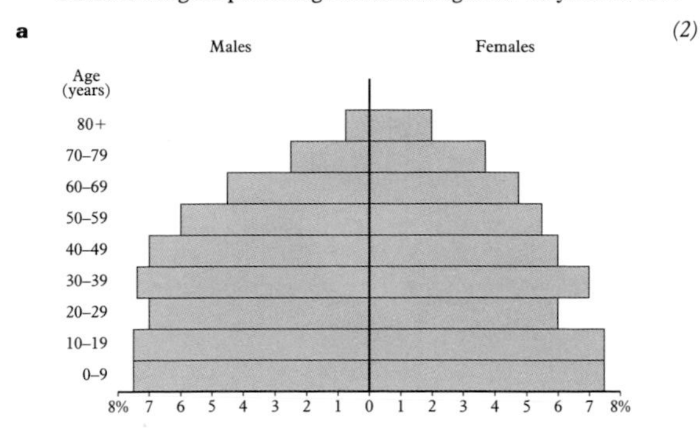

b $7.5 + 7.5 = 15$ *(2)*
c $70 - 79$ *(1)*
d $7 + 7.4 + 6 + 7 = 27.4$ *(2)*
e Fewer males in each group *(1)*

Test Yourself

1 **a** 10
b 7
c 17

2

	Men	Women	Total
Hardback	26	12	38
Paperback	31	51	82
Total	57	63	120

3 *Andrew's scores*

Stem	Leaf
4	4
5	2 2
6	1 3 3
7	5 6 7 8
8	4 6 8 8 9 9

Where 7|5 means 75

4 **a**

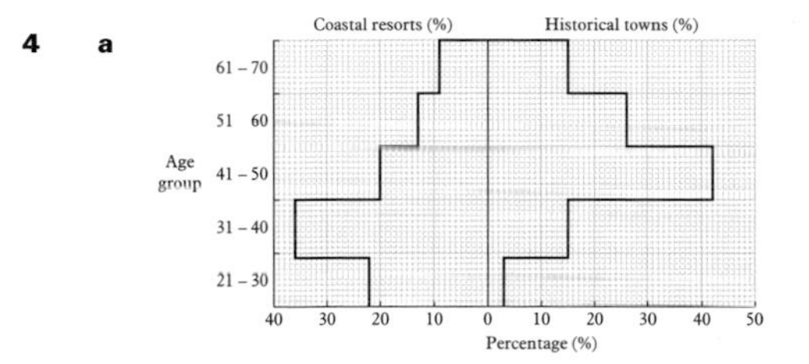

b 58%
c 41%
d Younger people prefer coastal resorts.
Older people prefer historical towns.

CHAPTER 6

Exercise 6:1

1

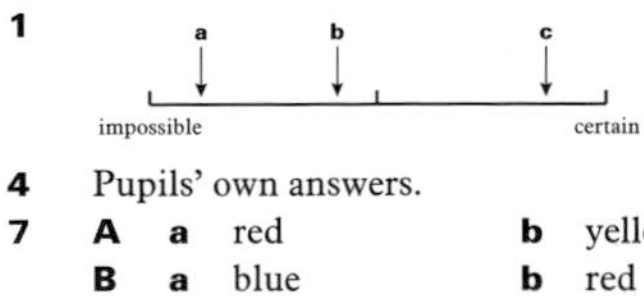

2 Pupils' own answers.

3 a c b — 0 0.5 1

4 Pupils' own answers.

5 Probability only lies between 0 and 1.

6 Probability only lies between 0 and 1.

7 **A** **a** red **b** yellow
B **a** blue **b** red
C **a** yellow **b** green
D **a** red **b** yellow

8 **a, c**

9 There are equal numbers of plain and milk chocolates in the box.

10 Unfair. Sue has 2 chances whereas Liam has 4 chances of winning.

11 **a** Each of the numbers 1 to 6 is equally likely.
b Jenny's numbers 2, 4 and 6 are all higher than Lisa's numbers 1, 3 and 5 if paired so Jenny should reach 24 more quickly.

Exercise 6:2

1 **a** $\frac{3}{8}$ **b** $\frac{5}{8}$ **2** **a** $\frac{1}{6}$ **b** $\frac{2}{6}\left(\frac{1}{3}\right)$ **c** $\frac{3}{6}\left(\frac{1}{2}\right)$ **d** $\frac{3}{6}\left(\frac{1}{2}\right)$

3 **a** $\frac{8}{100}\left(\frac{2}{25}\right)$ **b** 0.08 **c** 8% **4** **a** $\frac{4}{52}\left(\frac{1}{13}\right)$ **b** $\frac{13}{52}\left(\frac{1}{4}\right)$ **c** $\frac{1}{52}$ **d** $\frac{8}{52}\left(\frac{2}{13}\right)$

5 $\frac{7}{8}$ **6** 60% **7** **a** $\frac{2}{8}\left(\frac{1}{4}\right)$ **b** 0.25 **c** 25% **8** $\frac{9}{15}\left(\frac{3}{5}\right)$

9 **a**

10 **a** $\frac{50}{120}\left(\frac{5}{12}\right)$ **b** $\frac{26}{120}\left(\frac{13}{60}\right)$ **c** $\frac{56}{120}\left(\frac{7}{15}\right)$ **d** $\frac{70}{120}\left(\frac{7}{12}\right)$ **e** $\frac{38}{120}\left(\frac{19}{60}\right)$ **f** $\frac{64}{120}\left(\frac{8}{15}\right)$

11 **a** $\frac{110}{200}\left(\frac{11}{20}\right)$ **b** $\frac{78}{200}\left(\frac{39}{100}\right)$ **c** $\frac{105}{200}\left(\frac{21}{40}\right)$

12 **a** 130 **b** $\frac{14}{130}\left(\frac{7}{65}\right)$ **c** $\frac{12}{130}\left(\frac{6}{65}\right)$ **d** $\frac{40}{130}\left(\frac{4}{13}\right)$ **e** $\frac{38}{130}\left(\frac{19}{65}\right)$

Exercise 6:3

1 **a**

		Spinner		
		R	G	B
Coin	H	H R	H G	H B
	T	T R	T G	T B

b $\frac{1}{6}$ **c** $\frac{1}{6}$

2 **a**

	R	R	B	Y
R	R R	R R	R B	R Y
R	R R	R R	R B	R Y
B	B R	B R	B B	B Y
Y	Y R	Y R	Y B	Y Y

b $\frac{4}{16}\left(\frac{1}{4}\right)$ **c** $\frac{6}{16}\left(\frac{3}{8}\right)$ **d** $\frac{4}{16}\left(\frac{1}{4}\right)$

3

4 **a**

		Ace			
		H	S	C	D
King	H	H H	H S	H C	H D
	S	S H	S S	S C	S D
	C	C H	C S	C C	C D
	D	D H	D S	D C	D D

b $\frac{1}{16}$ **c** $\frac{2}{16}\left(\frac{1}{8}\right)$ **d** $\frac{12}{16}\left(\frac{3}{4}\right)$ **e** $\frac{4}{16}\left(\frac{1}{4}\right)$

Exercise 6:4

1 **a** $\frac{1}{4}$ **b** $\frac{1}{3}$ **2** **a** $\frac{3}{7}$ **b** $\frac{3}{6}\left(\frac{1}{2}\right)$

3 **a** $\frac{3}{100}$ **b** $\frac{3}{99}\left(\frac{1}{33}\right)$ **c** $\frac{2}{98}\left(\frac{1}{49}\right)$ **4** **a** $\frac{7}{12}$ **b** $\frac{6}{11}$

5 **a** $\frac{7}{16}$ **b** $\frac{7}{15}$ **c** $\frac{4}{14}\left(\frac{2}{7}\right)$

Exercise 6:5

1 **a** (1) $\frac{156}{1000}\left(\frac{39}{250}\right)$ (2) $\frac{500}{1000}\left(\frac{1}{2}\right)$ **b** $\frac{344}{500}\left(\frac{86}{125}\right)$ **c** $\frac{281}{437}$

2 **a** (1) $\frac{191}{600}$ (2) $\frac{194}{600}$ **b** $\frac{59}{194}$ **c** $\frac{274}{409}$

3 **a** (1) $\frac{43}{69}$ (2) $\frac{17}{69}$ **b** $\frac{14}{31}$ **c** $\frac{14}{43}$

4 **a** 74 **c** 12 **e** 31 **g** $\frac{12}{26}\left(\frac{6}{13}\right)$
b 29 **d** 17 **f** $\frac{26}{74}\left(\frac{13}{37}\right)$ **h** $\frac{12}{29}$

5 **a** 37 **b** (1) 11 (2) 5 (3) 11 (4) 18 (5) 5 (6) 10 (7) 2
c $\frac{3}{37}$ **d** $\frac{3}{11}$ **e** $\frac{5}{24}$ **f** $\frac{2}{3}$

6

7 **a**

b 4

8 **a**

b 5

Exercise 6:6

1 **a** (1) $\frac{34}{100}\left(\frac{17}{50}\right)$ (2) $\frac{66}{100}\left(\frac{33}{100}\right)$
b No. You would expect the relative frequencies to be closer to $\frac{50}{100}\left(\frac{1}{2}\right)$ for both.

2 **a** 100 **b** (1) $\frac{32}{100}$ (2) 0.32 (3) 32% **c** (1) $\frac{11}{100}$ (2) 0.11 (3) 11%
d red **e** $\frac{23}{100}$ **f** Collect more data.

3 **a** $\frac{31}{100}$ **b** $\frac{69}{100}$ **c** $\frac{6}{100}\left(\frac{3}{50}\right)$ **d** $\frac{63}{100}$

Exercise 6:7

1 **a, b**

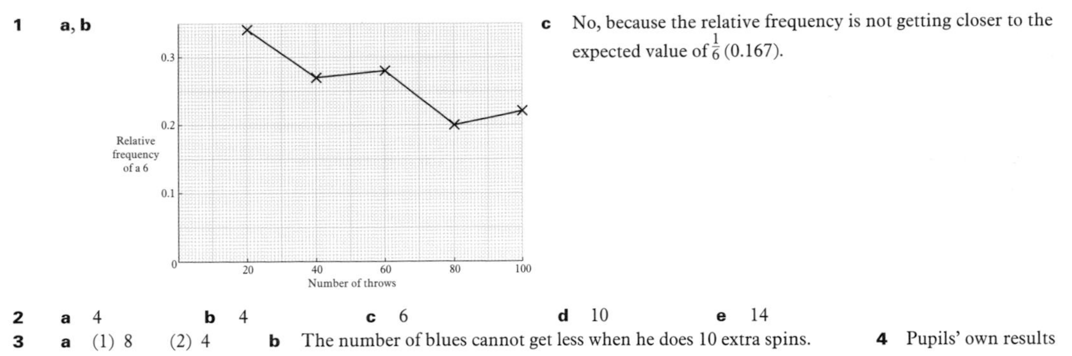

c No, because the relative frequency is not getting closer to the expected value of $\frac{1}{6}$ (0.167).

2 **a** 4 **b** 4 **c** 6 **d** 10 **e** 14

3 **a** (1) 8 (2) 4 **b** The number of blues cannot get less when he does 10 extra spins. **4** Pupils' own results

Exercise 6:8

1 190

2 **a** 576 **b** No because 589 is close to the expected number of faulty cans and the expected number is only an estimate.

3 **a** 2838 **b** more **4** 500 **5** **a** 6 (5.71) **b** £143 **6** **a** $\frac{32}{400}\left(\frac{2}{25}\right)$ **b** $\frac{368}{400}\left(\frac{23}{25}\right)$

Exercise 6:9

1 Method 1 **3** Method 1 **5** Method 2 **7** Method 2

2 Method 3 **4** Method 2 **6** Method 1 **8** Method 1

Exercise 6:10

1 Allocate the numbers 1 to 4 to the members of the pop group.

2 Allocate the numbers 1 to 20 and use random number tables.

3 Allocate the numbers from a dice: 1 and 2 to one colour, 3 and 4 to a second colour and 5 and 6 to the third colour.

Exercise 6:11

2 Allocate the numbers 1 to 8: 12, 2 to fish, 3 to vegetarian and 4–8 to meat. Use a table of random numbers.

3 **c** The second answer should be closer because the experiment has been repeated more times.

4 **a** There are 37 numbers between 00 and 36.

b Use 2-digit random number tables.

c

Category	Percentage	Numbers allocated
Male adult	21	00–20
Female adult	38	21–58
Male child	16	59–74
Female child	25	75–99

5 Use a table of 2-digit random numbers. Allocate 00 to 57 for games won, 58–76 for games lost and 77–99 for games drawn.

Questions

1 **a** $\frac{1}{4}$ *(2)* **b** **i** $\frac{20}{78}\left(\frac{10}{39}\right)$ *(2)* **ii** $\frac{18}{78}\left(\frac{3}{13}\right)$ *(2)*

2 **a** $\frac{1}{6}$ *(1)* **b** $\frac{1}{6}$ *(1)* **c** 4 *(1)* **d** $\frac{2}{6}\left(\frac{1}{3}\right)$ *(1)*

e **i**

First Throw	2	3	4	5	6
Second Throw	6	5	4	3	2

(2)

ii $\frac{5}{36}$ *(3)*

3 **a** Relative frequencies: 0.4, 0.3, 0.2, 0.2, 0, 0.2, 0.2, 0.5, 0.35, 0.4 *(2)* **b** $\frac{38}{140} = 0.27$ *(3)*

4 **a** 5 *(1)*

b Point B which gives 4 heads in 20 throws, which contradicts A (5 out of 10) and C (15 out of 30) *(2)*

c Coordinates – (50, 0.46) *(2)*

5 **a** $\frac{12}{50} = 0.24$ *(2)* **b** $\frac{23}{50} = 0.46$ *(2)* **c** $\frac{13}{23} = 0.57$ *(2)*

6 **a** Heads 15, Tails 5 *(1)* **b** 12 *(1)* **c** $25 - 12 = 13$ *(2)* **d** $\frac{1}{4} = 0.25$ *(1)*

e 0.75 *(1)* **f** Yes. Probability of head or tail should be close to 0.5 *(2)*

7 **a** $\frac{1}{6}$ *(1)* **b** $\frac{1}{6}$ *(1)* **c** $\frac{1}{6} \times \frac{1}{6} = \frac{1}{36}$ *(2)*

d

First Throw	6	5	5
Second Throw	6	5	6
Third Throw	6	6	5

(3)

e One triple product $\frac{1}{6} \times \frac{1}{6} \times \frac{1}{6}$

Triple addition $3 \times \frac{1}{216} = \frac{1}{72}$ *(3)*

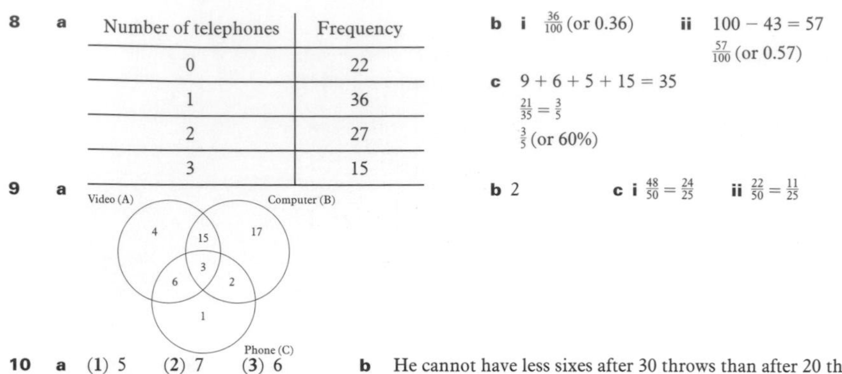

8 **a**

Number of telephones	Frequency
0	22
1	36
2	27
3	15

b **i** $\frac{36}{100}$ (or 0.36) **ii** $100 - 43 = 57$
$\frac{57}{100}$ (or 0.57)

c $9 + 6 + 5 + 15 = 35$
$\frac{21}{35} = \frac{3}{5}$
$\frac{3}{5}$ (or 60%)

9 **a**

b 2 **c** **i** $\frac{48}{50} = \frac{24}{25}$ **ii** $\frac{22}{50} = \frac{11}{25}$

10 **a** (1) 5 (2) 7 (3) 6 **b** He cannot have less sixes after 30 throws than after 20 throws.
c You would expect a relative frequency of $\frac{1}{6} = 0.17$.
The data shows higher relative frequencies so the dice is probably biased.

Test Yourself

1 $\frac{7}{20}$

2 **a**

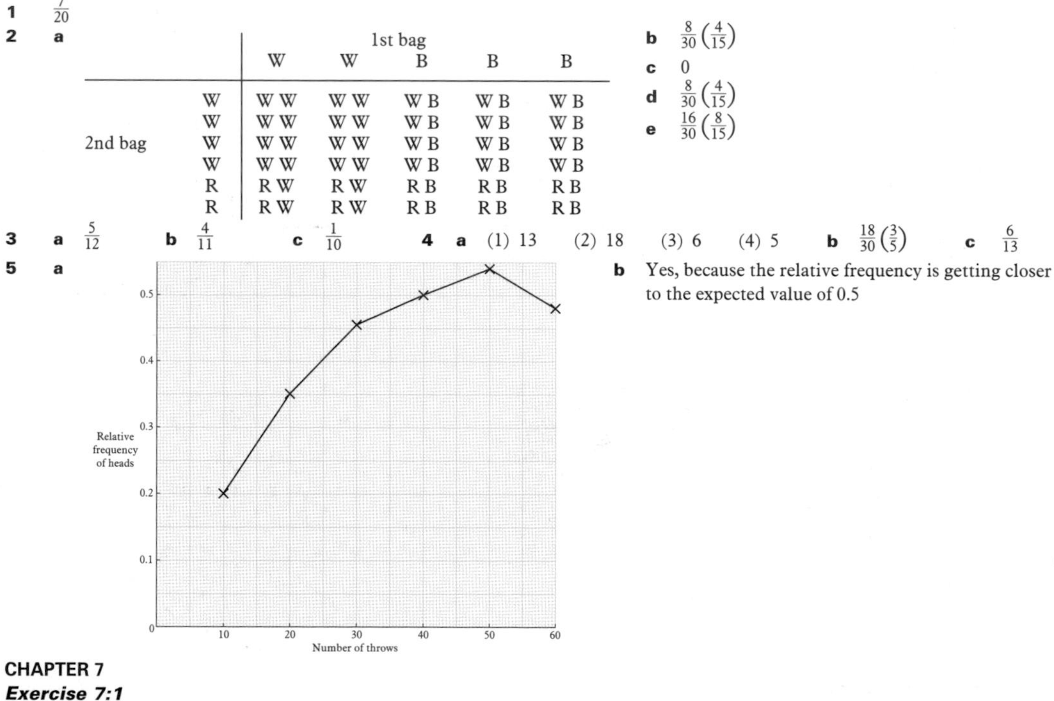

		1st bag				
		W	W	B	B	B
2nd bag	W	W W	W W	W B	W B	W B
	W	W W	W W	W B	W B	W B
	W	W W	W W	W B	W B	W B
	W	W W	W W	W B	W B	W B
	R	R W	R W	R B	R B	R B
	R	R W	R W	R B	R B	R B

b $\frac{8}{30}\left(\frac{4}{15}\right)$
c 0
d $\frac{8}{30}\left(\frac{4}{15}\right)$
e $\frac{16}{30}\left(\frac{8}{15}\right)$

3 **a** $\frac{5}{12}$ **b** $\frac{4}{11}$ **c** $\frac{1}{10}$

4 **a** (1) 13 (2) 18 (3) 6 (4) 5 **b** $\frac{18}{30}\left(\frac{3}{5}\right)$ **c** $\frac{6}{13}$

5 **a**

b Yes, because the relative frequency is getting closer to the expected value of 0.5

CHAPTER 7

Exercise 7:1

1

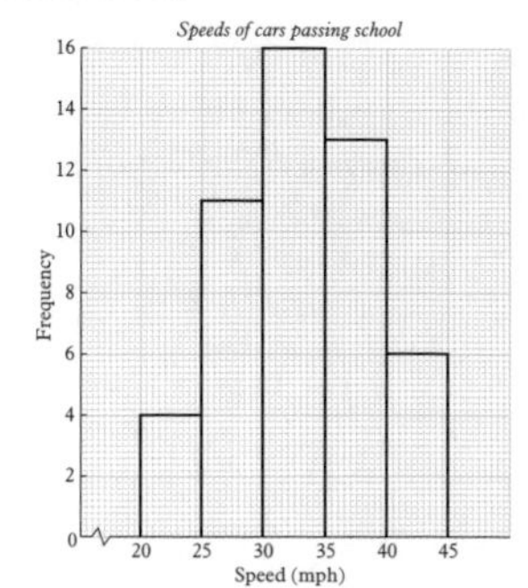

3

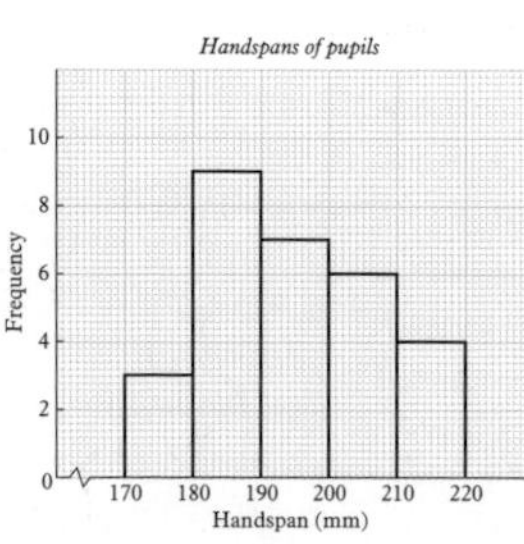

2

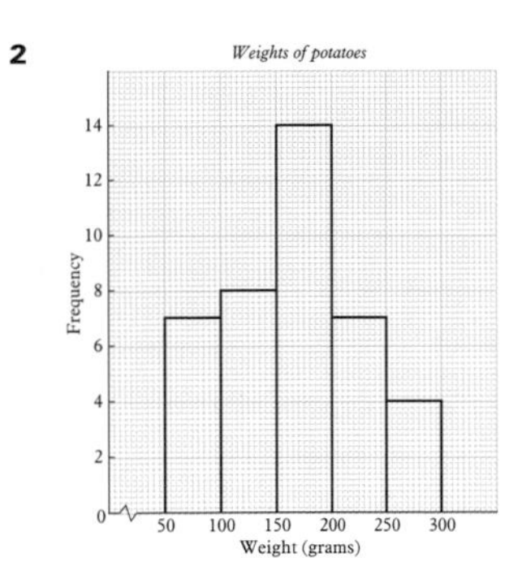

4

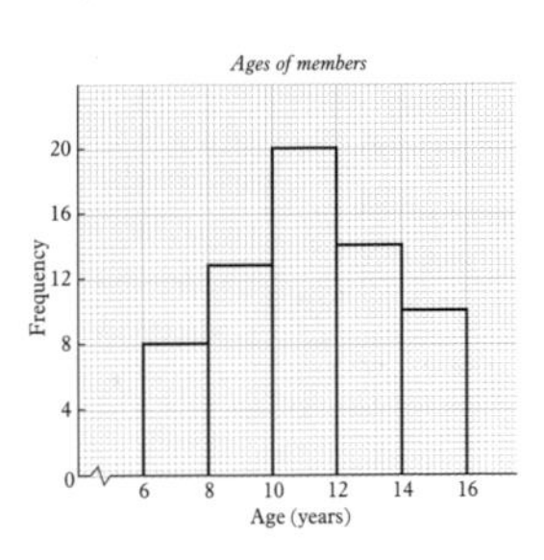

Exercise 7:2

1 **a**

Time in minutes (t)	Frequency (f)	Class width	Frequency density
$0 \leqslant t < 10$	4	10	$4 \div 10 = 0.4$
$10 \leqslant t < 15$	8	5	$8 \div 5 = 1.6$
$15 \leqslant t < 20$	8	5	$8 \div 5 = 1.6$
$20 \leqslant t < 30$	10	10	$10 \div 10 = 1$
$30 \leqslant t < 45$	3	15	$3 \div 15 = 0.2$

b

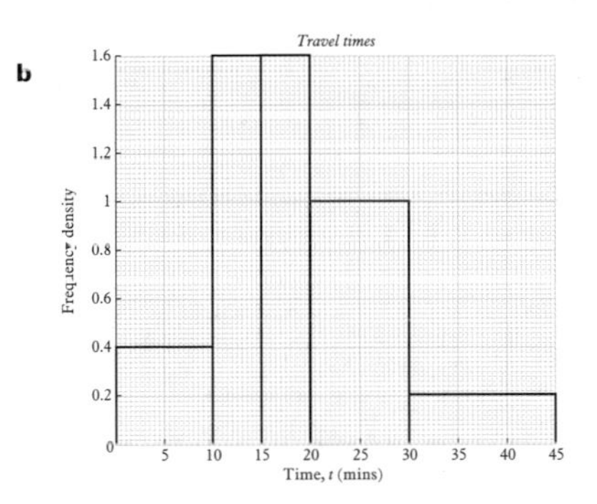

2 **a**

Height in cm (h)	Frequency (f)	Class width	Frequency density
$120 \leqslant h < 124$	4	4	$4 \div 4 = 1$
$124 \leqslant h < 128$	6	4	$6 \div 4 = 1\frac{1}{2}$
$128 \leqslant h < 130$	2	2	$2 \div 2 = 1$
$130 \leqslant h < 136$	6	6	$6 \div 6 = 1$
$136 \leqslant h < 140$	2	4	$2 \div 4 = \frac{1}{2}$

b

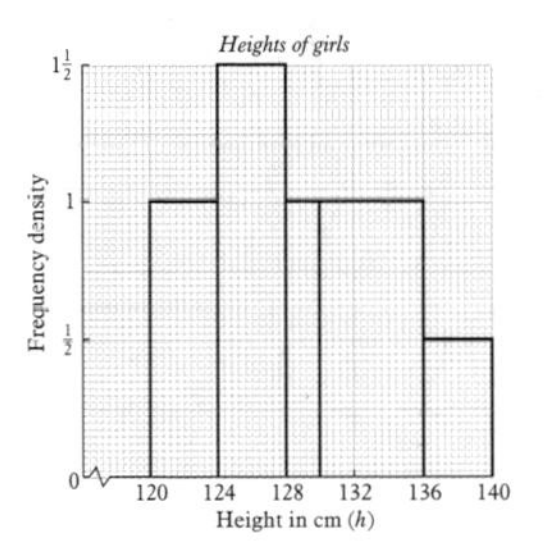

3 **a** 0.4, 0.3, 1.2, 1.0, 1.2, 0.4, 0.2

b

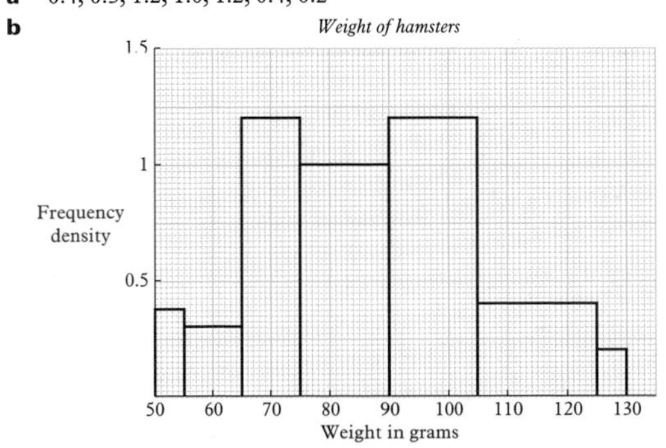

4 a

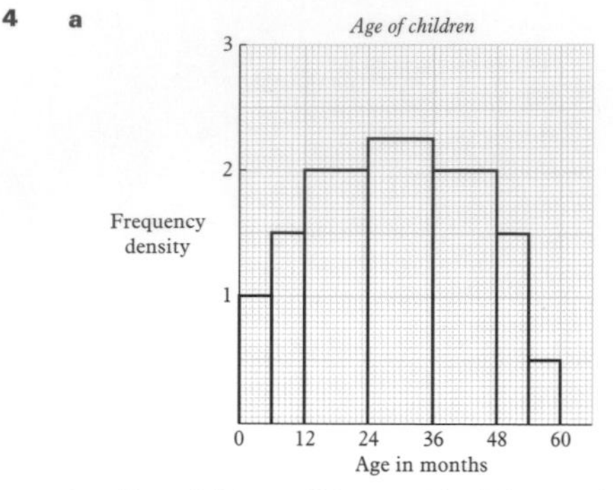

b Many children will be attending Primary school at this age.

5 a

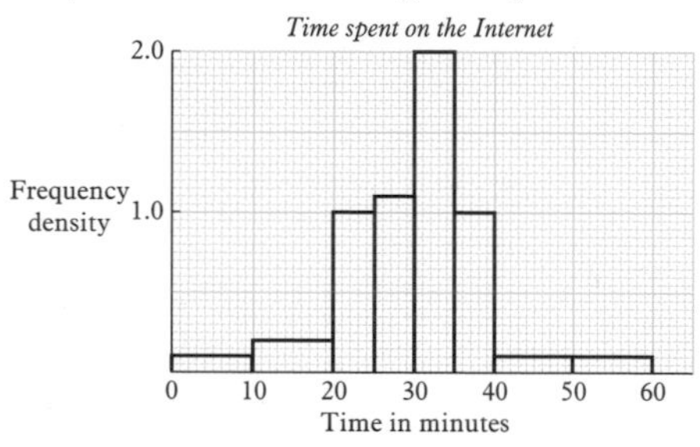

b To show the times more clearly. $40 \leqslant t < 50$ has only 1 pupil. There is no point in having 2 smaller groups here.

Exercise 7:3

1 a

Time (to the nearest min)	2–4	5–7	8–10	11–13
Time (minutes)	$1\frac{1}{2}$–$4\frac{1}{2}$	$4\frac{1}{2}$–$7\frac{1}{2}$	$7\frac{1}{2}$–$10\frac{1}{2}$	$10\frac{1}{2}$–$13\frac{1}{2}$
Frequency	6	4	2	1

b

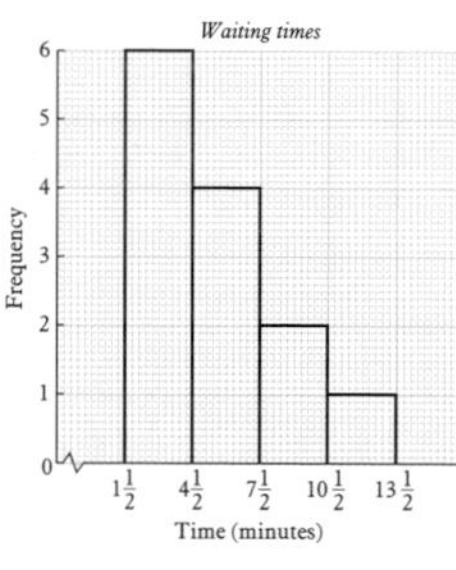

2 a

Weight (to the nearest kg)	2–3	4–5	6–8	9–10
Weight (kg)	$1\frac{1}{2}$–$3\frac{1}{2}$	$3\frac{1}{2}$–$5\frac{1}{2}$	$5\frac{1}{2}$–$8\frac{1}{2}$	$8\frac{1}{2}$–$10\frac{1}{2}$
Frequency	2	7	9	5
Class widths	2	2	3	2
Frequency density	1	3.5	3	2.5

b

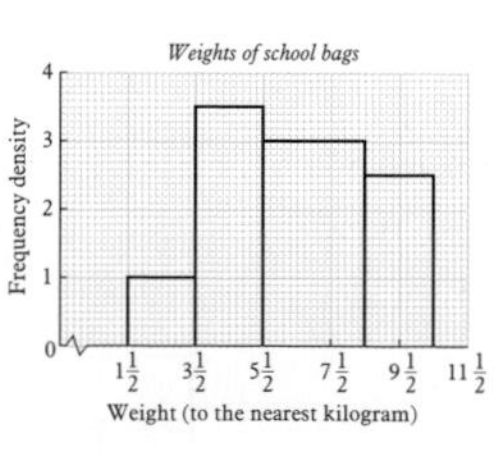

Exercise 7:4

1 a

Time (to the nearest min)	10–14	15–19	20–24	25–29
Time (minutes)	$9\frac{1}{2}$–$14\frac{1}{2}$	$14\frac{1}{2}$–$19\frac{1}{2}$	$19\frac{1}{2}$–$24\frac{1}{2}$	$24\frac{1}{2}$–$29\frac{1}{2}$
Frequency	2	8	5	3

b

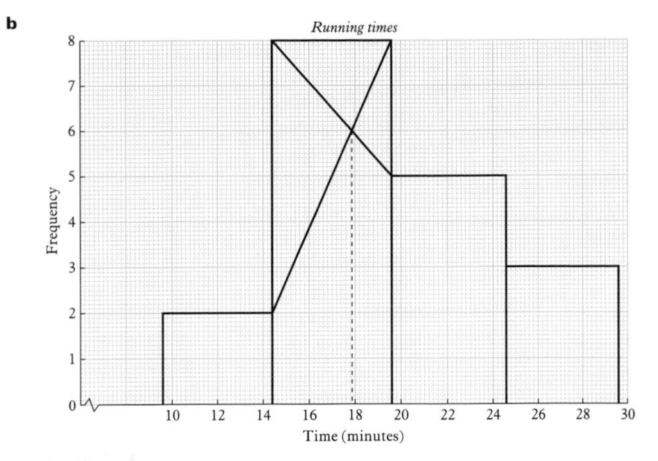

c $14\frac{1}{2}$–$19\frac{1}{2}$ minutes

d 17.8 minutes

2

Distance (to the nearest km)	30–39	40–49	50–59	60–69
Distance (km)	$29\frac{1}{2}$–$39\frac{1}{2}$	$39\frac{1}{2}$–$49\frac{1}{2}$	$49\frac{1}{2}$–$59\frac{1}{2}$	$59\frac{1}{2}$–$69\frac{1}{2}$
Frequency	4	8	10	4

a

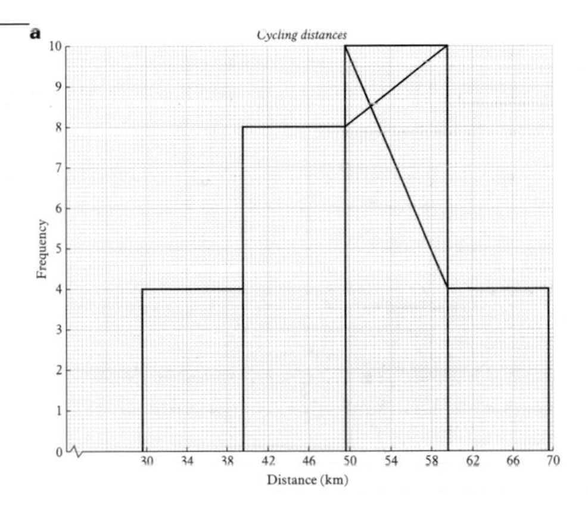

b $49\frac{1}{2}$–$59\frac{1}{2}$ km

c 52 km

3

Weight (to the nearest pound)	2–3	4–5	6–7	8–9	10–11
Weight (pounds)	$1\frac{1}{2}$–$3\frac{1}{2}$	$3\frac{1}{2}$–$5\frac{1}{2}$	$5\frac{1}{2}$–$7\frac{1}{2}$	$7\frac{1}{2}$–$9\frac{1}{2}$	$9\frac{1}{2}$–$11\frac{1}{2}$
Frequency	4	29	42	20	5

a

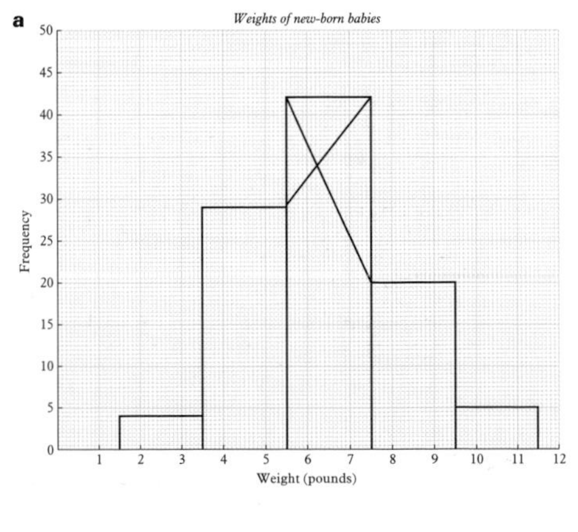

b $5\frac{1}{2}$–$7\frac{1}{2}$ pounds

c $6\frac{1}{4}$ pounds

Exercise 7:5

1 **a** Positive skew **b** Median **c** Interquartile range

2 **a** Negative skew **b** Median **c** Interquartile range

3 **a** Symmetrical **b** Mean **c** Standard deviation

4 **a** Positive skew **b** Median **c** Interquartile range

5 **a** Symmetrical **b** Mean **c** Standard deviation

6 **a** Bimodal **b** Mean **c** Standard deviation

Questions

1 a

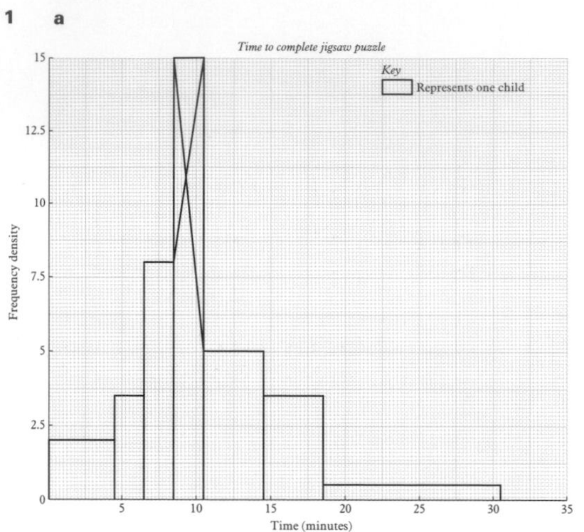

Width	$4\frac{1}{2}$	2	2	2	4	4	12
Frequency density	2	$3\frac{1}{2}$	8	15	5	3	$\frac{1}{2}$

(6)

b $9.1 \leqslant \text{answer} < 9.5$ *(2)*

2 a

Playing time of tracks on a disc

Key: Represents one disc

Frequency density; Time (seconds)

Width	40	10	5	5	5	5	30
Frequency density	0.3	1.3	2.4	3.0	3.4	2.8	0.7

(6)

b Mode = 242 seconds *(2)*

3 a

Waiting time at a doctors' surgery

Key: Represents one patient

Frequency density; Waiting time (minutes)

Width	10	5	5	10	10	20
Frequency distribution	1.3	4.4	3.6	1.5	0.9	0.4

(6)

b Mode = 14 minutes *(2)*

c 54 or 55 patients *(2)*

4 a Positive skew *(1)*

b Modal time = 33.5 minutes *(2)*

c 10 squares represent one worker, 30–35 group has 200 squares. Number of workers = 20. *(2)*

d Median is journey time for 46th worker equivalent to 460 squares. Median is 41 minutes. *(5)*

5 **a** 30, 35, 17, 12, 6 *(3)*

b **i** $\bar{x} = 5.3$, standard deviation $= 6.6$ *(6)*

ii Female, lower mean, lower variation. *(2)*

c **i** Male 1.06, Female 0.91. *(3)*

ii Positive skew in both cases. *(2)*

d Longer periods would place too big a weight on memory. *(1)*

e Repeat at regular intervals. *(2)*

6 **a**

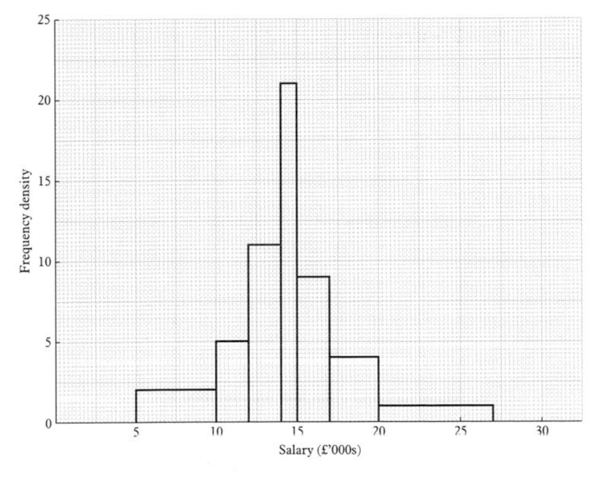

Width	5	2	2	1	2	3	7
Frequency density	2	5	11	21	9	4	1

(6)

b Modal class: $14 \leqslant x < 15$ *(1)*

c $\left(\frac{32}{100}\right)^2 = 0.102$ or $\frac{32}{100} \times \frac{31}{99} = 0.100$ *(3)*

7 **a** (1) No label on vertical axis.

(2) Has not used frequency density – areas not proportional to frequency. *(3)*

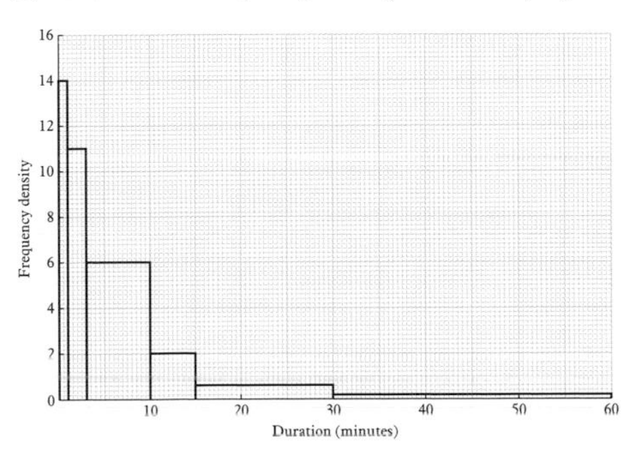

Width	1	2	7	5	15	30
Frequency density	14	11	6	2	0.6	0.1

b Distribution is more skewed. *(6)*

c Modal class $= 0 \leqslant x < 1$ *(1)*

8 **a**

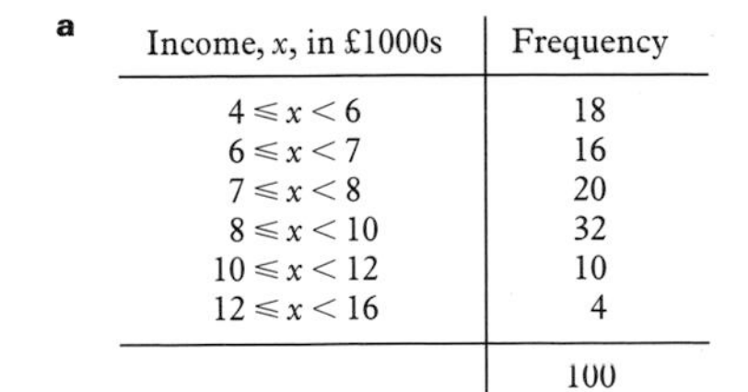

Income, x, in £1000s	Frequency
$4 \leqslant x < 6$	18
$6 \leqslant x < 7$	16
$7 \leqslant x < 8$	20
$8 \leqslant x < 10$	32
$10 \leqslant x < 12$	10
$12 \leqslant x < 16$	4
	100

b positive

c $7 \leqslant x < 8$

Test Yourself

1

Height (to the nearest cm)	140–144	145–149	150–154	155–159	160–164
Height (cm)	$139\frac{1}{2}$–$144\frac{1}{2}$	$144\frac{1}{2}$–$149\frac{1}{2}$	$149\frac{1}{2}$–$154\frac{1}{2}$	$154\frac{1}{2}$–$159\frac{1}{2}$	$159\frac{1}{2}$–$164\frac{1}{2}$
Frequency	13	10	7	6	4

a

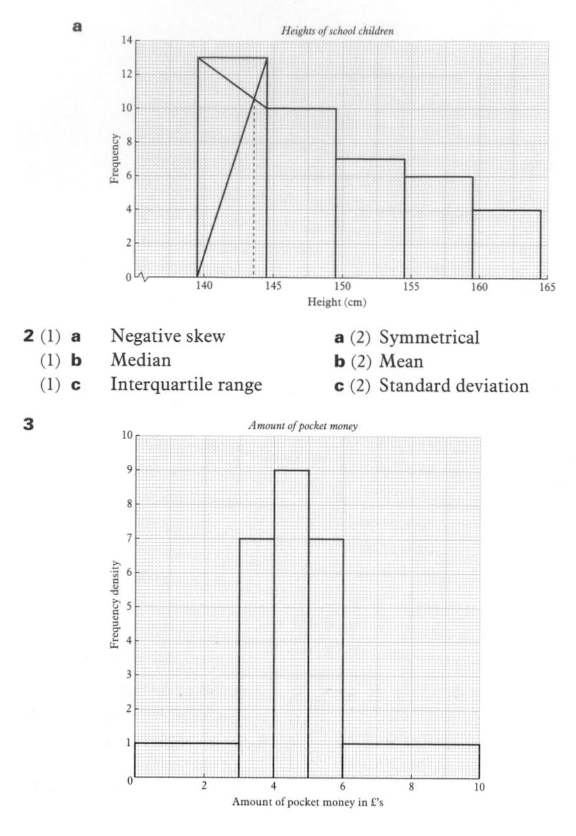

b $139\frac{1}{2}$–$144\frac{1}{2}$ cm
c 143 cm
d Positive skew

2 (1) **a** Negative skew
(1) **b** Median
(1) **c** Interquartile range
a (2) Symmetrical
b (2) Mean
c (2) Standard deviation

3

CHAPTER 8

Exercise 8:1

1

110 120 130 140 150 160 170 180 190

2 **a** Median = 100 minutes
b Lower quartile = 70 minutes
c Upper quartile = 120 minutes
d

50 60 70 80 90 100 110 120 130 140 150 160 170 180

3 **a** Median = 21.8 seconds
b Lower quartile = 20.8 seconds
c Upper quartile = 24.2 seconds
d

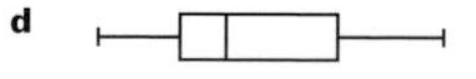

4 **a** Median = 9 minutes
b Lower quartile = 5 minutes
c Upper quartile = 11 minutes
d

0 5 10 15 20

Exercise 8:2

1 **a** (1) LQ = 12 (2) m = 15 (3) UQ = 23
b m − LQ = 3 UQ − m = 8
c Positive skew

2 **a** (1) LQ = 140 (2) m = 220 (3) UQ = 250
b m − LQ = 80 UQ − m = 30
c Negative skew

3 **a** (1) LQ = 12 (2) m = 23 (3) UQ = 34
b m − LQ = 11 UQ − m = 11
c Symmetrical

4 **a** (1) LQ = 10 (2) m = 15 (3) UQ = 65
b m − LQ = 5 UQ − m = 50
c Positive skew

Exercise 8:3

1 **a** Anna's times are negatively skewed. Brian's times are positively skewed.
b Brian's range of times is greater than Anna's.
c The interquartile range for Anna is greater than the interquartile range for Brian.
d Anna's median time is greater than Brian's.

2 **a** Classical CDs
(1) LQ = 2.8 minutes, m = 6.3 minutes, UQ = 11.1 minutes (2)

b Pop CDs
(1) LQ = 2.5 minutes, m = 5.5 minutes, UQ = 7.1 minutes (2)

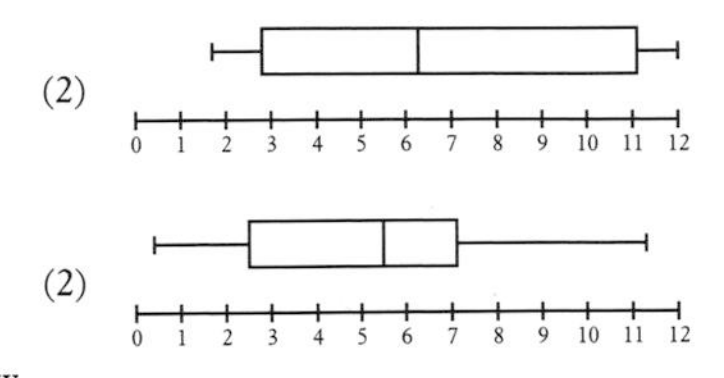

c (1) Classical times have a positive skew. Pop times have a negative skew.
(2) Classical tracks have a slightly lower range than Pop tracks.
(3) Interquartile range for Classical tracks is much higher than for Pop tracks.
(4) Median time for Classical tracks is higher than for Pop tracks.

3 **a** (1) Lower quartile = 8.9
Median = 9.8
Upper quartile = 11.2
b (1) Lower quartile = 6.7
Median = 7.2
Upper quartile = 9.1
(2)

North East
5.0 5.5 6.0 6.5 7.0 7.5 8.0 8.5 9.0 9.5 10.0 10.5 11.0 11.5 12.0 12.5 13.0 % Unemployment
North West

c South East: 4.5, 5.7, 7.4
South West: 4.0, 5.2, 6.5

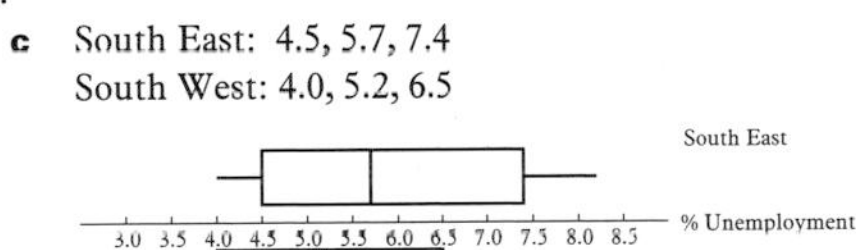

d For North East and North West the spread is very similar. The median is much greater in North East. For South East and South West spread is about the same but less than that for the North East and North West. Medians are about the same for South East and South West.

Exercise 8:4

1 **a** m = 49.5 min, LQ = 46 min, UQ = 57 min
b Interquartile range = 11 min
c No small outliers
d 75 and 79 are large outliers.
e

40 44 48 52 56 60 64 68 72 76 80

2 **a** m = 29°C, LQ = 27°C, UQ = 30°C
b Interquartile range = 3°C
c 20°, 21°, 22° are small outliers
d 35°, 36° are large outliers.
e

20 22 24 26 28 30 32 34 36

3 **a** m = 36, **b** LQ = 32 **c** UQ = 38 **d** 22, 49, 50

4 **A** **a** (1) m = 5 (2) LQ = 4 (3) UQ = 8 (4) 15 **b** positive
B **a** (1) m = 38 (2) LQ = 34 (3) UQ = 40 (4) 22, 23, 50, 51 **b** negative

Exercise 8:5

1 **A** **a** 24 **b** 24 **c** 24 **d** **B** **e** **A**
B **a** 16 **b** 16 **c** 16

Exercise 8:6

1 **a** 2 **b** $-1\frac{1}{2}$ **c** $1\frac{1}{2}$ **d** −2
2 **a** 1 **b** $1\frac{1}{2}$ **c** $-1\frac{1}{2}$ **d** $-1\frac{1}{4}$
3 **a** 75 **b** 78 **c** 69 **d** 63
4 **a** $\frac{1}{2}$ **b** −1 **c** $1\frac{1}{2}$ **d** −2
5 **a** 2 **b** 1.5 **c** −1.5 **d** −2
6 **a** (1) 0.917 (2) −0.792 **b** (1) 11.8 cm (2) 6.59 cm

Exercise 8:7

1 81.5% 2 81.5% 3 68% 4 81.5%

Exercise 8:8

1 **a** 0.8 **b** 0.67

c The standardised score in French is higher than in German. Performance in French was better than in German even though the French score was slightly lower.

2 **a** 1.33 **b** 1.0

c Yes. Her standardised score in Physics is higher than that for Geography.

3 For the first tree 7 cm leaves have a standardised score of +2. For the second tree 7 cm leaves have a standardised score of −0.67 Mr. Cross is correct because there will be more 7 cm leaves for the second tree.

4 For the first beach the standardised score is 1.43. For the second beach the standardised score is 1.25. Her claim is correct.

Questions

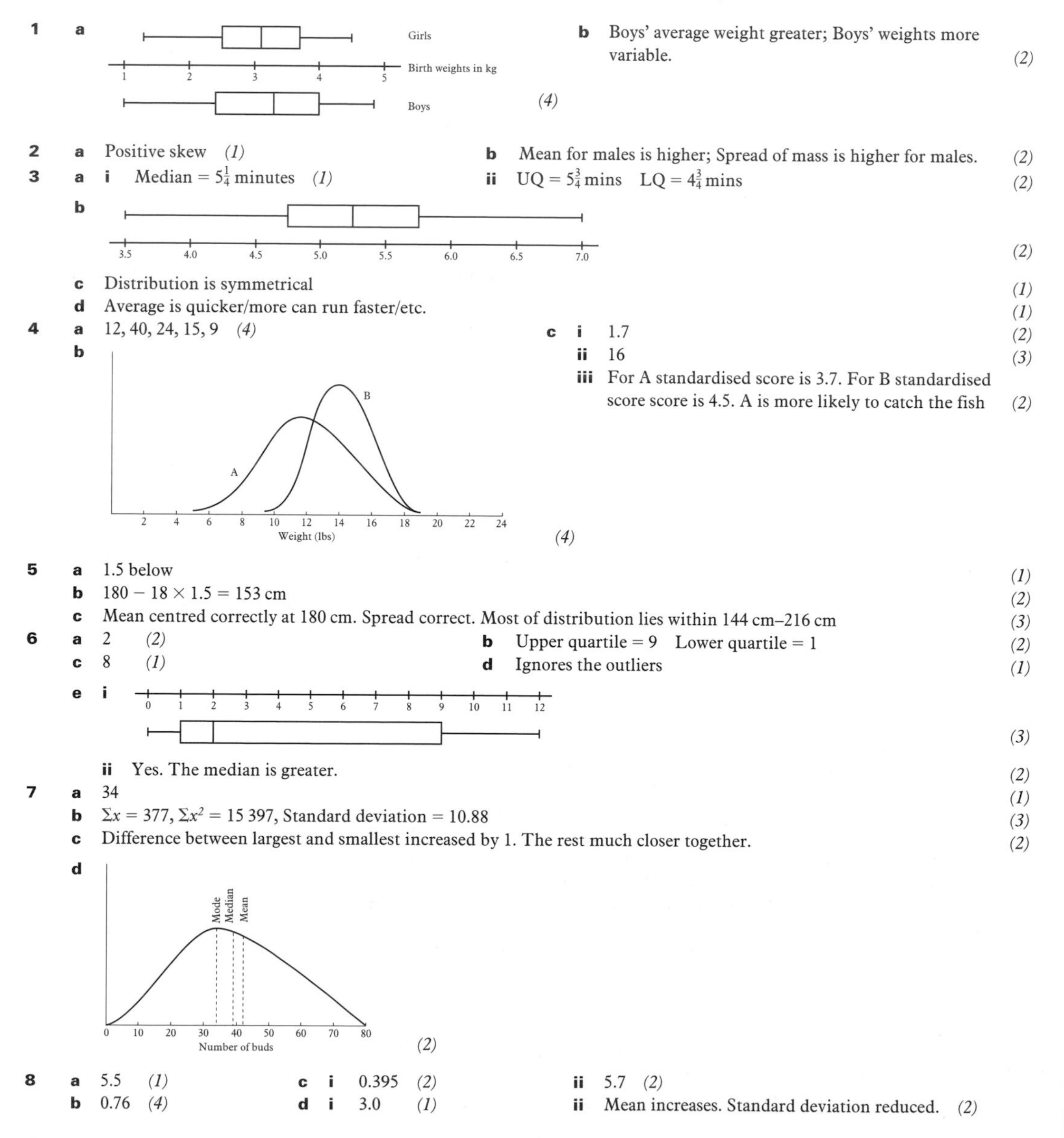

1 **a** (4)

b Boys' average weight greater; Boys' weights more variable. *(2)*

2 **a** Positive skew *(1)* **b** Mean for males is higher; Spread of mass is higher for males. *(2)*

3 **a** **i** Median = $5\frac{1}{4}$ minutes *(1)* **ii** UQ = $5\frac{3}{4}$ mins LQ = $4\frac{3}{4}$ mins *(2)*

b *(2)*

c Distribution is symmetrical *(1)*

d Average is quicker/more can run faster/etc. *(1)*

4 **a** 12, 40, 24, 15, 9 *(4)*

b *(4)*

c **i** 1.7 *(2)*

ii 16 *(3)*

iii For A standardised score is 3.7. For B standardised score score is 4.5. A is more likely to catch the fish *(2)*

5 **a** 1.5 below *(1)*

b $180 - 18 \times 1.5 = 153$ cm *(2)*

c Mean centred correctly at 180 cm. Spread correct. Most of distribution lies within 144 cm–216 cm *(3)*

6 **a** 2 *(2)* **b** Upper quartile = 9 Lower quartile = 1 *(2)*

c 8 *(1)* **d** Ignores the outliers *(1)*

e **i** *(3)*

ii Yes. The median is greater. *(2)*

7 **a** 34 *(1)*

b $\Sigma x = 377$, $\Sigma x^2 = 15\,397$, Standard deviation = 10.88 *(3)*

c Difference between largest and smallest increased by 1. The rest much closer together. *(2)*

d *(2)*

8 **a** 5.5 *(1)* **c** **i** 0.395 *(2)* **ii** 5.7 *(2)*

b 0.76 *(4)* **d** **i** 3.0 *(1)* **ii** Mean increases. Standard deviation reduced. *(2)*

9 **a**

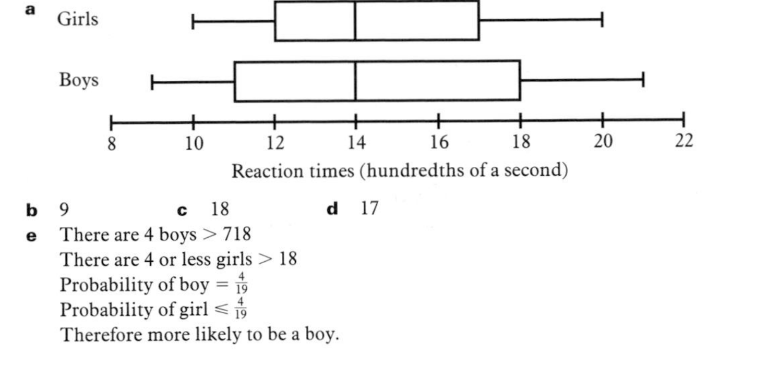

b 9 **c** 18 **d** 17

e There are 4 boys > 718
There are 4 or less girls > 18
Probability of boy $= \frac{4}{19}$
Probability of girl $\leqslant \frac{4}{19}$
Therefore more likely to be a boy.

Test Yourself

1 **a** m = 6, LQ = 4, UQ = 10 **b**

0 2 4 6 8 10 12 14 16 18 20 22

c m − LQ = 2, UQ − m = 4 **d** Positive skew
e Interquartile range = 6 **f** 21 is a large outlier.

2 **a** −0.5 **b** −1 **c** Her results in Biology are better.

3 **a** 0.167 **b** 32 minutes **c** Chloe is most likely to have won as her standardised score is closer to the mean.

CHAPTER 9

Exercise 9:1

1

	a	b	c	d	e	f
lower bound	7.285	85.5	38.05	9365	350	5500
upper bound	7.295	86.5	38.15	9375	450	6500
absolute error	0.005	0.5	0.05	5	50	500

2 **a** 48.5 cm **b** No, because the minimum frame size would be 484.5 mm (48.45 cm) and $48.45 < 48.5$

3 **a** 79.5 cm **b** $12 \times 80.5 = 966$ cm **c** 14.65 m (1465 cm) **d** 19 ($1465 \div 79.5$ rounded up)

4 **a** 5.85 m **b** 10.24 m^2 **c** 0.38 m^2

5 **a** 1.285 m **b** 0.005 m (0.5 cm) **c** 335 000 cm^3 or 0.335 m^3 **d** 8300 cm^3 or 0.0083 m^3

6 **a** 48.5 seconds **b** 0.5 m **c** 1.57 m/sec **d** 0.027 m/sec

7 **a** 7.335 kg **b** $7.345 \times 60 = 440.7$ kg **c** 1833.75 kg

Exercise 9:2

1 **a** (1) 2 cm (2) 0.44% **b** (1) 2 cm (2) 8.7%

c The percentage error is less when the actual measurement is larger because you are dividing by a larger number.

2

		(1)	(2)	(3)	(4)
a	Absolute error	0.05 cm	0.5 sec	5 g	0.5 million
b	Relative error	0.0033	0.01	0.0088	0.0125
c	Percentage relative error	0.33%	1.04%	0.88%	1.25%

3 Mary's absolute error = 0.5 cm, William's absolute error = 5 cm.
Mary's % error = $0.5 \times 100/67 = 0.75\%$, William's % error = $5 \times 100/460 = 1.09\%$.
Therefore William has the greater percentage error.

4 **a** 9294.75 cm^2 **b** 0.011 **c** 1.1% **5** **a** 625026 mm^3 **b** 1.8%

Exercise 9:3

1 **a**

Month	June	July	Aug.	Sept.
Sales (£1000s)	22	24	26	28

b

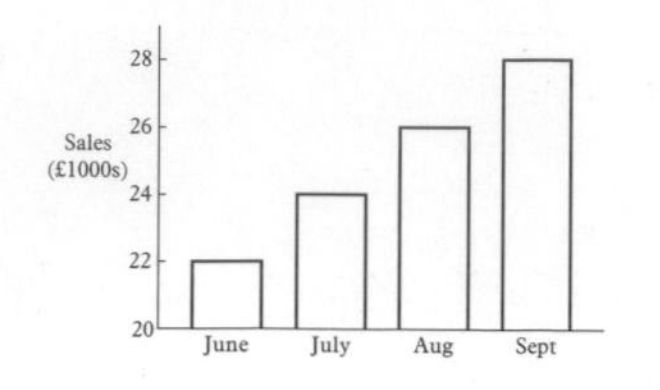

c The chart drawn for question **b** because it apparently shows a higher rate of growth.

2

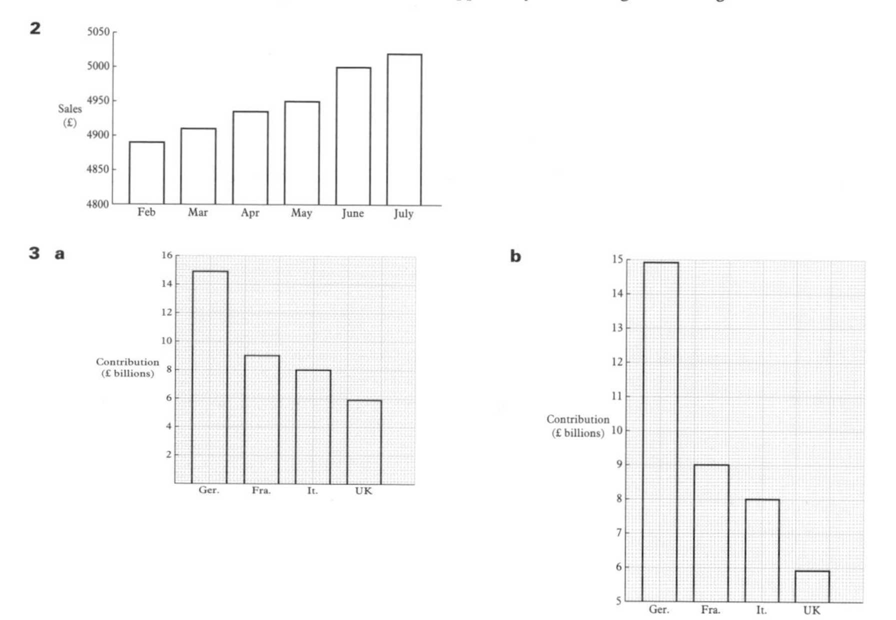

4 Vertical axis does not start at zero.
5 Sloping scale gives the impression that profits are rising faster than they actually are.
6 The gaps between the years on the horizontal scale are different.
7 Thick band prevents precise interpretation.
8 Bars have different widths, gives false impression of growth.
9 Actual numbers show no relation to shape of graph, no scale on vertical axis.

Exercise 9:4

1 Both length and width have been doubled in the diagram. This gives the impression that Chester sales are 4 times those in Swansea.
2 Length, width and height of the tea packet diagram have been multiplied by 3. This gives the impression that sales of Earl Grey are 27 times higher than those of Lapsang Souchong.
3 Because the diameter of the Paris diagram is twice that of the London diagram the impression is that Paris has 4× number of sunny days rather than 2×.
4 In the diagrams both length and diameter of the cylinders have been doubled. This gives the impression that sales have increased by a factor of 8 rather than 2.
5 **a** Diagram for Basic machines is 3× bigger in length, width and height than the diagram for Premium machines. This gives the impression that Basic machines sell 27 times the number of Premium machines, rather than 3 times.
b Height must be $\sqrt[3]{3} \times 50$ cm = 72.1 cm.

Questions

1 **a** 0.5 cm *(1)* **b** 49.5 cm *(1)* **2** **a** 0.5 cm *(1)* **b** 443.5 cm *(1)* **c** 4725 cm *(3)*
3 **a** 5 seconds *(1)* **b** Minimum = 7.5 litres × 95 seconds = 712.5 litres *(3)*
4 **a** 10.5 cm *(1)* **b** 1157.625 cm^3 *(1)* **c** 0.16 (2 sf) *(3)*

5 **a** ±0.5 m/sec *(1)* **b** $0.5/61 = 0.82\%$ *(2)* **c** $61.5 \times 2.75 \simeq 169$ m *(3)*

6 **a** Scale on vertical axis does not begin at zero. Labelling on vertical axis is incomplete – 'millions' left out. Title is misleading. *(2)*

b **i** Should say 'bottled water'. *(1)*

ii Scale on horizontal axis is not regular – the changes between years are different. *(1)*

7 Misleading title; False zero; sloping lines; unequal distances between numbers of horizontal scale. *(2)*

8 **a** **i** Time scale not linear *(1)* **ii** 5 *(1)* **b** **i** T *(1)* **ii** T *(1)*

9 Two from – Growth looks bigger than it really is; Slanted time axis; No title; thickness of blocks/gaps different *(2)*

10 **a** Ratio of areas in diagrams is 1 : 4; should be 1 : 2 so length and height should increase by $\sqrt{2}$, not 2. *(1)*

b $4 \times \sqrt[3]{2} = 5.04$ cm *(3)*

Test Yourself

1 **a** 0.5 cm **b** 18.5 cm **c** $25.5 \times 19.5 = 497.25\ \text{cm}^2$

d $497.25 - 25 \times 19 = 22.25\ \text{cm}^2$ **e** $25.5 \times 60.5 \times 19.5 = 30\,083.625\ \text{cm}^3$ **f** $\dfrac{30\,083.625 - 25 \times 60 \times 19}{25 \times 60 \times 19} = 0.056$

2 Vertical scale does not start at 0. Vertical scale is not labelled. Vertical scale not linear. Shading of bars is misleading. Figure of cat climbing misleads the eye.

CHAPTER 10

Exercise 10:1

1 **a** Seasonal **b** Cyclical **c** Random

2 **a** Seasonal with rising trend **b** 15 000 **c** 7000 **d** 22 000 (12 000 + 10 000) **e** 80%

3 **a, b, c** **d** (1) 22 000 (2) 30 000

e 2002, because it is outside the range of the data

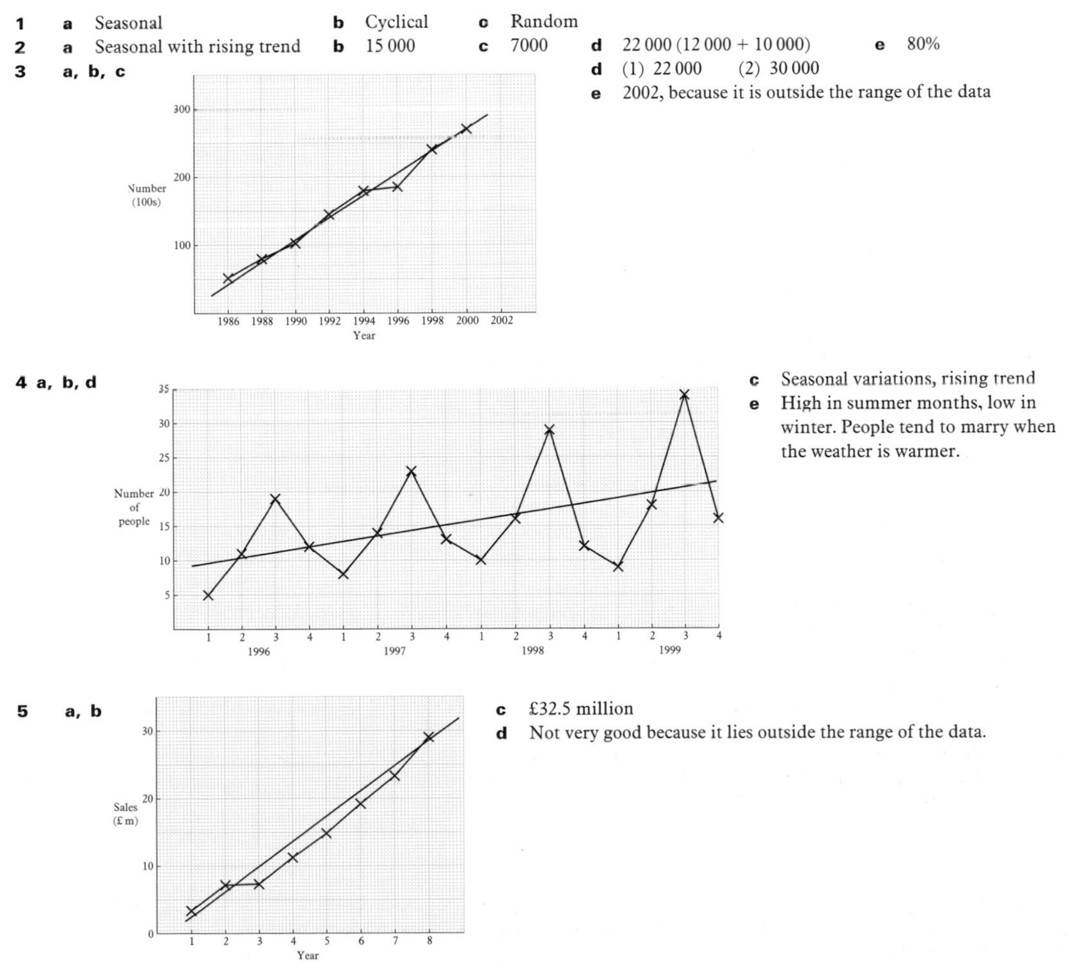

4 **a, b, d**

c Seasonal variations, rising trend

e High in summer months, low in winter. People tend to marry when the weather is warmer.

5 **a, b**

c £32.5 million

d Not very good because it lies outside the range of the data.

Exercise 10:2

1 7, 9.$\dot{3}$, 11, 12, 11 **2** 10, 12, 15, 19, 22 **3** 5, 6, 8, 11, 12

4 7, 7, 8, 9, 10, 11, 10, 11, 10 **5** 42, 40.6, 43.6, 41.4, 44, 45.6, 48.2, 47.8, 47.8, 49.2, 48.6

6 **a** 18, 20, 21.5, 23.5, 25, 24.3, 19.3, 17.8, 18, 21.3, 24.3, 23, 21.8, 16.5, 15, 18.3, 23

b 18.4, 22.6, 22.4, 23, 23.4, 22, 19.4, 18.6, 20.4, 22, 22.4, 21.8, 19.2, 17, 17.6, 21.8

7 162.75, 193.25, 195, 196.75, 181.25, 167, 168.50, 159.75, 157.50

Exercise 10:3

1 **a, e**

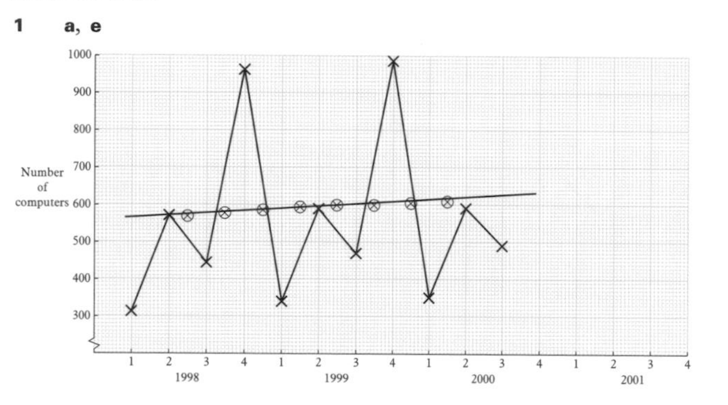

b

Year	Quarter	Computers Sold	4-Point Moving Average
1998	1	315	
	2	571	573.75
	3	446	580.00
	4	963	584.75
1999	1	340	590.75
	2	590	597.25
	3	470	600.00
	4	989	600.50
2000	1	351	605.75
	2	592	
	3	491	

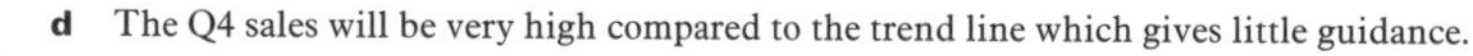

d The Q4 sales will be very high compared to the trend line which gives little guidance.

2 **a, c**

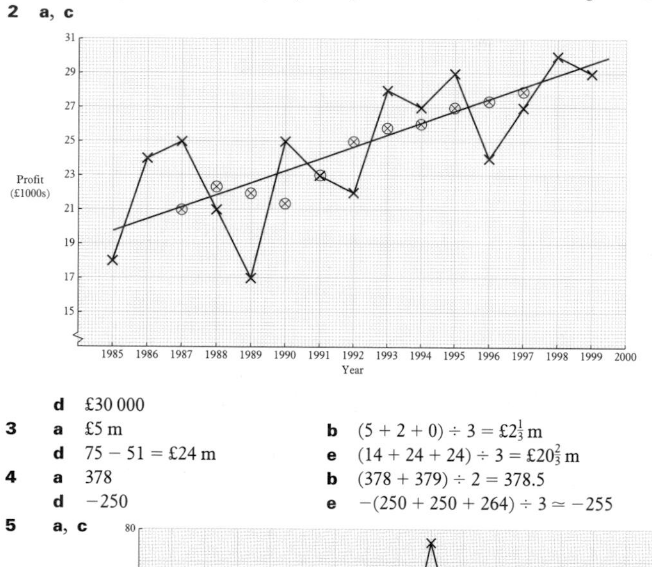

b

Year	Profit (£1000s)	5-Point Moving Average
1985	18	
1986	24	
1987	25	21
1988	21	22.4
1989	17	22.2
1990	25	21.6
1991	23	23.0
1992	22	25.0
1993	28	25.8
1994	27	26.0
1995	29	27.0
1996	24	27.4
1997	27	27.8
1998	30	
1999	29	

d £30 000

3 **a** £5 m **b** $(5 + 2 + 0) \div 3 = £2\frac{1}{3}$ m **c** $£62\frac{1}{3}$ m

d $75 - 51 = £24$ m **e** $(14 + 24 + 24) \div 3 = £20\frac{2}{3}$ m **f** $61 + 20\frac{2}{3} = £81\frac{2}{3}$ m

4 **a** 378 **b** $(378 + 379) \div 2 = 378.5$ **c** $630 + 378.5 \simeq 1009$

d -250 **e** $-(250 + 250 + 264) \div 3 \simeq -255$ **f** $640 - 255 = 385$

5 **a, c**

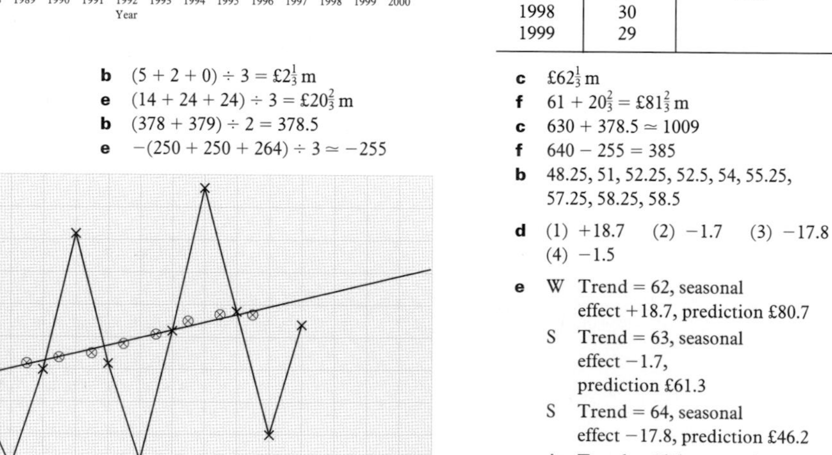

b 48.25, 51, 52.25, 52.5, 54, 55.25, 57.25, 58.25, 58.5

d (1) +18.7 (2) −1.7 (3) −17.8 (4) −1.5

e W Trend = 62, seasonal effect +18.7, prediction £80.7

S Trend = 63, seasonal effect −1.7, prediction £61.3

S Trend = 64, seasonal effect −17.8, prediction £46.2

A Trend = 65.5, seasonal effect −1.5, prediction £64

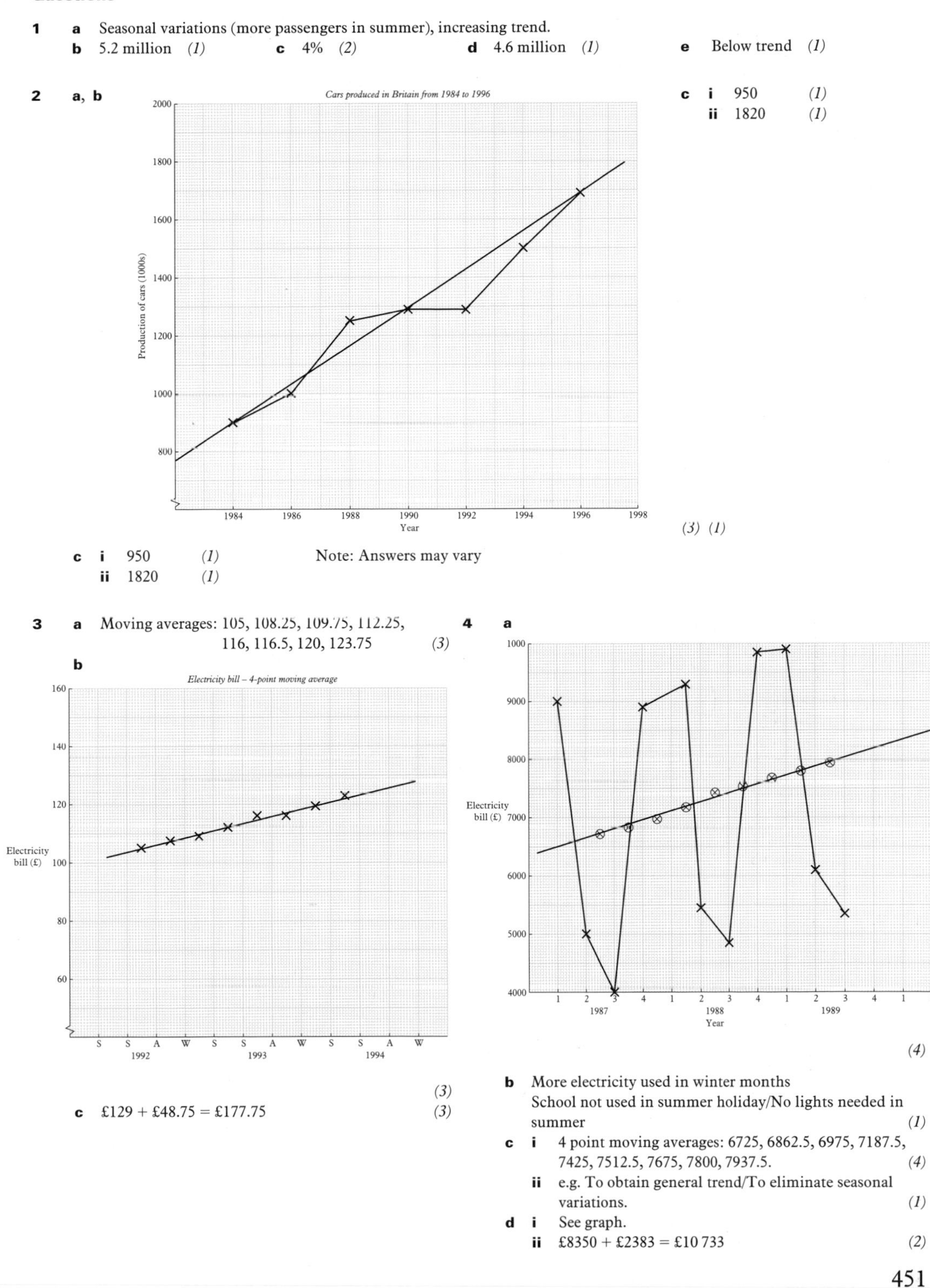

Questions

1 **a** Seasonal variations (more passengers in summer), increasing trend.

b 5.2 million *(1)* **c** 4% *(2)* **d** 4.6 million *(1)* **e** Below trend *(1)*

2 **a, b**

c **i** 950 *(1)*
ii 1820 *(1)*

(3) (1)

c **i** 950 *(1)*
ii 1820 *(1)*

Note: Answers may vary

3 **a** Moving averages: 105, 108.25, 109.75, 112.25, 116, 116.5, 120, 123.75 *(3)*

b

(3)

c £129 + £48.75 = £177.75 *(3)*

4 **a**

(4)

b More electricity used in winter months
School not used in summer holiday/No lights needed in summer *(1)*

c **i** 4 point moving averages: 6725, 6862.5, 6975, 7187.5, 7425, 7512.5, 7675, 7800, 7937.5. *(4)*

ii e.g. To obtain general trend/To eliminate seasonal variations. *(1)*

d **i** See graph.

ii £8350 + £2383 = £10 733 *(2)*

5 **a** *(2)* **d** *(2)* **e** *(1)*

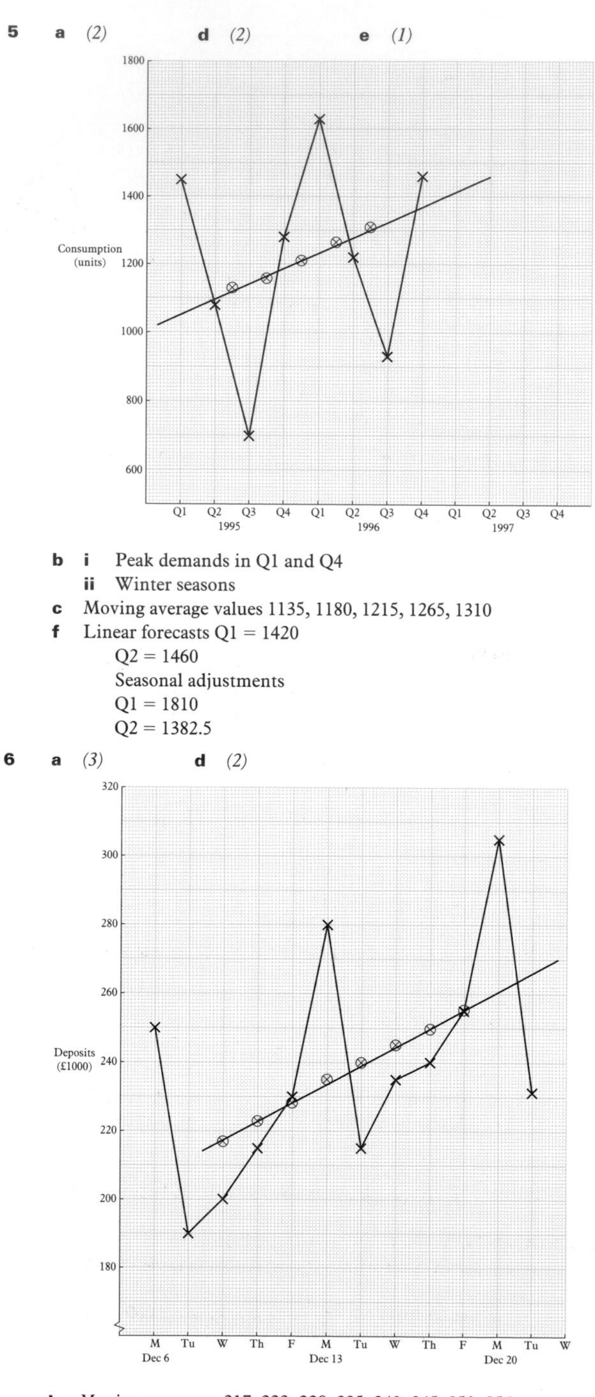

b **i** Peak demands in Q1 and Q4 *(1)*

ii Winter seasons *(1)*

c Moving average values 1135, 1180, 1215, 1265, 1310 *(3)*

f Linear forecasts Q1 = 1420

Q2 = 1460

Seasonal adjustments

Q1 = 1810

Q2 = 1382.5 *(4)*

6 **a** *(3)* **d** *(2)*

b Moving averages: 217, 223, 228, 235, 240, 245, 250, 256 *(4)*

c Weekly takings (or shape of graph). *(1)*

e $-\dfrac{(17+10)}{2} = -13.5$ (× £1000) *(2)*

f 256 to 263 (× £1000) *(2)*

Test Yourself

1 **a** Sales are higher in Q2 and Q3 (Spring and Summer) which are the main gardening seasons.

b **c, d**

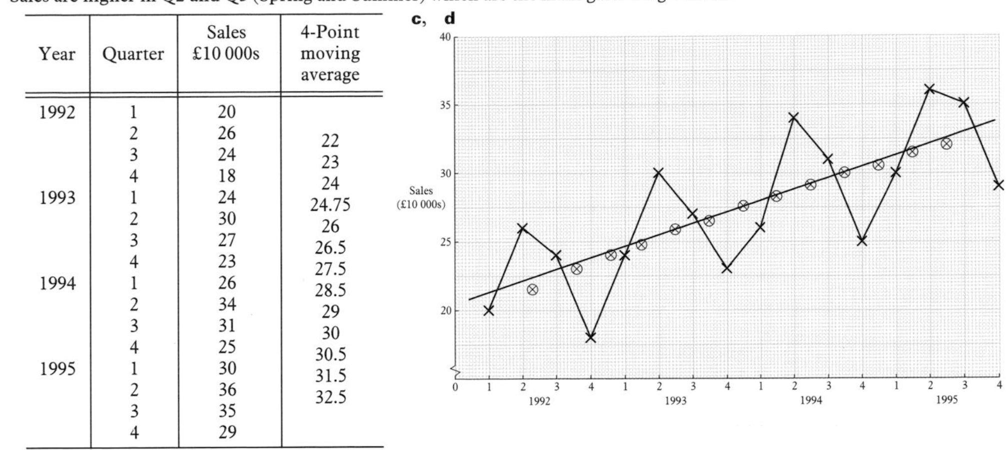

Year	Quarter	Sales £10 000s	4-Point moving average
1992	1	20	
	2	26	22
	3	24	23
	4	18	24
1993	1	24	24.75
	2	30	26
	3	27	26.5
	4	23	27.5
1994	1	26	28.5
	2	34	29
	3	31	30
	4	25	30.5
1995	1	30	31.5
	2	36	32.5
	3	35	
	4	29	

CHAPTER 11

Exercise 11:1

1 **a**

Year	1996	1997	1998	1999
Index	100	104	109	112

b 105 (or 104.8)

2

Year	1997	1998	1999
Index	100	70	35

3 **a** 118 **b** 18%
4 **a** 111 **b** 119, 123
5 **a** 109.5, 95 **b** 9.5% **c** 1.095 **d** 13.2%
6 **a** 108 **b** 5783 **c** 105

Exercise 11:2

1 **a** £9.24 **b** There has been a 5% increase compared to 1999.
2 **a** (1) £132 (2) £174 **b** 10%
c 31.8% **d** The increase between 1998 and 1999 was much higher than between 1997 and 1998.
3 **a** £97 500 **b** £330 400
c There was no change in the cost of raw materials between 1990 and 1995.
d Raw materials rose by 25% but wages rose more steeply by 52%.
4 **a** Between 1994 and 1998 **b** £36.50 **c** £42.71 **d** 17%
5 **a** £144 **b** £156.96 **c** £147.20
d 7% **e** Living costs rose by 15% but salary only rose by 9%. **f** 2.83%
6 **a** Between 1996 and 1997 **b** £2640 **c** In 1996 **d** £2688 **e** 10%
7 **a** 1997 1998 1999 2000 2001
100 115.5 104.5 91.2 73.1
b The value increased by 15.5% during 1998 and by 4.5% during 1999. It then decreased by 8.8% during 2000 and by 26.9% during 2001.
8 Hadley's profits increased for each year. They increased by 8% in 2000, 10% in 2001 and 7% in 2002.
Marchmont's profits also iincreased for each year but had a much greater increase in 2000. The profits increased by 56% in 2000, 9% in 2001 and 18% in 2002.
Carmend's profits also increased for each year, but only by a small amount. The profits increased by 2% in 2000, 4% in 2001 and 3% in 2002.
9 **a**

Year	1998	1999	2000	2001	2002
Chain	100	110.7	111.3	105.8	119.2

b

Year	1998	1999	2000	2001	2002
Value (£)	72 000	94 320	92 160	90 000	90 720

c

Year	1998	1999	2000	2001	2002
	100	131	97.7	97.7	100.8

d The value of the newsagents increased each year but the value of the corner shop decreased in 2000 and 2001.

Exercise 11:3

1 **a** (1) 64.1 (2) 67.7 **b** Danny **c** 82 **2** **a** 5.8 **b** 6.1
3 **a** (1) £390 926 (2) £380 412 **b** 97.3 **c** Lower number of employees in production, which is the most costly type of work.
4 108.8
5 **a** Teachers and support staff are the major costs for schools. **b** 110.1
c Because the weighting for salaries is so much higher than for repairs or books.
6 **a** Bread 114.5 **b** (1) 106.9 (2) 95.8
Steak 116.5
Baked Beans 39.1
Eggs 105.6
Apples 111.4
c Williams eat less steak (high index) and more baked beans (low index).

Exercise 11:4

1 117.8
2 **a** 115.2 **b** Because housing is not included, which has both a high index and a high weighting (or similar reason).
3 112.2 **4** **a** 111.1 **b** 110.8 **c** 112.1 **d** 120.3
5 **a** 117.5 **b** It has fallen because motoring has a high index and a high weighting, and this has been removed.
6 **a** 117.6 **b** It has fallen because fares and travel have a high index and have been removed. Because the weighting is low, the fall is less than in **5b**.
7 118.3 **8** 117.9 **9** **a** 113.5 **b** 118.1 **c** 118.4 **d** 118.5

Exercise 11:5

1 4.4 per 1000
2 **a** (1) 7.1 per 1000 (2) 14.6 per 1000 **b** (1) 11.5 per 1000 (2) 10.6 per 1000
c Sally's town is likely to have a lower average age than Ken's town.
3 **a** 8.6 per 100 **b** 1299 (or 1292 if 8.6 is used)
4 **a** 394 **b** 524
c Weather and other factors (e.g. flu epidemic, change in population, might be younger etc.) would affect the death rate.
5 **a** (1) 136 (2) 141 **b** Bears will become extinct in this area.

Exercise 11:6

1 0–25, 206; 26–50, 432 (431); 51–75, 272; 76–100, 90.
2 **a** 1991 – Under 16, 203; 16–39, 355(354); 40–64, 287; 65–79, 120; 80 and over, 36.
2051 – Under 16, 176; 16–39, 280; 40–64, 303; 65–79, 149; 80 and over, 92.
b There are more older people in 2051, more younger people in 1991.
3 **a** 13.9 per 1000 **b** 13.1 **4** **a** 6.69 **b** 7.77 **c** Harpdon
5 Barton because it has the highest death rate.
6 **a** A, 2.77; B, 6.99 **b** Area A – which has a much lower standardised death rate.
c Areas A and B have very different population profiles, the standardised death rate takes this into account. **7** **b** 89.0

Questions

1 **a** Cost is 5% higher than in 1994. *(1)* **b** **i** £32.40 *(3)* **ii** £2.40 *(1)* **c** 116 *(2)*
2 **a** Better use of data; more representative of data. *(1)* **b** £500 *(2)* **c** 117.25 *(3)*
3 **a** 122.6 *(2)* **b** 127 *(2)*
4 **a** $(2 \times 2.5) + (4 \times 1) + (12 \times 4) + (1 \times 5) =$ £62.00 *(2)*
b $(2 \times 3) + (4 \times 1.5) + (12 \times 4.5) + (1 \times 6) =$ £72.00 *(2)* **c** 116.1 *(2)*
d 33.3% *(2)* **e** Labour costs rising at a much higher rate *(2)*
5 **a** Total wage bill: 1990 = £98 590, 1995 = £130 480, Index = 132.3 *(4)*
b Total wage bill: 1990 = £98 590, 1995 = £157 240 for same number of employees, Index = 159.5 *(4)*
c Change in weighting procedure; also change in distribution of numbers of employees in each category. *(2)*
6 **a** 120.9 *(3)* **b** Groups 11 and 12 show a more significant increase. *(1)*
c **i** 114.57 *(4)* **ii** Reduction in all groups index. *(1)*
7 **a** 13.4 per thousand. *(3)* **b** Town Y has more older people or fewer younger people. *(1)*
c 15.8 *(4)* **d** Can compare towns with differing age distributions. *(1)*
8 **a** 2.94 *(4)* **b** The standardised death rate takes into account the ages of the population. *(1)*
c Dormingly because it has a lower standardised death rate. *(1)*
9 **a** 0–4.0, 20–2.0, 40–15.0, 60–50 *(4)* **b** Westhope: 13.6, Martrent: 19.1 *(4)*
c Yes. Westhope has the lower standardised death rate, even though the percentage of the population 60 – is much higher in Westhope ($\simeq$28%) than in Martrent (5%). *(2)*

10 **a** 125.92 **b** 71.8 (3 sf)

11 **a** **i** 167 (3 sf) **ii** 125 **b** 4320p, 480p **c** 150

12 **a** $69 - 6 = 63$ years **b** 19

c $\frac{19}{750} \times 100 = 2.53$ deaths per 100 of population or 25.3 deaths per 1000 of population.

d 22 per 1000

Population $= \frac{7}{22} \times 1000$

$= 318$

Test Yourself

1 **a** $\dfrac{72.9}{41.2} \times 100 \simeq 177$ **b** $41.2 \times \dfrac{128}{100}\text{p} \simeq 52.7\text{p}$ **c** The price has risen by 28%.

2 **a** $\dfrac{2 \times 70 + 3 \times 45 + 5 \times 40}{2 + 3 + 5} = 47.5$ **b** $\dfrac{2 \times 40 + 3 \times 50 + 5 \times \text{m}}{10} = 40.5$

$80 + 150 + 5\text{m} = 405$

$5\text{m} = 175$

$\text{m} = 35$

Charles scored 35

3 **a** $\dfrac{19 + 104 + 77 + 121 + 192}{2800 + 8300 + 8900 + 7400 + 4900} \times 1000 = 15.9$

b $\dfrac{19}{2800} \times 240 + \dfrac{104}{8300} \times 210 + \dfrac{77}{8900} \times 190 + \dfrac{121}{7400} \times 200 + \dfrac{192}{4900} \times 160 \simeq 15.4$

CHAPTER 12

Exercise 12:1

1 Independent **2** Dependent **3** Independent **4** Independent **5** Dependent

Exercise 12:2

1 $\frac{2}{20} = \frac{1}{10}$ **2** **a** $\frac{1}{9}$ **b** $\frac{2}{9}$ **3** **a** 0.27 **b** 0.07

4 **a** $\frac{3}{8}$ **b** $\frac{9}{64}$ **5** **a** 0.3 **b** 0.7 **c** 0.21

6 $\frac{1}{8}$ **7** $\frac{16}{625}$ **8** **a** 0.0034 (2 s.f.) **b** 0.61 **9** **a** $\frac{1}{4}$ **b** $\frac{1}{64}$ **c** $\frac{1}{1024}$

10 0.0000036 (or 1/279936)

Exercise 12:3

1 **a** Mutually exclusive **b** Not mutually exclusive **c** Mutually exclusive **d** Not mutually exclusive

2 **a** $\frac{1}{4}$ **b** $\frac{9}{20}$ **c** $\frac{11}{20}$ **d** $\frac{3}{4}$

3 **a** $\frac{1}{12}$ **b** $\frac{1}{2}$ **c** $\frac{1}{6}$ **d** $\frac{1}{3}$ **e** $\frac{7}{12}$ **f** $\frac{5}{12}$

4 **a** Exhaustive **b** Not exhaustive **c** Exhaustive

Exercise 12:4

1 **a** 0.28 **b** 0.12 **c** 0.54 **d** 0.18 **2** **a** 0.83 **b** 0.17

3 **a** $\frac{1}{4}$ **b** $\frac{1}{5}$ **c** $\frac{1}{20}$ **d** $\frac{1}{10}$

4 **a** 0.375 **b** 0.3375 **c** 0.3375 **d** 0.0125

5 **a** 0.3 **b** 0.53

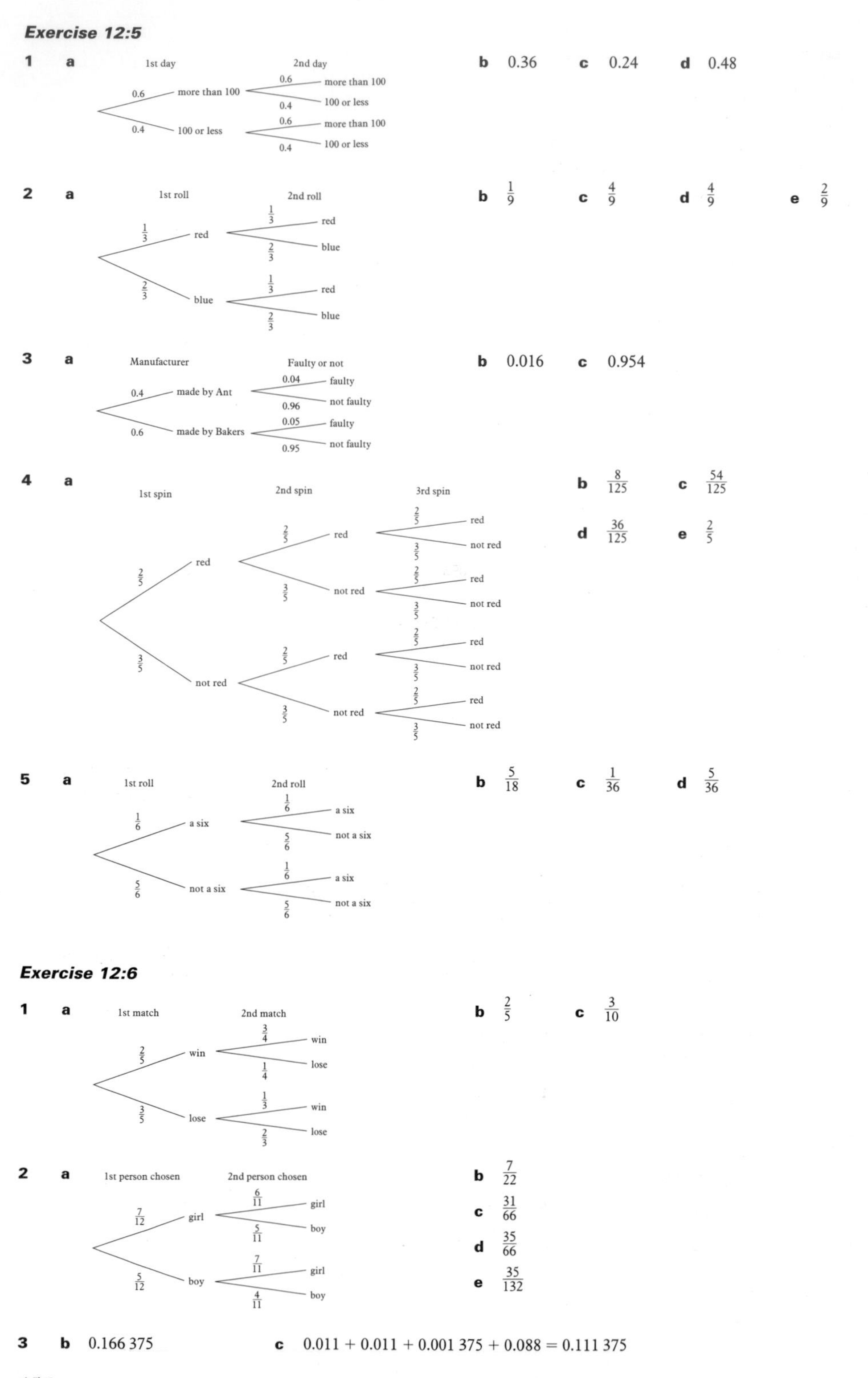

Exercise 12:5

1 **a**

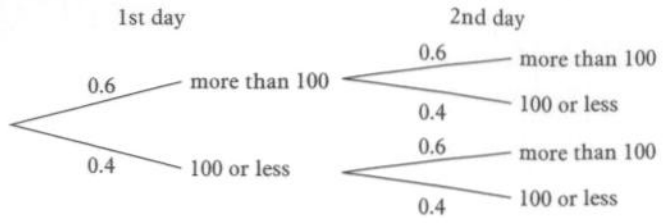

b 0.36 **c** 0.24 **d** 0.48

2 **a**

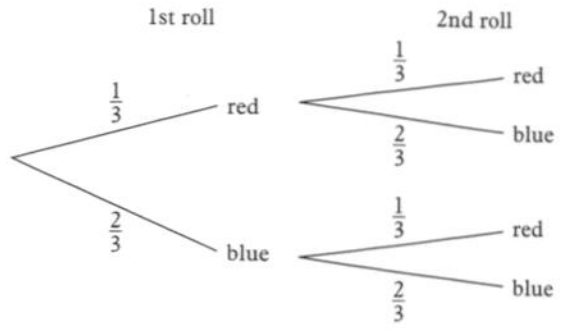

b $\frac{1}{9}$ **c** $\frac{4}{9}$ **d** $\frac{4}{9}$ **e** $\frac{2}{9}$

3 **a**

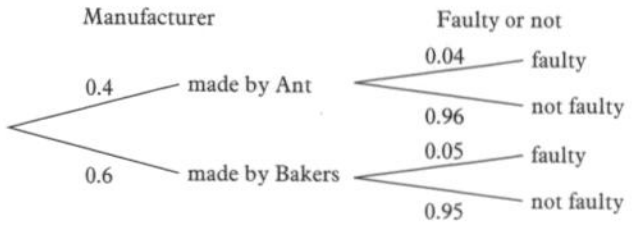

b 0.016 **c** 0.954

4 **a**

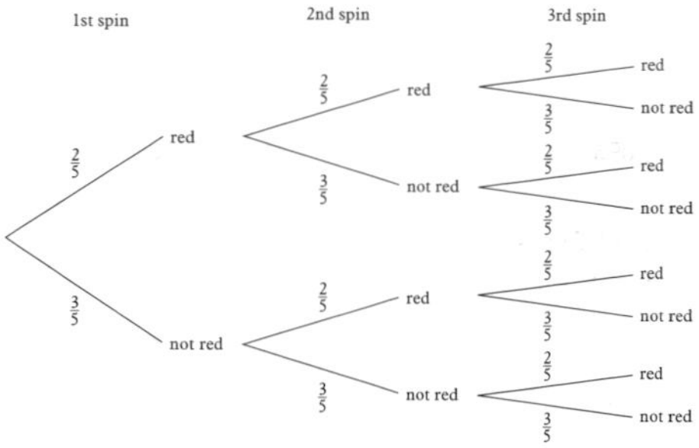

b $\frac{8}{125}$ **c** $\frac{54}{125}$

d $\frac{36}{125}$ **e** $\frac{2}{5}$

5 **a**

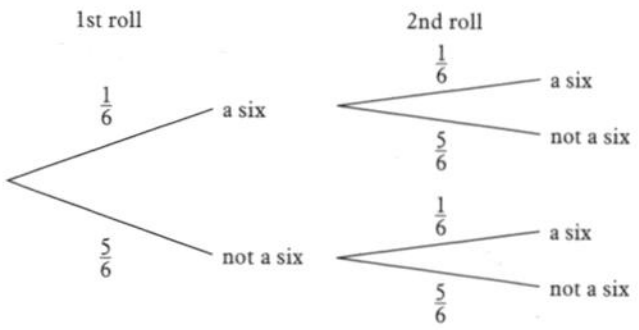

b $\frac{5}{18}$ **c** $\frac{1}{36}$ **d** $\frac{5}{36}$

Exercise 12:6

1 **a**

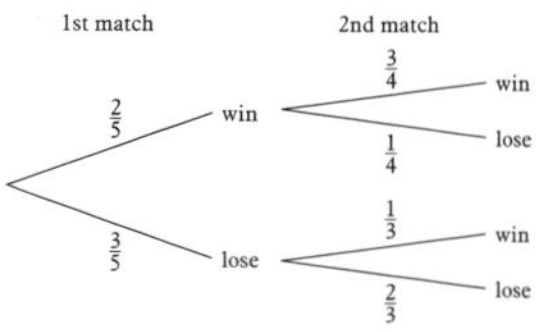

b $\frac{2}{5}$ **c** $\frac{3}{10}$

2 **a**

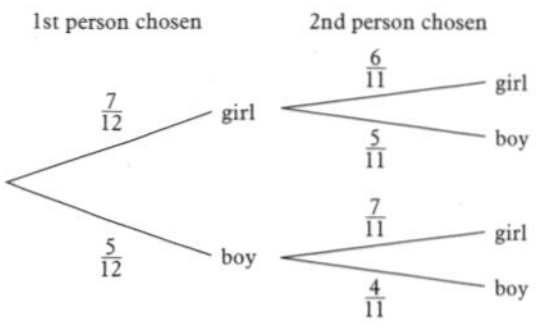

b $\frac{7}{22}$

c $\frac{31}{66}$

d $\frac{35}{66}$

e $\frac{35}{132}$

3 **b** 0.166 375 **c** $0.011 + 0.011 + 0.001\,375 + 0.088 = 0.111\,375$

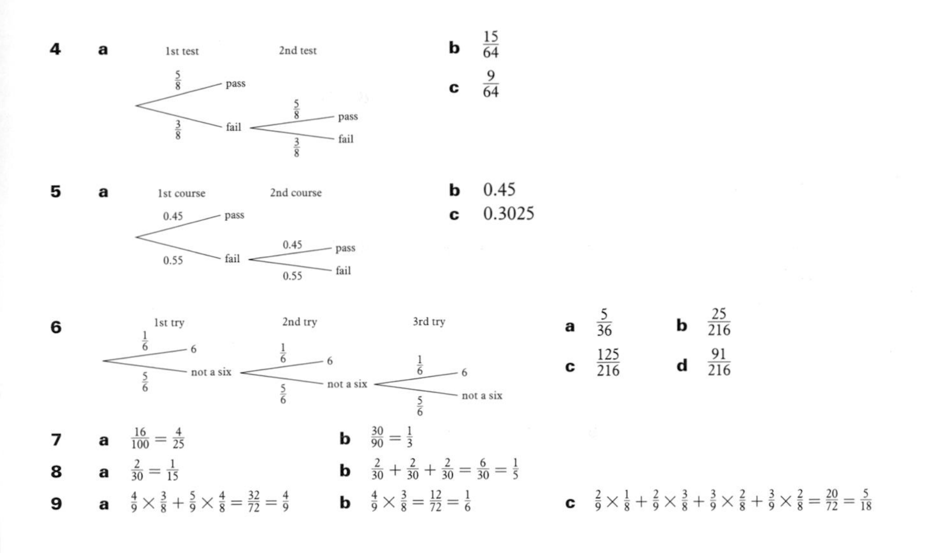

4 **a**

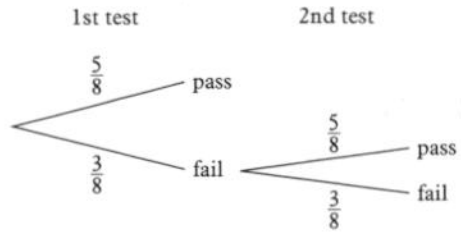

b $\frac{15}{64}$

c $\frac{9}{64}$

5 **a**

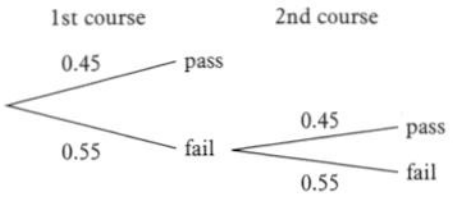

b 0.45

c 0.3025

6

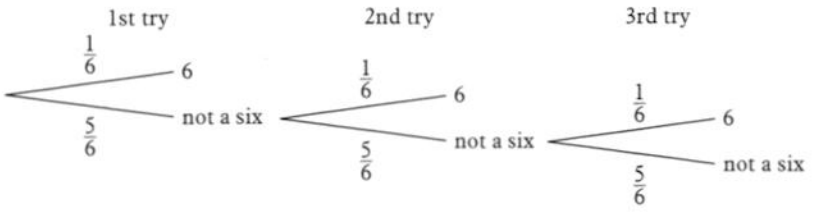

a $\frac{5}{36}$ **b** $\frac{25}{216}$

c $\frac{125}{216}$ **d** $\frac{91}{216}$

7 **a** $\frac{16}{100} = \frac{4}{25}$ **b** $\frac{30}{90} = \frac{1}{3}$

8 **a** $\frac{2}{30} = \frac{1}{15}$ **b** $\frac{2}{30} + \frac{2}{30} + \frac{2}{30} = \frac{6}{30} = \frac{1}{5}$

9 **a** $\frac{4}{9} \times \frac{3}{8} + \frac{5}{9} \times \frac{4}{8} = \frac{32}{72} = \frac{4}{9}$ **b** $\frac{4}{9} \times \frac{3}{8} = \frac{12}{72} = \frac{1}{6}$ **c** $\frac{2}{9} \times \frac{1}{8} + \frac{2}{9} \times \frac{3}{8} + \frac{3}{9} \times \frac{2}{8} + \frac{3}{9} \times \frac{2}{8} = \frac{20}{72} = \frac{5}{18}$

Exercise 12.7

1 $1 - \frac{1}{4} = \frac{3}{4}$ 2 $1 - \frac{8}{27} = \frac{19}{27}$

3 **a** 0.512 **b** 0.384 **c** 0.096 **d** $1 - 0.512 = 0.488$

4 **a** 0.001 **b** $1 - 0.001 = 0.999$

5 **a** 0.1 **b** 0.216 **c** $1 - 0.216 = 0.784$

6 **a** $\frac{27}{1000} + \frac{8}{1000} + \frac{125}{1000} = \frac{160}{1000} = \frac{4}{25}$ **b** $1 - \frac{4}{25} = \frac{21}{25}$

Questions

1 **a** $\frac{13}{20}$ *(2)* **b** $\frac{7}{20}$ *(2)* **c** 0 *(1)* **d** 15 *(2)* **e** 10 *(2)*

2 **a** **i** $\frac{3}{4}$ *(1)* **ii** $\frac{1}{4}$ *(1)* **b** $\frac{9}{16}$ *(2)* **c** **i** $\frac{81}{256}$ *(2)* **ii** Unlikely *(1)*

3 **a**

	Large	Small	Total
Leaded	272	880	1152
Unleaded	68	580	648
Total	340	1460	1800

(4)

b **i** $\frac{880}{1800} = 0.489$ *(1)* **ii** $\frac{1152}{1800} + \frac{580}{1800} = 0.962$ *(2)* **iii** $\frac{580}{1460} = 0.397$ *(2)*

4 **a** $\frac{1}{5}$ *(1)* **b** $\frac{4}{5}$ *(1)*

c

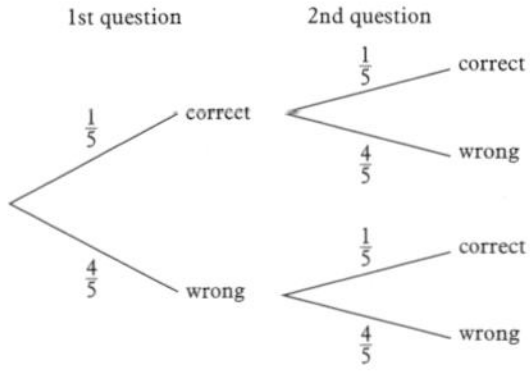

(3)

d $\frac{1}{25}$ *(2)*

e $\frac{8}{25}$ (0.32) *(2)*

f $30 \times \frac{1}{5} = 6$ *(2)*

5 **a**

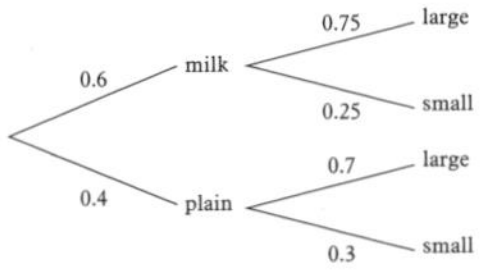

(3)

b **i** 0.15 *(2)* **ii** 0.27 *(3)*

c $\frac{0.15}{0.27} \times 270 = 150$ *(3)*

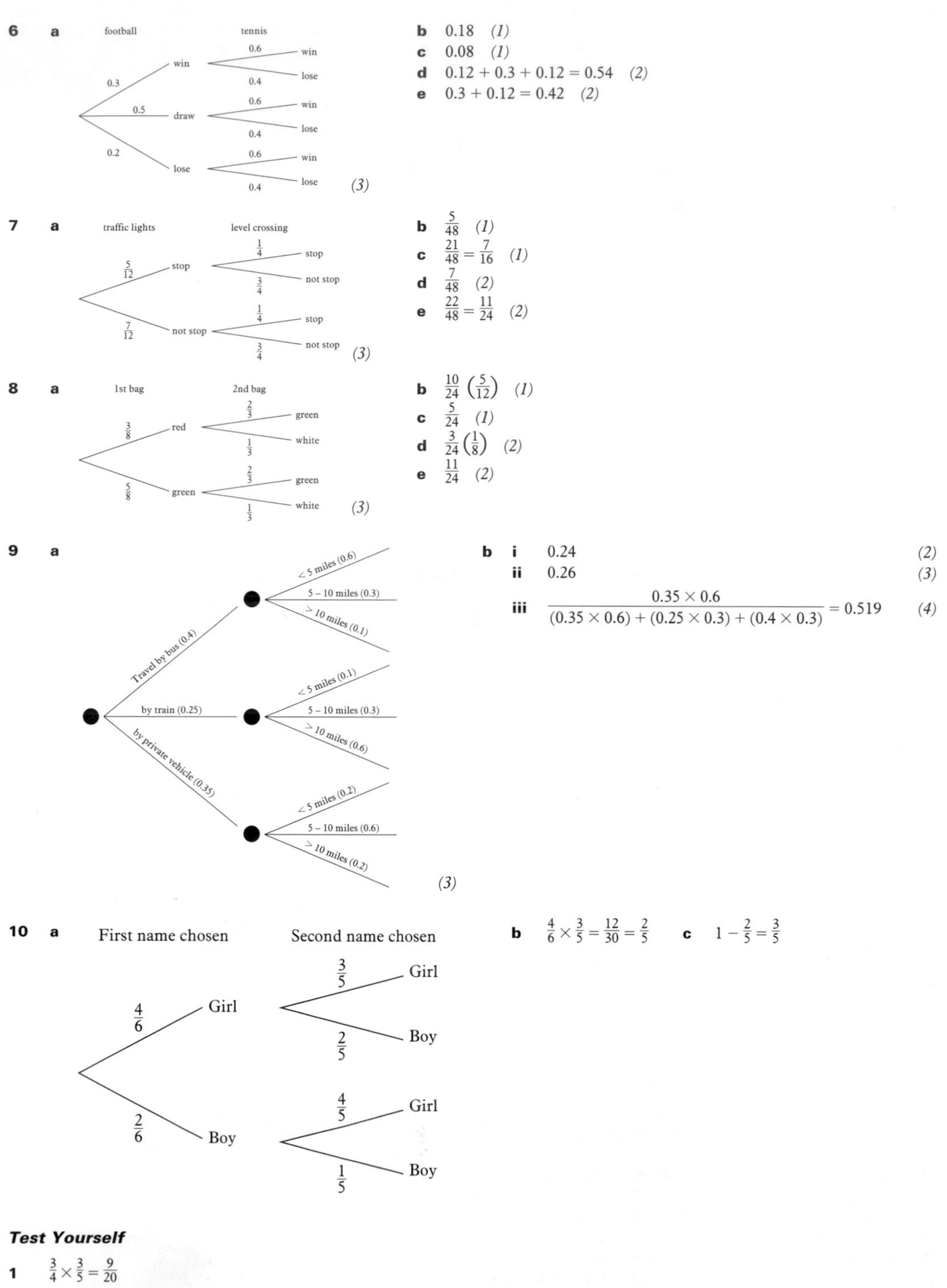

6 **b** 0.18 *(1)*
c 0.08 *(1)*
d $0.12 + 0.3 + 0.12 = 0.54$ *(2)*
e $0.3 + 0.12 = 0.42$ *(2)*

7 **b** $\frac{5}{48}$ *(1)*
c $\frac{21}{48} = \frac{7}{16}$ *(1)*
d $\frac{7}{48}$ *(2)*
e $\frac{22}{48} = \frac{11}{24}$ *(2)*

8 **b** $\frac{10}{24}\left(\frac{5}{12}\right)$ *(1)*
c $\frac{5}{24}$ *(1)*
d $\frac{3}{24}\left(\frac{1}{8}\right)$ *(2)*
e $\frac{11}{24}$ *(2)*

9 **b** **i** 0.24 *(2)*
ii 0.26 *(3)*
iii $\dfrac{0.35 \times 0.6}{(0.35 \times 0.6) + (0.25 \times 0.3) + (0.4 \times 0.3)} = 0.519$ *(4)*

10 **b** $\frac{4}{6} \times \frac{3}{5} = \frac{12}{30} = \frac{2}{5}$ **c** $1 - \frac{2}{5} = \frac{3}{5}$

Test Yourself

1 $\frac{3}{4} \times \frac{3}{5} = \frac{9}{20}$

2 **a** $0.02 \times 0.02 = 0.0004$
b $0.02 \times 0.98 + 0.98 \times 0.02 = 0.0392$
c $0.98 \times 0.98 = 0.9604$
d $0.02 \times 0.98 = 0.0196$

3 **a**

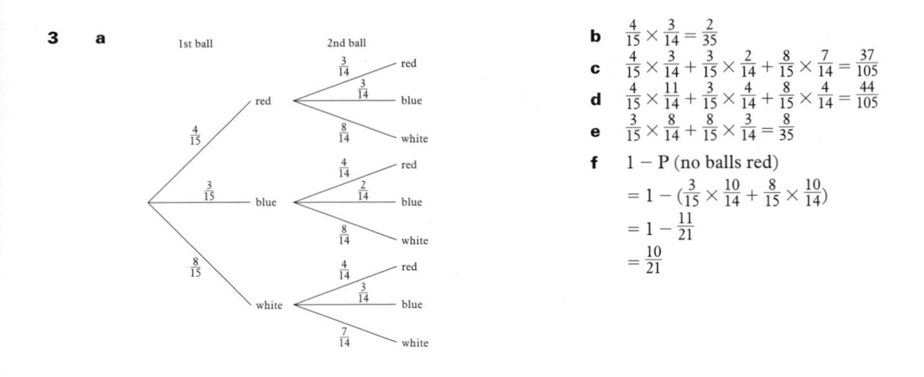

b $\frac{4}{15} \times \frac{3}{14} = \frac{2}{35}$

c $\frac{4}{15} \times \frac{3}{14} + \frac{3}{15} \times \frac{2}{14} + \frac{8}{15} \times \frac{7}{14} = \frac{37}{105}$

d $\frac{4}{15} \times \frac{11}{14} + \frac{3}{15} \times \frac{4}{14} + \frac{8}{15} \times \frac{4}{14} = \frac{44}{105}$

e $\frac{3}{15} \times \frac{8}{14} + \frac{8}{15} \times \frac{3}{14} = \frac{8}{35}$

f $1 - \text{P (no balls red)}$

$= 1 - (\frac{3}{15} \times \frac{10}{14} + \frac{8}{15} \times \frac{10}{14})$

$= 1 - \frac{11}{21}$

$= \frac{10}{21}$

CHAPTER 13

Exercise 13:1

1 **a** 5040 **b** 40 320 **c** 144 **d** 240 **e** 3600 **f** 1440

2 **a** 239 500 800 **c** 1440 **e** 950 400 **g** 3360 **i** 1 965 600

b 8 709 120 **d** 1 814 400 **f** 192 **h** 168

3 **a** $2.5x$ **b** $7x$ **c** $90x$ **d** x **e** $x + 1$ **f** $x^2 + 3x + 2$ or $(x + 2)(x + 1)$

Exercise 13:2

1 $6! = 720$ **2** $12! = 479\,001\,600$ **3** $7! = 5040$ **4** $5! = 120$

5 $4! = 24$ **6** **a** $1 \times 7! = 5040$ **b** $1 \times 6! \times 1 = 720$ **7** $5! \times 1 = 120$

8 $1 \times 6! = 720$ **9** $5! \times 1 = 120$ **10** $2 \times 1 \times 5! \times 1 = 240$ **11** $2 \times 1 \times 8! \times 1 = 80\,640$

Exercise 13:3

1 $\frac{1}{42}$ **2** **a** $\frac{1}{5}$ **b** $\frac{2}{5}$ **3** **a** $\frac{1}{7}$ **b** $\frac{2}{7}$ **4** $\frac{1}{720}$

Exercise 13:4

1 15 **2** 252 **3** 924 **4** 70 **5** 1820 **6** 3003 **7** 20

Exercise 13:5

1 **a, b** $\frac{216}{625}$ **2** **a** $\frac{45}{512}$ **b** $\frac{405}{1024}$ **3** **a** 0.000643 **b** 0.0536

4 **a** 0.000408 **b** 0.149 **5** **a** 0.02835 **b** 0.3087 **6** 0.0576

7 0.2001 **8** **a** 0.23 **b** 0.31 **9** 0.046

10 0.270 **11** 0.0027 **12** 0.9536

13 0.0547 **14** 0.68256

Questions

1 **a** 5! *(1)* **b** 4! = 24 *(1)* **c** 2 × 3! = 12 *(2)*

2 **a** **i** 6 *(1)* **ii** $\frac{1}{3}$ *(2)* **b** 6! = 720 *(2)* **3** 10 *(2)*

4 **a** 0.5404 *(2)* **b** 0.3413 *(2)* **c** 0.0988 *(2)* **d** 0.0196 *(2)*

5 **a** 0.3456 *(3)* **b** 0.2592, 0.0778 *(3)* **c** 0.0105 *(4)*

6 **a** Fixed number of girls/only two outcomes/fixed probability of being left handed *(2)*

b **i** 0.302 *(2)* **ii** 0.268 *(2)* **iii** 0.323 *(3)*

7 **a** 8! = 40320 *(2)* **b** 6720 *(2)* **c** 56 *(2)*

8 **a** 6! = 720 *(2)* **b** 5! = 120 *(2)* **c** 15 *(2)*

9 **a** 479 001 600 *(2)* **b** 39 916 800 *(2)* **c** 792 *(2)* **d** 0.417 *(3)*

10 **a** 0.015 *(2)* **b** 0.386, 0.116, 0.015, 0.001 *(3)* **c** **i** 0.261 *(2)* **ii** 0.004 *(2)*

11 **a** 7.29×10^{-10} *(2)* **b** 0.012 *(2)* **c** 0.988 *(2)* **d** 0.999 *(2)*

e The probability of two light bulbs is very small. *(1)*

Test Yourself

1 **a** 3 628 800 **b** 241 920 **c** 59 875 200 **d** 33 **e** 30 **f** 17 280

2 $7! = 5040$ **3** $5! = 120$ **4** $\dfrac{1 \times {}^8C_2}{{}^{10}C_4} = \dfrac{2}{15}$

5 $2 \times 2! \times 6! = 2880$ **6** ${}^{20}C_6 = 38\,760$ **7** ${}^6C_3 = 20$

8 ${}^5C_3 \times 0.8^3 \times 0.2^2 = 0.2048$ **9** $({}^5C_3 \times 0.3^3 \times 0.7^2) + ({}^5C_4 \times 0.3^4 \times 0.7^1) + ({}^5C_5 \times 0.3^5 \times 0.7^0) = 0.16308$

CHAPTER 14

Exercise 14:1

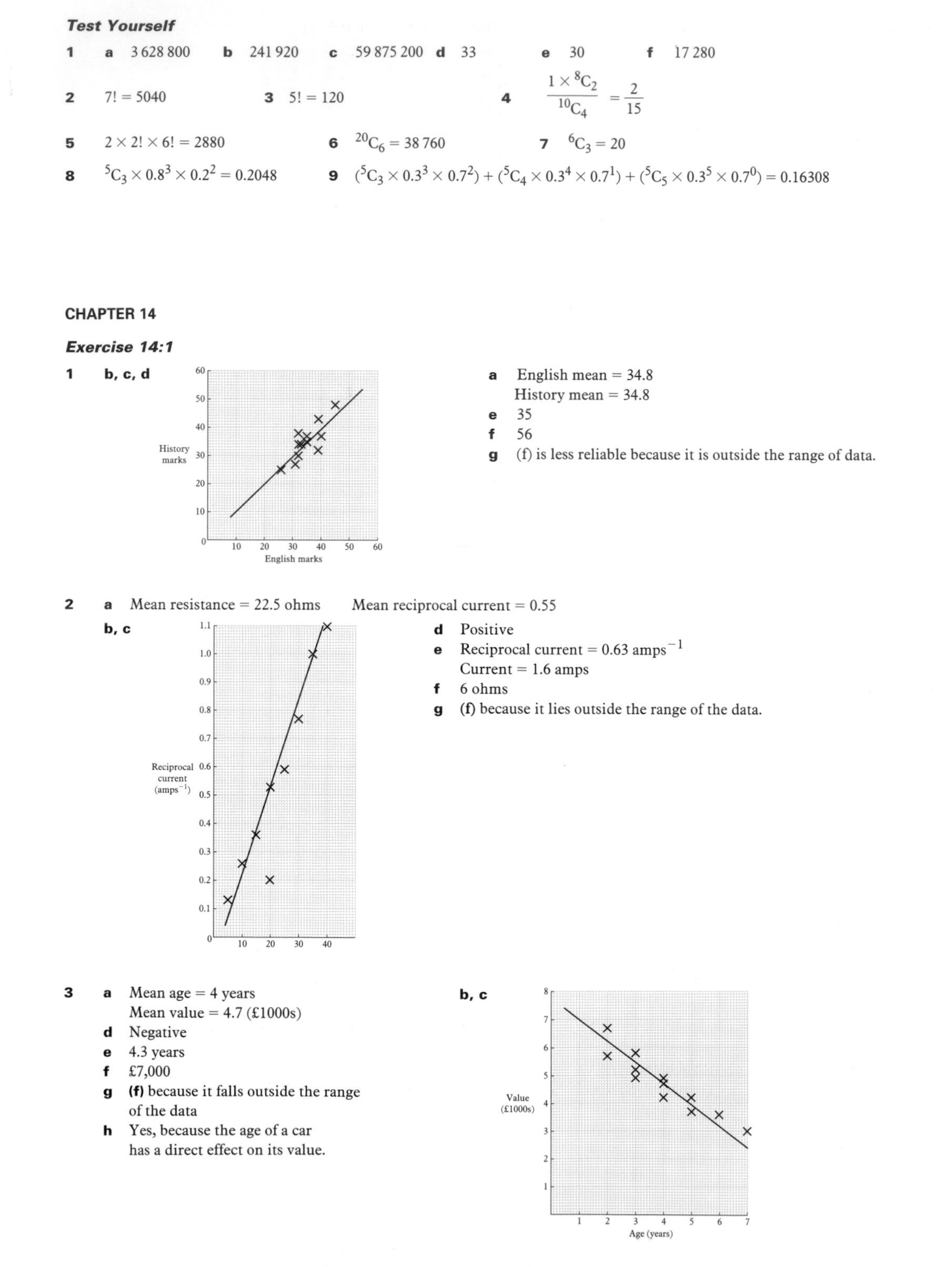

1 **b, c, d**

a English mean = 34.8
History mean = 34.8
e 35
f 56
g (f) is less reliable because it is outside the range of data.

2 **a** Mean resistance = 22.5 ohms Mean reciprocal current = 0.55
b, c
d Positive
e Reciprocal current = 0.63 amps^{-1}
Current = 1.6 amps
f 6 ohms
g (f) because it lies outside the range of the data.

3 **a** Mean age = 4 years
Mean value = 4.7 (£1000s)
b, c
d Negative
e 4.3 years
f £7,000
g **(f)** because it falls outside the range of the data
h Yes, because the age of a car has a direct effect on its value.

4 **a** 100 kg **b** 170 cm **c** height = 175 cm, weight = 85 kg

5 **a** **b** **c** **d**

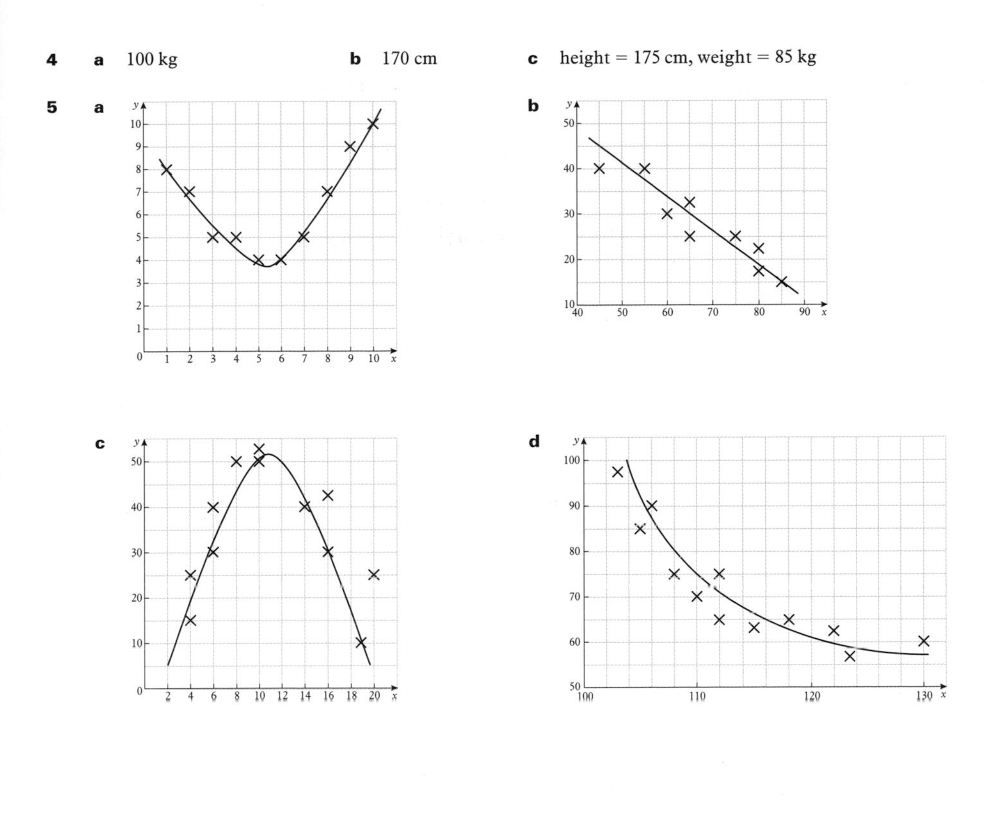

Exercise 14:2

1 **a** $w = 0.3h + 10$ **b** $t = 0.4s + 5$ **c** $v = -100a + 1000$

2 **a** Mean mass = 450 g Mean length = 18.75 cm

b, c

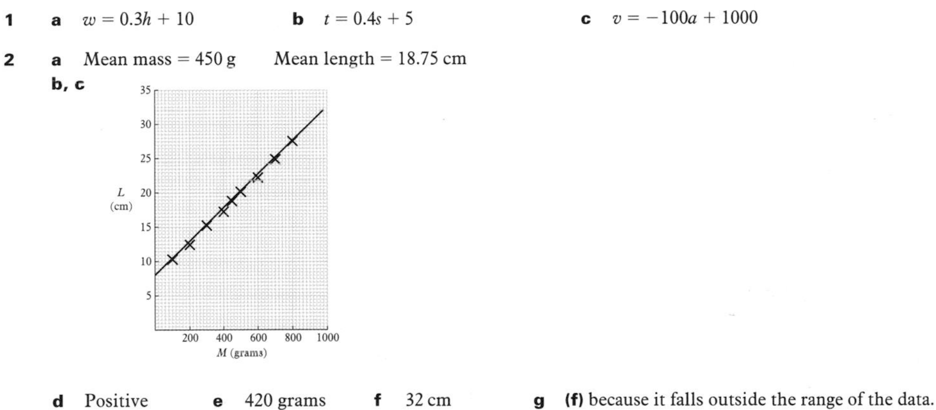

d Positive **e** 420 grams **f** 32 cm **g** **(f)** because it falls outside the range of the data.

h $L = 0.023\text{ m} + 8.5$

i $c = 8.5$ which is the length of the spring with no load.

$m = 0.023$ which is the amount in cm, that the spring stretches for each gram of mass added.

3 **a** Mean number = 73.6 Mean cost = £242

b, c

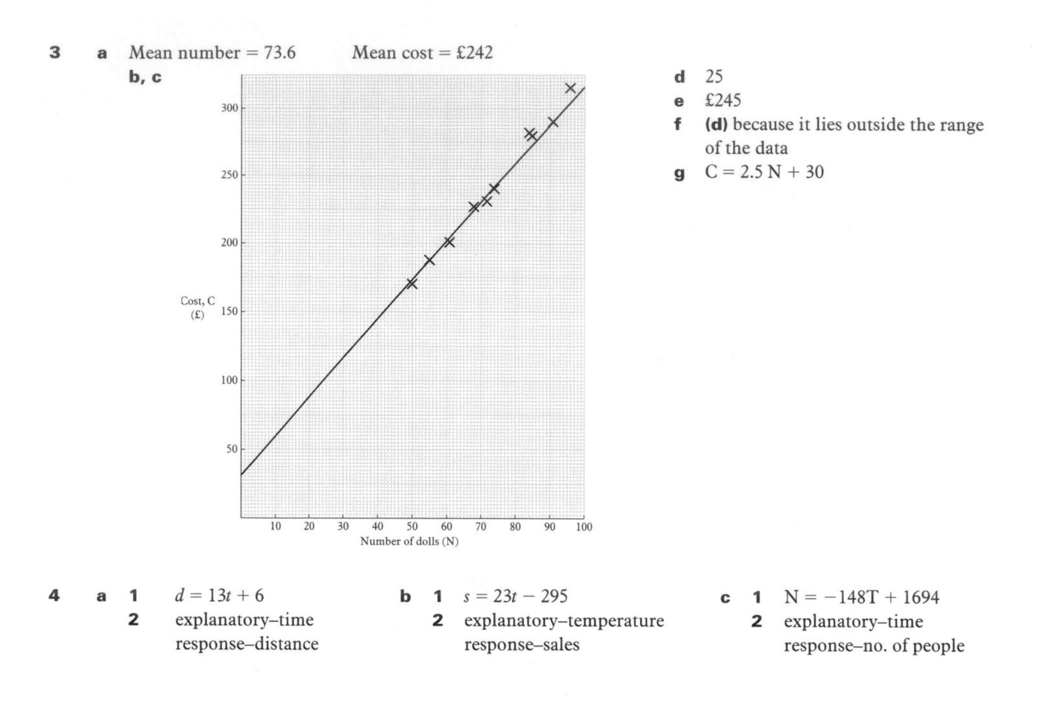

d 25

e £245

f **(d)** because it lies outside the range of the data

g C = 2.5 N + 30

4 **a** **1** $d = 13t + 6$
2 explanatory–time
response–distance

b **1** $s = 23t - 295$
2 explanatory–temperature
response–sales

c **1** N = −148T + 1694
2 explanatory–time
response–no. of people

Exercise 14:3

Pupil's answers should be close to the following.

1 **a** $y = 3x^2 + 4$ **b** $y = \frac{1}{2}x^2 + 1$ **2** **a** $y = 3\sqrt{x} + 1$ **b** $y = 2\sqrt{x} + 0.3$

3 $y = \dfrac{2}{x} + 3$

Exercise 14:4

1 **a** G **b** B, D **2** 0.07

3 **a** $\rho = 0.89$ **b** There is a good positive correlation, so a good level of agreement.

4 **a**

Pupil	A	B	C	D	E	F	G	H
Maths	1	7	6	2	5	4	3	8
Science	1	6	7	2	5	3	4	8

b $\rho = 0.95$

c There is a strong positive correlation between rankings in Maths and Science.

5 **a** −0.9 **b** 0.3 **c** 0.03

6 **a**

Jam	A	B	C	D	E	F	G	H
Judge 1	$5\frac{1}{2}$	2	1	3	8	7	4	$5\frac{1}{2}$
Judge 2	8	5	1	7	6	2	3	4

b $\rho = 0.24$

c There is only a weak positive correlation between the two judges' ratings (although they agreed on the 1st place) because ρ is closer to 0 than 1.

7 **a** $\rho = -0.86$

b There is a strong negative correlation between the two sets of data because Spearman's coefficient is negative and close to −1.

8 **a** $\rho = 0.88$

b There is a strong positive correlation between the two sets of data because Spearman's coefficient is positive and close to +1.

Questions

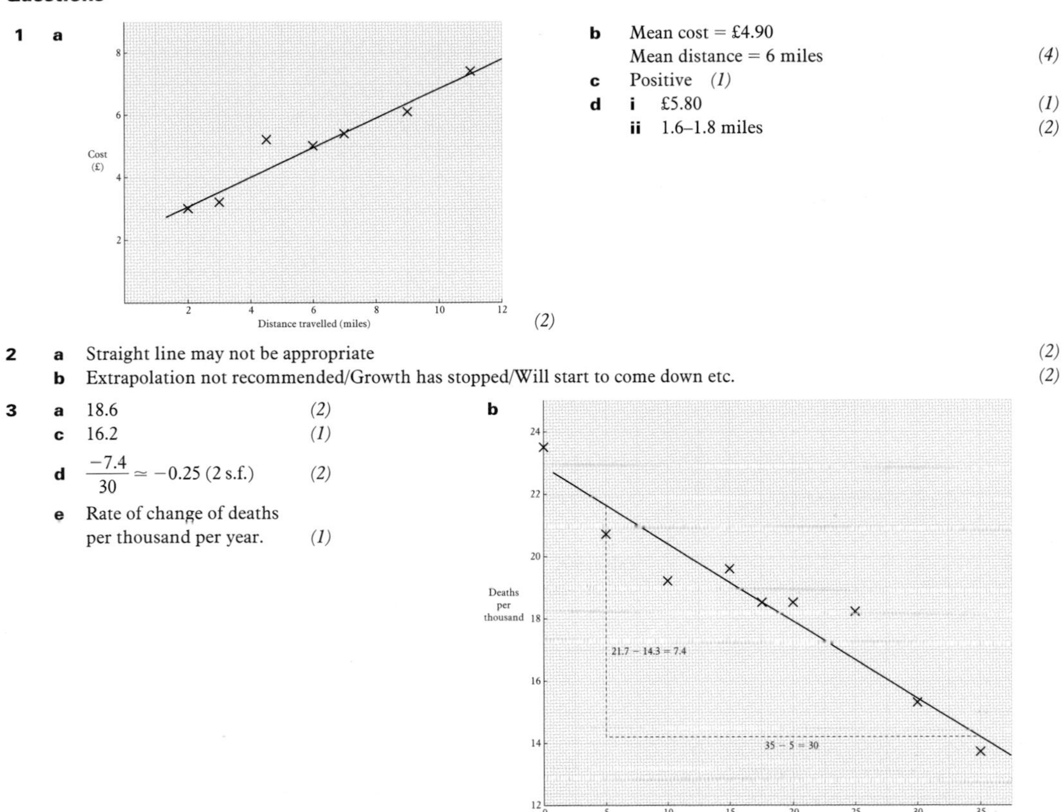

1 **a** (2)

b Mean cost = £4.90
Mean distance = 6 miles (4)

c Positive (1)

d **i** £5.80 (1)
ii 1.6–1.8 miles (2)

2 **a** Straight line may not be appropriate (2)

b Extrapolation not recommended/Growth has stopped/Will start to come down etc. (2)

3 **a** 18.6 (2)

b (2)

c 16.2 (1)

d $\frac{-7.4}{30} \simeq -0.25$ (2 s.f.) (2)

e Rate of change of deaths per thousand per year. (1)

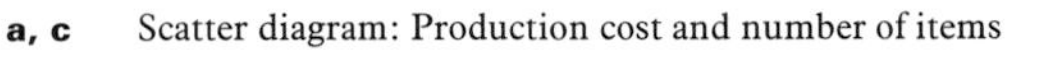

4 **a, c** Scatter diagram: Production cost and number of items

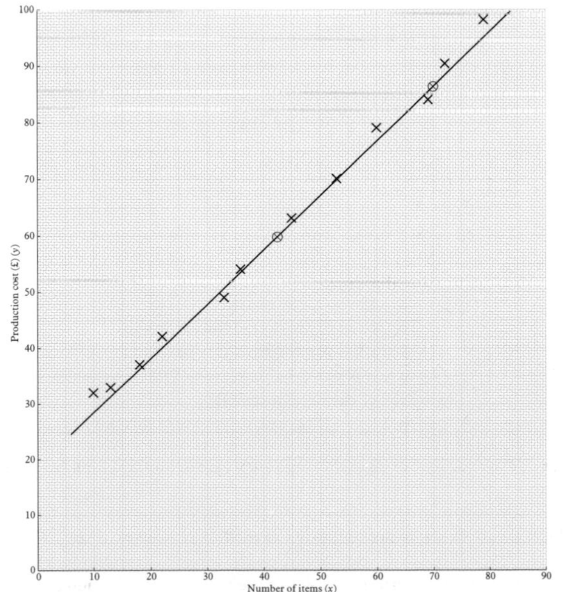

(3) (2)

b Mean $x = 42.5$ Mean $y = 61$ (2)

d **i** £83 (2) **ii** £102 (2)

e (i) is more reliable because it is interpolation (ii) is extrapolation, so less reliable (2)

f Gradient $= \frac{27}{27.5} = 0.982$ (2)

g Rate of change of production cost as number of items changes. (1)

5 **a**

Year	1990	1989	1988	1987	1986	1985	1984	1983	1982	1981
d	2	6	1	1	4	−4	−1	2	−2	−9
d^2	4	36	1	1	16	16	1	4	4	81

$\Sigma d^2 = 164$

$\rho = 1 - \dfrac{6 \times 164}{10 \times 99} = 0.006$ *(4)*

b No correlation – no evidence to suggest older wines taste better because 0.006 is very close to zero. *(1)*

c Strong negative correlation – as beer ages, quality decreases *(1)*

6 *Rankings*

a

Sunshine (hrs)	*Temp* (°F)	d	d^2
4	6	−2	4
6	10	−4	16
(8) 8.5	9	−0.5	0.25
10	8	2	4
(8) 8.5	2	6.5	42.25
5	(4) 4.5	0.5	0.25
3	7	−4	16
2	3	−1	1
7	1	6	36
1	(4) 4.5	−3.5	12.25

$\Sigma d^2 = 132$ *(3)*

b $\rho = 1 - \dfrac{6 \times 132}{10 \times 99} = 1 - 0.8 = 0.2$ *(4)*

c Very low positive correlations (almost no correlation) since 0.2 is quite close to zero. *(2)*

d **i** Saturday – no correlation *(2)*

ii Sunday – strong negative correlation – inverse relationship between temperature and sunshine hours *(3)*

7 **a**

Ranks			
Sales	*Grants*	d	d^2
1	1	0	0
2	4	2	4
3	6	3	9
4	2	2	4
5	7	2	4
6	5	1	1
7	8	1	1
8	3	5	25
9	9	0	0
10	10	0	0

$\Sigma d^2 = 48$ *(2)*

b $\rho = 1 - \dfrac{6 \times 48}{10 \times 99} = 1 - 0.2909 = +0.709$ *(3)*

c Fairly strong positive correlation – grants are roughly proportional to sales *(1)*

d £125 − 17.5 = 107.5. It can be reduced up to but less than £107.5 million, but no lower. *(2)*

8 **a**

Team	A	B	C	D	E	F	G	H
Ranking points	1	2	3	4	5	7	6	8
Ranking goals conceded	5	8	7	4	2	6	1	3
d	4	6	4	0	3	1	5	5
d^2	16	36	16	0	9	1	25	25

$\Sigma d^2 = 128$

$\rho = 1 - \dfrac{6 \times 128}{8 \times 63}$

$= 1 - 1.524 = -0.524$ *(6)*

b **i** goals conceded, because the Spearman coefficient is numerically higher. *(1)*

9 **a** temperature **b** times **c** typing ability, age, both groups should reflect the same ranges, etc.

10 **a** aspirin **b** life of flowers **c** same flowers, kept under same external conditions (ie temperature) etc.

Test Yourself

1 **a** Mean retail price (£) $= \dfrac{3385}{9} = £376$ Mean sale price (£) $= \dfrac{2545}{9} = £283$

b, c

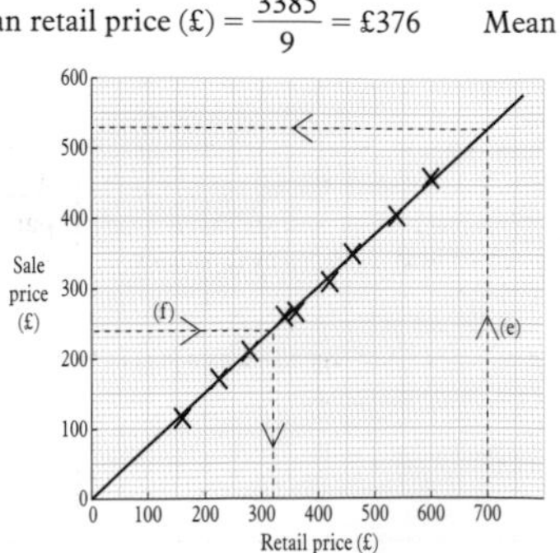

d Positive correlation

e £530 (from dotted line on graph)

f £320 (from dotted line on graph)

g **e** is less reliable because £700 lies outside the range of the data in the table

h ($y = 0.77x$) sale price = 0.77× retail price.

2 **a**

Student	A	B	C	D	E	F
Paper 1 (rank)	3	6	2	5	4	1
Paper 2 (rank)	5	6	1	3	4	2
d	2	0	1	2	0	1
d^2	4	0	1	4	0	1

$\Sigma d^2 = 10$

$$\rho = 1 - \frac{6 \times 10}{6(36-1)} = 0.71 \text{ (2 sf)}$$

b $\rho = 0.71$ shows a fairly strong positive correlation between the rankings for the two papers. A student high in the ranking of one paper tended to come high in the ranking of the other paper.

CHAPTER 15

This chapter is aimed at stimulating discussion. The answers are only to be used as guidelines. There are too many possible answers to be listed here.

Exercise 15:1

1 **a** (1) What is your shoe size? (2) What is your height? (3) What is your age?

b (1) What was the age of your mother when the oldest person in your family was born?
(2) What was the age of your father when the oldest person in your family was born?
(3) When was your grandmother born? What was her age when her first child was born?

c (1) How much television do you watch each day? (2) Do you have a television in your bedroom?
(3) Does your family only have terrestrial television?

2 **a** (1) The average shoe size for boys aged 11 is 6
(2) There is a positive correlation between shoe sizes and heights of boys.
(3) The shoe sizes of boys increases until they are 16 years old.

b (1) The average age of a mother of a pupil in school, at the birth of her first child, was 24 years.
(2) The average age of a father of a pupil in school, at the birth of his first child, was 28 years.
(3) The average age of a mother, at the birth of her first child, has increased in the last fifty years.

c (1) Pupils watch the least television on Mondays.
(2) Most students have a television in their bedroom.
(3) Most families in school only have terrestrial television.

Exercise 15:2

1 Population — All the students in school

Method of sampling — Random sampling — Use alphabetical year lists and a table of random numbers

Size of sample — About 10 students from each of years 7–11

Collecting the data — Collect primary data
Write a questionnaire
Conduct a pilot survey
Send out questionnaires

Recording the results — Make a clear table of the required information.

Name	Year in school	Age of your mother when her first child was born	Date of birth of maternal grandmother	Age of maternal grandmother at birth of her first child	Date of birth of paternal grandmother	Age of paternal grandmother at birth of her first child

2 Population — All the students in Year 7 and Year 11

Method of sampling — Systematic sampling from the registers

Size of sample — The year groups are of equal size so choose 30 pupils in each year.

Collecting the data — Collect primary data
Write a questionnaire
Conduct a pilot survey
Interview each member of the sample

Results — Make a clear table of the required information.

Name	Year group	Day on which student watches most television	Favourite television programme

Exercise 15:3

1 For both **(a)** and **(b)** each of the following could be looked at. The diagrams would then allow for comparisons to be made.
Continuous data
Diagrams: Histogram, Cumulative frequency polygon, Stem and leaf, Box plot
Calculations: Mean, Median, Mode, Range, Interquartile range, Standard deviation

2 **a** *Continuous data*
Diagrams: Cumulative frequency diagrams for each day's viewing. Use this to find medians and quartiles.
Box plot for each day. This allows comparisons to be made.
Calculations: Mean number of hours viewing per day
Standard deviation of the number of hours viewing each day
Median and interquartile range.

b *Discrete data*
Diagrams: Bar chart
Pairs of data: For both years rank the top ten programmes. Draw a scatter graph. Calculate Spearman's rank correlation coefficient.

RANDOM NUMBER TABLE

65 23 68 00	77 82 58 14	10 85 11 85	57 11 73 74	45 25 50 46
09 56 76 51	04 73 94 30	16 74 69 59	04 38 83 98	30 20 87 85
55 99 98 60	01 33 06 93	85 13 232 17	25 51 92 04	52 31 38 70
72 82 45 44	09 53 04 83	03 83 98 41	67 41 01 38	66 83 11 99
04 21 28 72	73 25 02 74	35 81 78 49	52 67 71 40	60 50 47 50
87 01 80 59	89 36 41 59	60 27 64 89	47 45 18 21	69 84 76 06
31 62 46 53	84 40 56 31	74 76 52 53	72 95 96 06	56 83 85 22
29 81 57 94	35 91 90 70	94 24 19 35	50 22 23 72	87 34 83 15
39 98 74 22	77 19 12 81	29 42 04 50	62 34 36 81	43 07 97 92
56 14 80 10	76 52 38 54	84 13 99 90	22 55 41 04	72 37 89 33
29 56 62 74	12 67 09 35	89 33 04 28	44 75 01 57	87 45 52 21
93 32 57 38	39 36 87 42	72 55 73 97	98 36 57 41	76 09 11 68
95 69 51 54	43 19 20 49	57 25 90 55	26 20 70 98	43 73 56 45
65 71 32 43	64 67 22 55	65 65 48 86	10 88 20 12	40 18 49 25
90 27 33 43	97 84 20 57	49 91 41 20	17 64 29 60	66 87 55 97
90 29 42 45	61 34 30 13	30 39 21 52	59 28 64 98	08 76 09 27
99 74 06 29	20 55 72 70	11 43 95 82	75 37 90 24	77 43 63 21
87 87 66 91	16 97 51 50	61 36 96 47	76 68 49 11	50 56 51 06
46 24 17 74	97 37 39 03	54 83 34 00	74 61 77 51	43 63 15 67
66 79 81 43	40 92 84 72	88 32 83 24	67 01 41 34	70 19 26 93
36 42 94 58	83 30 92 39	18 40 03 00	12 90 32 37	91 65 48 15
07 66 25 08	99 27 69 48	85 32 16 46	19 31 85 02	86 36 22 96
93 10 05 72	18 26 36 67	68 48 31 69	68 58 93 49	45 86 99 29
49 50 63 99	26 71 47 94	32 71 72 91	34 18 74 06	32 14 40 80
20 75 58 89	39 04 42 73	37 93 11 07	28 77 91 36	60 47 82 62
02 40 62 09	00 71 09 37	80 44 50 37	32 70 20 38	71 86 75 34
59 87 21 38	29 78 72 67	42 83 65 21	54 79 66 42	47 86 31 15
48 08 99 66	43 38 28 13	50 25 47 93	11 15 07 84	28 30 19 07
54 26 86 75	44 15 20 39	20 03 58 54	80 29 62 53	06 97 71 51
35 35 58 45	23 58 63 66	09 62 80 92	14 55 81 41	21 48 87 34
73 84 90 49	01 21 90 29	57 06 68 73	51 10 51 95	63 08 57 99
34 64 78 00	92 59 67 74	58 48 92 09	42 20 40 37	63 80 58 93
68 56 87 47	63 06 24 71	41 98 79 06	07 18 58 29	16 49 67 37
72 47 05 42	88 07 27 55	58 74 82 08	42 28 26 48	25 32 00 31
44 44 96 75	89 57 12 60	42 38 77 36	45 69 21 68	32 70 04 96
28 11 57 47	61 57 89 88	62 18 93 67	57 32 96 72	21 17 13 54
87 22 38 88	91 99 16 08	17 76 27 47	52 14 98 86	35 68 23 85
44 93 14 59	67 40 24 10	11 63 40 47	07 56 14 22	62 74 93 39
81 84 37 25	90 43 56 62	94 58 49 03	84 22 57 22	47 98 86 37
09 75 35 21	04 47 54 08	98 44 08 16	44 86 69 71	20 52 64 94
77 65 05 04	22 18 20 10	81 87 05 69	43 70 96 76	42 05 21 10
19 06 51 61	34 03 61 55	98 58 83 50	01 48 99 85	08 67 15 91
52 91 87 07	19 62 32 28	04 91 42 48	65 24 86 09	87 68 55 51
52 47 25 14	93 91 75 51	49 26 49 41	20 83 30 30	43 22 69 08
52 67 87 40	63 41 91 86	10 47 80 70	56 87 25 86	89 94 21 42
66 25 71 73	78 60 50 62	91 04 95 97	64 16 71 31	32 80 19 61
29 97 56 42	56 90 16 75	74 95 99 26	01 63 25 16	54 18 54 46
15 25 03 68	92 45 53 00	06 29 46 43	46 66 27 12	85 05 22 44
82 08 65 67	64 13 51 14	38 28 24 30	39 62 20 35	23 90 57 36
81 35 03 25	87 24 83 59	04 67 51 52	26 21 69 75	87 28 6 50

INDEX